D0948918

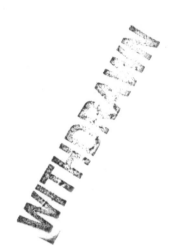

DATE DUE		
DEC 10 '86'S		
JAN 2 7 '88'S		
OCT 31 '91'S		
OCT 3 1 1994		
ILL 370 9388		
MAR 1 8 1995 S		

Texts and Monographs in Computer Science

Editor
David Gries

Texts and Monographs in Computer Science

Computational Geometry
An Introduction

Franco P. Preparata
Michael Ian Shamos

With 231 Illustrations

Springer-Verlag
New York Berlin Heidelberg Tokyo

Franco P. Preparata
Coordinated Science Laboratory
University of Illinois at Urbana-Champaign
Urbana, IL 61801
U.S.A.

Michael Ian Shamos
Department of Computer Science
Carnegie-Mellon University
Pittsburgh, PA 15213
U.S.A.

Series Editor

David Gries
Department of Computer Science
Cornell University
Upson Hall
Ithaca, NY 14853
U.S.A.

Library of Congress Cataloging in Publication Data
Preparata, Franco P.
  Computational geometry.
  (Texts and monographs in computer science)
  Bibliography: p.
  Includes index.
  1. Geometry—Data processing.   I. Shamos, Michael.
II. Title.   III. Series.
QA447.P735   1985   516′.028′54   85-8049

Typeset by Asco Trade Typesetting Ltd., Hong Kong.
Printed and bound by R. R. Donnelley & Sons, Harrisonburg, Virginia.
Printed in the United States of America.

9 8 7 6 5 4 3 2 1

ISBN 0-387-96131-3 Springer-Verlag New York Berlin Heidelberg Tokyo
ISBN 3-540-96131-3 Springer-Verlag Berlin Heidelberg New York Tokyo

To Paola and Claudia
and
To Julie

Preface

The objective of this book is a unified exposition of the wealth of results that have appeared—mostly in the past decade—in Computational Geometry. This young discipline—so christened in its current connotation by one of us, M. I. Shamos—has attracted enormous research interest, and has grown from a collection of scattered results to a mature body of knowledge. This achieved maturity, however, does not prevent computational geometry from being a continuing source of problems and scientific interest.

As the research endeavor reaches a level of substantial sophistication, with the emergence of powerful techniques and the beneficial interaction with combinatorial and algebraic geometry, there is increasing demand for a pedagogically organized presentation of the existing results. The demand issues both from the classroom, where experimental graduate courses are being taught from papers and notes, and from the professional environment, where several applied fields—such as computer-aided design, computer graphics, robotics, etc.—are mature for the transfer.

This book is intended to respond to this need. Basically conceived as an early graduate text, it is also directed to the professional in the applied fields mentioned above. It must be pointed out, however, that this book is not exactly a catalog of readily usable techniques, although some algorithms are eminently practical. Rather, the philosophical outlook is inherited from the discipline known as design and analysis of algorithms (or "algorithmics"), which aims at the characterization of the difficulty of individual problems. Our analyses are mainly concerned with the worst-case behavior of algorithms; moreover, they achieve full validity for problems of sufficiently large size (asymptotic analysis). These two features should be carefully considered in selecting an algorithm, because the technique most suitable for small sizes is not necessarily the asymptotically optimal one, and because an algorithm inferior in the worst case may be superior in its average-case behavior.

We have attempted to offer a rather detailed panoramic view of the "continuous brand" of computational geometry, as distinct from its "discretized" counterpart. However, our principal objective has been a coherent discourse rather than a meticulous survey. We have attempted to partially remedy this deficiency with 'Notes and Comments' sections at the end of each chapter. Our apologies go to the numerous authors whose work has not been described or barely mentioned.

The initial core of this book was M. I. Shamos' doctoral dissertation. When in 1981 R. L. Graham suggested to one of us, (F. P. Preparata) to develop the original manuscript into a textbook, none of us had really a clear appraisal of the effort involved. Fortunately many of our colleagues and friends generously assisted in this task by patiently reading successive drafts, suggesting improvements, and catching errors. We are greatly indebted to them; we would like to explicitly thank (in alphabetical order) D. Avis, J. L. Bentley, B. Bhattacharya, B. M. Chazelle, D. P. Dobkin, M. Dyer, S. C. Eisenstat, Dan Hoey, M. Kallay, D. T. Lee, J. van Leeuwen, W. Lipski, Jr., A. J. Perlis, J. O'Rourke, L. C. Rafsky, R. Seidel, M. H. Schultz, G. T. Toussaint, and D. Wood. We also acknowledge the partial assistance of NSF grant MCS 81-05552 (for the work of F. P. Preparata), the Office of Naval Research, IBM Corporation, Carnegie-Mellon University, and Bell Laboratories (for the work of M. I. Shamos) and the enthusiastic cooperation of the staff at Springer-Verlag.

Finally, our thanks go to our respective wives, Rosa Maria and Julie, for their patience and support during the long hours that have lead to this finished product.

May 1985   F. P. Preparata
                                                                  M. I. Shamos

Contents

CHAPTER 3
Convex Hulls: Basic Algorithms   89

CHAPTER 4
Convex Hulls: Extensions and Applications   144

CHAPTER 8
The Geometry of Rectangles   315

CHAPTER 1
Introduction

1.1 Historical Perspective

Egyptian and Greek geometry were masterpieces of applied mathematics. The original motivation for geometric problems was the need to tax lands accurately and fairly and to erect buildings. As often happens, the mathematics that developed had permanence and significance that far transcend the Pharaoh's original revenue problem, for Geometry is at the heart of mathematical thinking. It is a field in which intuition abounds and new discoveries are within the compass (so to speak) of nonspecialists.

It is popularly held that Euclid's chief contribution to geometry is his exposition of the axiomatic method of proof, a notion that we will not dispute. More relevant to this discussion, however, is the invention of *Euclidean construction*, a schema which consists of an algorithm *and its proof*, intertwined in a highly stylized format. The Euclidean construction satisfies all of the requirements of an algorithm: it is unambiguous, correct, and terminating. After Euclid, unfortunately, geometry continued to flourish, while analysis of algorithms faced 2000 years of decline. This can be explained in part by the success of *reductio ad absurdum*, a technique that made it easier for mathematicians to prove the existence of an object by contradiction, rather than by giving an explicit construction for it (an algorithm).

The Euclidean construction is remarkable for other reasons as well, for it defines a collection of allowable instruments (ruler and compass) and a set of legal operations (primitives) that can be performed with them. The Ancients were most interested in the closure of the Euclidean primitives under finite composition. In particular, they wondered whether this closure contained all conceivable geometric constructions (e.g., the trisection of an angle). In modern terms, this is a computer science question—do the Euclidean primitives suffice to perform all geometric "computations"? In an attempt to answer this question, various alternative models of computation were considered by allowing the primitives and the instruments themselves to vary. Archimedes proposed a (correct) construction for the trisector of a 60-degree angle with the following addition to the set of primitives: Given two circles, A and B, and a point P, we are allowed to mark a segment MN on a straightedge and position it so that the straightedge passes through P, with M on the boundary of A and N on the boundary of B. In some cases, restricted sets of instruments were studied, allowing compasses only, for example. Such ideas seem almost a premonition of the methods of automata theory, in which we examine the power of computational models under various restrictions. Alas, a proof of the insufficiency of the Euclidean tools would have to await the development of Algebra.

The influence of Euclid's *Elements* was so profound that it was not until Descartes that another formulation of geometry was proposed. His introduction of coordinates enabled geometric problems to be expressed algebraically, paving the way for the study of higher plane curves and Newton's calculus. Coordinates permitted a vast increase in computational power, bridged the gulf between two great areas of Mathematics, and led to a renaissance in constructivist thinking. It was now possible to produce new geometric objects by solving the associated algebraic equations. It was not long before computability questions arose once again. Gauss, now armed with algebraic tools, returned to the problem of which regular polygons with a prime number of sides could be constructed using Euclidean instruments, and solved it completely. At this point a close connection between ruler and compass constructions, field extensions, and algebraic equations became apparent. In his doctoral thesis, Gauss showed that every algebraic equation has at least one root (Fundamental Theorem of Algebra). Abel, in 1828, went on to consider the same problem *in a restricted model of computation*. He asked whether a root of every algebraic equation could be obtained using only arithmetic operations and the extraction of nth roots, and proved that the answer was negative. While all constructible numbers were known to be algebraic, this demonstrated that not all algebraic numbers are constructible. Shortly thereafter, he characterized those algebraic equations which *can* be solved by means of radicals, and this enabled him to discuss the feasibility of specific geometric problems, such as the trisection of the angle.

1.1.1 Complexity notions in classical geometry

Euclidean constructions for any but the most trivial of problems are very complicated because of the rudimentary primitives that are allowed. An apparently frequent pastime of the post-Euclidean geometers was to refine his constructions so that they could be accomplished in fewer "operations." It was not until the twentieth century, however, that any quantitative measure of the complexity of a construction problem was defined. In 1902, Emile Lemoine established the science of *Geometrography* by codifying the Euclidean primitives as follows [Lemoine (1902)]:

1. Place one leg of the compass at on a given point.
2. Place one leg of the compass on a given line.
3. Produce a circle.
4. Pass the edge of the ruler through a given point.
5. Produce a line.

The total number of such operations performed during a construction is called its *simplicity*, although Lemoine recognized that the term "measure of complication" might be more appropriate. This definition corresponds closely to our current idea of the *time complexity* of an algorithm, although in Lemoine's work there is no functional connection between the size of the input (number of given points and lines) in a geometric construction and its simplicity. Indeed, Lemoine's interest was in improving Euclid's original constructions, not in developing a theory of complexity. At the former he was remarkably successful—Euclid's solution to the Circles of Apollonius problem requires 508 steps, while Lemoine reduced this to fewer than two hundred [Coolidge (1916)]. Unfortunately, Lemoine did not see the importance of proving, or perhaps was unable to prove, that a certain number of operations were *necessary* in a given construction, and thus the idea of a lower bound eluded him.

Hilbert, however, appreciated the significance of lower bounds. Working in a restricted model, he considered only those constructions performable with straightedge and *scale*, an instrument which is used only to mark off a segment of fixed length along a line. Not all Euclidean constructions can be accomplished with this set of instruments. For those that can, we may view the coordinates of the constructed points as a function F of the given points. Hilbert gave a necessary and sufficient condition for F to be computable using exactly n square root operations, one of the earliest theorems in algebraic computational complexity [Hilbert (1899)].

Further evidence suggests that many of our present-day techniques for analyzing algorithms were anticipated by the geometers of previous centuries. In 1672, Georg Mohr showed that any construction performable with ruler and compass can be accomplished with compass alone, insofar as the given and required objects are specified by points. (Thus, even though a straight line cannot be drawn with compass alone, two points on the line can each be specified by intersecting two circular arcs.) What is notable about Mohr's proof is that it is a *simulation*, in which he demonstrates that any operation in which the ruler participates can be replaced by a finite number of compass operations. Could one ask for a closer connection with automata theory? Along similar lines is the result that the ruler used in any construction may have any positive length, however small, and yet be able to simulate a ruler of arbitrary length.

While Lemoine and others were occupied with the time complexity of Euclidean constructions, the question of the amount of *space* needed for such constructions was also raised. While the measure of space that was used does not coincide with our current definition as the amount of memory used by an algorithm, it comes remarkably close and is quite natural: the area of the plane needed to perform the construction. The space used depends, in general, on the area of the convex hull of the given loci and on the size of the required result, as well as on the size of any intermediate loci that need to be formed during the construction [Eves (1972)]. Our point here is that time and space notions are not entirely foreign to Geometry.

When the impossibility of certain Euclidean constructions was demonstrated by Galois, it was realized that this prevented the *exact* trisection of an angle but said nothing about the feasibility of an approximate construction. In fact, asymptotically convergent procedures for the quadrature of the circle and duplication of the cube were known to the ancient Greeks [Heath (1921)]. The history of iterative algorithms is indeed long.

1.1.2 The theory of convex sets, metric and combinatorial geometry

Geometry in the nineteenth century progressed along many lines. One of these, promulgated by Klein, involved a comprehensive study of the behavior of geometric objects under various transformations, and projective geometry formed an important offshoot (see Section 1.3.2). While research on finite projective planes leads to fascinating questions in both combinatorial theory and discrete algorithms, this aspect of geometry will not be pursued in this book.

The growth of real analysis had a profound effect on geometry, resulting in formal abstraction of concepts that had previously been only intuitive. Two such developments, metric geometry and convexity theory, provide the principal mathematical tools that aid in the design of fast algorithms.

Distance is an essential notion of geometry. The *metric*, its generalization, was able to draw geometric concepts and insights into analysis, where the idea of the "distance" between functions gave rise to function spaces and other powerful constructs. Unfortunately, much of what developed withdrew toward the nonconstructive. Function spaces by their nature are not computational objects.

The significance of convexity theory is that is deals analytically with *global* properties of objects and enables us to deal with extremal problems. Unfortunately, many questions in convexity are cumbersome to formulate algebraically, and the subject tends to encourage nonconstructive methods.

Combinatorial geometry is much closer in spirit to our goal of algorithmic geometry. It is based on characterizations of geometric objects in terms of properties of *finite* subsets. For example, a set is convex if and only if the line segment determined by every *pair* of its points lies entirely in the set. The inadequacy of combinatorial geometry for our purposes lies in the fact that for most sets of interest the number of finite subsets is itself infinite, which precludes algorithmic treatment. Recent work on geometric algorithms aimed at remedying these deficiencies and at producing mathematics conducive to good algorithms.

1.1.3 Prior related work

Algorithms of a geometric nature have been developed in several different contexts, and the phrase "Computational Geometry" has been already used in at least two other connotations. We will now try to place these related endeavors in a proper perspective and contrast them to the nowadays prevalent connotation:

1. Geometric modeling by means of spline curves and surfaces, a topic that is closer in spirit to numerical analysis than it is to geometry, has been dealt with expertly by Bézier, Forrest and Riesenfeld. We should note that Forrest refers to his discipline as "Computational Geometry" [Bézier (1972); Forrest (1971); Riesenfeld (1973)].
2. In an elegant and fascinating book entitled *Perceptrons* (of which the subtitle is also "Computational Geometry"), Minsky and Papert (1969) deal with the complexity of predicates that recognize certain geometric properties, such as convexity. The intent of their work was to make a statement about the possibility of using large retinas composed of simple circuits to perform pattern recognition tasks. Their theory is self-contained and does not fall within the algorithmic scope of this book.
3. Graphics software and geometric editors are undoubtedly targets for many of the algorithms presented in this book. However, they raise issues that are oriented more toward implementation details and the user interface than toward analysis of algorithms. Included in the same class are numerical control software for machine tool support, programs for graphic plotters, map-drawing systems, and software for architectural design and civil engineering.
4. Finally, the phrase "Computational Geometry" may sound to some people as the activity of proving geometry theorem by means of computers. While this is a fascinating study, it reveals much more about our theorem-proving heuristics and understanding of proof procedures than it does about geometry *per se*, and will thus not be treated here.

1.1.4 Toward computational geometry

A large number of applications areas have been the incubation bed of the discipline nowadays recognized as Computational Geometry, since they provide inherently geometric problems for which efficient algorithms have to be developed. These problems include the Euclidean traveling salesman, minimum spanning tree, hidden line, and linear programming problems, among hosts of others. In order to demonstrate the broad scope of computational geometry in a convincing way, we will defer presenting background material on such problems until they occur in the text.

Algorithmic studies of these and other problems have appeared in the past century in the scientific literature, with an increasing intensity in the past two decades. Only very recently, however, systematic studies of geometric algorithms have been undertaken, and a growing number of researchers have been attracted to this discipline, christened "Computational Geometry" in a paper by M. I. Shamos (1975a).

The philosophical outlook and the methodology of computational geometry will hopefully emerge from the detailed case studies presented in this book. One fundamental feature of this discipline is the realization that classical characterizations of geometric objects are frequently not amenable to the design of efficient algorithms. To obviate this inadequacy, it is necessary to identify the useful concepts and to establish their properties, which are conducive to efficient computations. In a nutshell, computational geometry must reshape—whenever necessary—the classical discipline into its computational incarnation.

1.2 Algorithmic Background

For the past fifteen years the analysis and design of computer algorithms has been one of the most thriving endeavors in computer science. The fundamental works of Knuth (1968; 1973) and Aho–Hopcroft–Ullman (1974) have brought order and systematization to a rich collection of isolated results, conceptualized the basic paradigms, and established a methodology that has become the standard of the field. Subsequent works [Reingold–Nievergelt–Deo (1977); Wirth (1976)] have further strengthened the theoretical foundations.

It is therefore beyond the scope of this work to review in detail the material of those excellent texts, with which the reader is assumed to be reasonably familiar. It is appropriate however—a least from the point of view of terminology—to briefly review the basic components of the language in which Computational Geometry will be described. These components are algorithms and data structures. Algorithms are programs to be executed on a suitable abstraction of actual "von Neumann" computers; data structures are ways to organize information, which, in conjunction with algorithms, permit the efficient and elegant solution of computational problems.

1.2.1 Algorithms: Their expression and performance evaluation

Algorithms are formulated with respect to a specific model of computation; as we shall see in Section 1.4, the model of computation is a comfortable abstraction of the physical machine used to execute programs. However, as pointed out by Aho–Hopcroft–Ullman (1974), it is neither necessary nor desirable to express algorithms in machine code. Rather, in the interest of clarity, effectiveness of expression, and conciseness, we shall normally[1] use a high-level language that has become the standard of the literature on the subject: *Pidgin Algol*. Pidgin Algol is an informal and flexible version of Algol-like languages; it is rather rigorous in its control structures but very loose in the format of its other statements, where the use of conventional mathematical notations alternates freely with the use of natural language descriptions. Of course a Pidgin Algol program can be routinely transformed into a formal higher level language program.

Following Aho–Hopcroft–Ullman, we briefly illustrate the constructs of Pidgin Algol. Formal declarations of data types are avoided, and the type of a variable is normally made evident by the context. Additionally, no special format is chosen for expressions and conditions.

A program is normally called **procedure**, and has the format

> **procedure** name (parameters): statement.

A statement can be rewritten (according to the terminology of phrase-structure grammars) as a string of two or more statements, in which case the "parentheses" "**begin** ... **end**" are used to bracket the string as follows:

> **begin** statement;
> ⋮
> statement
> **end**.

In turn, a statement can be specialized either as a natural language sentence or as one of the following specific more formal instances.

1. Assignment:

> variable := source

The "source" is a definition of the computation generating the value of the

---

[1] Occasionally, algorithms may be presented in an entirely narrative form.

variable; the computation is in general an *expression* on a set of variables. (Some of these variables may in turn be expressed as "functions"; a **function** is itself a special form of program, as shown below.)

2. Conditional:

> **if** condition **then** statement (**else** statement)

where the **else** portion is optional.

3. Loop. This appears in one of three formats:

> 3a. **for** variable := value **until** value **do** statement
> 3b. **while** condition **do** statement
> 3c. **repeat** statement **until** condition

The **while** and **repeat** constructs are distinguished by the fact that in the while-loop the condition is tested *before* the execution of the statement, whereas the opposite occurs in the repeat-loop.

4. Return:

> **return** expression

A statement of this sort must appear in a function type program, which has the format

> **function** name (parameters): statement.

The expression that is the argument of **return** becomes the source of an assignment statement, as indicated in 1 above.

Frequently, Pidgin Algol algorithms contain comments designed to aid the understanding of the action. In this text the typical format of a comment will be: (*natural-language sentence*).

The time used by a computation—the execution of an algorithm—is the sum of the times of the individual operations being executed (see also Section 1.4). As we mentioned earlier, a Pidgin Algol program can be transformed in a straightforward (although tedious) manner to a program in the machine code of a specific computer. This, in principle, provides a method for evaluating the program's running time. However, this approach is not only tedious but also scarcely illuminating, since it makes reference to a specific machine, while we are essentially interested in the more general functional dependence of computation time upon problem size (i.e., how fast the computation time grows with the problem size). Thus, it is customary in the field of algorithmic analysis and design to express running time—as well as any other measure of performance—modulo a multiplicative constant. This is normally done by counting only certain "key operations" executed by the algorithm (which is readily done by analyzing the high-level language version of the algorithm). Such approach is totally legitimate when establishing lower bounds to running time, since any unaccounted-for operation can only increase it; when dealing with upper bounds, however, we must ensure that the selected key operations account for a constant fraction of all operations executed by the algorithm. Knuth has popularized a notational device that distinguishes nicely between upper and lower bounds, which we will adopt [Knuth (1976)]:

$O(f(N))$ denotes the set of all functions $g(N)$ such that there exist positive constants C and N_0 with $|g(N)| \leq Cf(N)$ for all $N \geq N_0$.

$\Omega(f(N))$ denotes the set of all functions $g(N)$ such that there exist positive constants C and N_0 with $g(N) \geq Cf(N)$ for all $N \geq N_0$.

$\theta(f(N))$ denotes the set of all functions $g(N)$ such that there exist positive constants C_1, C_2, and N_0 with $C_1 f(N) \leq g(N) \leq C_2 f(N)$ for all $N \geq N_0$.

$o(f(N))$ denotes the set of all functions $g(N)$ such that *for all positive constants* C there is an N_0 with $g(N) \leq Cf(N)$ for all $N \geq N_0$ (or, equivalently $\lim_{n \to \infty} g(N)/f(N) = 0$).

Thus $O(f(N))$ is used to indicate functions that are *at most as large* as some constant times $f(N)$, the concept one needs to describe upper bounds; conversely, $\Omega(f(N))$ is used to indicate functions *at least as large* as some constant times $f(N)$, the analogous concept for lower bounds. Finally $\theta(f(N))$ is used to indicate functions of the same order as $f(N)$, the concept one needs for "optimal" algorithms.

The preceding discussion focuses on the computation time of an algorithm. Another important measure of performance is space, usually identified with the amount of memory used by the algorithm. Indeed, *space* and *time* complexity as functions of the problem size are the two fundamental performance measures of algorithmic analysis.

The central objective of this book is essentially the presentation of algorithms for geometric problems and the evaluation of their *worst-case complexity*. Worst-case complexity is the maximum of a measure of performance of a given algorithm over all problem instances of a given size, and is to be contrasted with *average-case* (or *expected*) *complexity*, which instead should give an estimate of the *observed* behavior of the algorithm. Unfortunately, average-case analysis is considerably more complicated than worst-case analysis, for a two-fold reason: first, substantial mathematical difficulties arise even when the underlying distribution is conveniently selected; second, there is frequently scarce consensus on a claim that the selected distribution is a realistic model of the situation being studied. This explains why the overwhelming majority of results concern worst-case analysis: correspondingly, this book will only occasionally discuss average-case results.

Another important item to be underscored is that the "order of" notation conceals multiplicative constants. Therefore a complexity result acquires its full validity only for sufficiently large problem sizes; for this reason this methodology is referred to as *asymptotic analysis*. It is therefore quite possible—and indeed not infrequent—that for small problem sizes the most suitable algorithm is not the asymptotically best algorithm. This *caveat* should never be ignored in the selection of an algorithm for a given application.

1.2.2 Some considerations on general algorithmic techniques

Efficient algorithms for geometric problems are frequently designed by re-sorting to general techniques of the discipline, such as divide-and-conquer, balancing, recursion, and dynamic programming. Excellent discussions of these techniques are available in the now classical texts of algorithm analysis and design (see, for example, [Aho–Hopcroft–Ullman (1974)]) and it would be superfluous to repeat them here.

There is however a technique that is suggested—uniquely and naturally—by the nature of some geometric problems. This technique is called *sweep*, and its most frequent instantiations are *plane sweep* (in two dimensions) and *space sweep* (in three dimensions). We shall now describe the main features of the plane sweep, its generalization to three dimensions being straightforward.

For concreteness, we illustrate the method in connection with a specific problem (discussed in full detail in Section 7.2.3): Given a set of segments in the plane, report all of their intersections. Consider a straight line *l* (assumed without loss of generality, to be vertical), which partitions the plane into a left and a right half-planes. Assume that each of these half-planes contains endpoints of the given segments. It is clear that the solution to our problem is the union of the solutions in each of the two half-planes; so, assuming we have already obtained the set of intersections to the left of *l*, this set is not going to be affected by segments lying to the right of *l*. We now observe that an intersection (to be reported) may occur only between two segments whose intersections with some vertical line are adjacent; so, if we generate *all* vertical cuts with the given set of segments, certainly will we discover all intersections. However, the (impossible) task of generating the (continuously) infinite set of all vertical cuts is avoided by realizing that the plane is partitioned into vertical strips, delimited either by segment endpoints or by segment intersections, in which the vertical order of the intercepts by a vertical cut is constant. Thus, all we need to do is to jump from the left boundary of one such strip to its right boundary, update the order of the intercepts and test for any new intersection among "adjacent" segments.

The previous discussion outlines the essential features of the plane-sweep technique. There is a vertical line that sweeps the plane from left to right, halting at special points, called "event points." The intersection of the sweep-line with the problem data contains all the *relevant* information for the continuation of the sweep. Thus we have two basic structures:

1. The *event point schedule*, which is a sequence of abscissae, ordered from left to right, which define the halting positions of the sweep-line. Notice that the event point schedule is not necessarily entirely extracted from the input data, but may be dynamically updated during the execution of the plane-sweep algorithm. Different data structures may be needed in different applications.
2. The *sweep-line status*, which is an adequate description of the intersection of the sweep-line with the geometric structure being swept. "Adequate" means that this intersection contains the information that is relevant to the specific application. The sweep-line status is updated at each event point, and a suitable data structure must be chosen in each case.

Examples of plane-sweep algorithms will be found in Section 2.2.2.

1.2.3 Data structures

Geometric algorithms involve the manipulation of objects which are not handled at the machine language level. The user must therefore organize these complex objects by means of the simpler data types directly representable by the computer. These organizations are universally referred to as *data structures*.

The most common complex objects encountered in the design of geometric algorithms are sets and sequences (ordered sets). Data structures particularly suited to these complex combinatorial objects are well described in the standard literature on algorithms to which the reader is referred [Aho–Hopcroft–Ullman (1974); Reingold–Nievergelt–Deo (1977)]. Suffice it here to review the classification of these data structures, along with their functional capabilities and computational performance.

Let S be a set represented in a data structure and let u be an arbitrary element of a universal set of which S is a subset. The fundamental operations occurring in set manipulation are:

1. MEMBER(u, S). Is $u \in S$? (YES/NO answer.)
2. INSERT(u, S). Add u to S.
3. DELETE(u, S). Remove u from S.

Suppose now that $\{S_1, S_2, \ldots, S_k\}$ is a collection of sets (with pairwise empty intersection). Useful operations on this collection are:

4. FIND(u). Report j, if $u \in S_j$.
5. UNION$(S_i, S_j; S_k)$. Form the union of S_i and S_j and call it S_k.

When the universal set is totally ordered, the following operations are very important:

6. MIN(S). Report the minimum element of S.

7. SPLIT(u, S). Partition S into $\{S_1, S_2\}$, so that $S_1 = \{v : v \in S$ and $v \leq u\}$ and $S_2 = S - S_1$.
8. CONCATENATE(S_1, S_2). Assuming that, for arbitrary $u' \in S_1$ and $u'' \in S_2$ we have $u' \leq u''$, form the ordered set $S = S_1 \cup S_2$.

Data structures can be classified on the basis of the operations they support (regardless of efficiency). Thus for ordered sets we have the following table.

Table I

Data Structure	Supported Operations
Dictionary	MEMBER, INSERT, DELETE
Priority queue	MIN, INSERT, DELETE
Concatenable queue	INSERT, DELETE, SPLIT, CONCATENATE

For efficiency, each of these data structures is normally realized as a height-balanced binary search tree (often an AVL or a 2-3-tree) [Aho–Hopcroft–Ullman (1974)]. With this realization, each of the above operations is performed in time proportional to the logarithm of the number of elements stored in the data structure; the storage is proportional to the set size.

The above data structures can be viewed abstractly as a linear array of elements (a *list*), so that insertions and deletions can be performed in an arbitrary position of the array. In some cases, some more restrictive modes of access are adequate for some applications, with the ensuing simplifications. Such structures are: *Queues*, where insertions occur at one end and deletions at the other; *Stacks*, where both insertions and deletions occur at one end (the stack-top). Clearly, one and two pointers are all that is needed for managing a stack or a queue, respectively. For brevity, we will use the notations "$\Rightarrow U$" and "$U \Leftarrow$" to indicate addition to or deletion from U, respectively, where U is either a queue or a stack.

Unordered sets can always be handled as ordered sets by artificially imposing an order upon the elements (for example, by giving "names" to the elements and using the alphabetical order). A typical data structure for this situation is the following.

Table II

Data Structure	Supported Operations
Mergeable heap	INSERT, DELETE, FIND, UNION, (MIN)

Each of the above operations can be executed in time $O(\log N)$, where N is the size of the set stored in the data structure, by deploying, as usual, height-balanced trees. If the elements of the set under consideration are represented as the integers from 1 to N, then a more sophisticated realization of the data structure enables the execution of N operations on a set of size N in time $O(N \cdot A(N))$, where $A(N)$ is an extremely slowly growing function related to a functional inverse of the Ackermann function (indeed for $N \leq 2^{2^{16}}$, or $\sim 10^{20,000}$, $A(N) \leq 5$).

The standard data structures reviewed above are used extensively in conjunction with the algorithms of Computational Geometry. However, the nature of geometric problems has led to the development of specific non-conventional data structures, two of which have proved of such general value that it is appropriate to present them in this introductory chapter. They are the Segment Tree and the Doubly-Connected-Edge-List.

1.2.3.1 The segment tree

The *Segment Tree*, originally introduced by J. L. Bentley [Bentley (1977)], is a data structure designed to handle intervals on the real line whose extremes belong to a *fixed* set of N abscissae. Since the set of abscissae is fixed, the segment tree is a *static* structure with respect to the abscissae, that is, one that does not support insertions or deletions of abscissae; in addition, the abscissae can be normalized by replacing each of them by its rank in their left-to-right order. Without loss of generality, we may consider these abscissae as the integers in the range $[1, N]$.

The *segment tree* is a rooted binary tree. Given integers l and r, with $l < r$, the segment tree $T(l, r)$ is recursively built as follows: It consists of a root v, with parameters $B[v] = l$ and $E[v] = r$ (B and E are mnemonic for "beginning" and "end," respectively), and if $r - l > 1$, of a left subtree $T(l, \lfloor (B[v] + E[v])/2 \rfloor)$ and a right subtree $T(\lfloor (B[v] + E[v])/2 \rfloor, r)$. (The roots of these subtrees are naturally identified as LSON[v] and RSON[v], respectively.) The parameters $B[v]$ and $E[v]$ define the *interval* $[B[v], E[v]] \subseteq [l, r]$ associated with node v. The segment tree $T(4, 15)$ is illustrated in Figure 1.1. The set of intervals $\{[B[v], E[v]] : v$ a node of $T(l, r)\}$ are the *standard intervals* of $T(a, b)$. The standard intervals pertaining to the leaves of $T(l, r)$ are called the *elementary intervals*.[2] It is straightforward to establish that $T(l, r)$ is balanced (all leaves belong to two contiguous levels) and has depth $\lceil \log_2 (r - l) \rceil$.

The segment tree $T(l, r)$ is designed to store intervals whose extremes belong to the set $\{l, l + 1, \ldots, r\}$, in a *dynamic* fashion (that is, supporting insertions and deletions). Specifically, for $r - l > 3$, an arbitrary interval $[b, e]$, with integers $b < e$, will be partitioned into a collection of at most $\lceil \log_2(r - l) \rceil + \lfloor \log_2(r - l) \rfloor - 2$ standard intervals of $T(l, r)$. The segmentation of interval $[b, e]$ is completely specified by the operation that stores (inserts) $[b, e]$ into the segment tree T, that is, by a call INSERT(b, e; root(T)) of the following primitive:

---

[2] Strictly speaking, the interval associated with v is the semiclosed interval $[B[v], E[v])$, except for the nodes of the rightmost path of $T(a, b)$, whose intervals are closed.

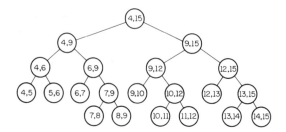

Figure 1.1   The segment tree T(4, 15).

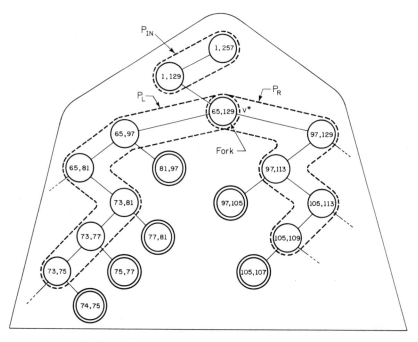

Figure 1.2   Insertion of interval [74, 107] into T(1,257). The allocation nodes are doubly-circled.

procedure INSERT($b, e; v$)
begin if $(b \leq B[v]]$ **and** $(E[v] \leq e)$ **then** allocate $[b, e]$ to v
   else begin if $(b < \lfloor (B[v] + E[v])/2 \rfloor)$ **then** INSERT(b, e; LSON[v]);
               if $(\lfloor (B[v] + E[v])/2 \rfloor < e)$ **then** INSERT(b, e; RSON[v])
       end
end.

The action of INSERT(b, e; root(T)) corresponds to a "tour" in T, having the following general structure (see Figure 1.2): a (possibly empty) initial path, called P_{IN}, from the root to a node v^*, called the *fork*, from which two (possibly empty) paths P_L and P_R issue. Either the interval being inserted is allocated entirely to the fork (in which case P_L and P_R are both empty), or all right-sons of nodes of P_L, which are not on P_L, as well as all left-sons of nodes of P_R, which are not on P_R, identify the fragmentation of $[b, e]$ (*allocation nodes*).

The allocation of an interval to a node v of T could take different forms, depending upon the requirements of the application. Frequently all we need to know is the cardinality of the set of intervals allocated to any given node v; this can be managed by a single nonnegative integer parameter $C[v]$, denoting this cardinality, so that the allocation of $[b, e]$ to v becomes

$$C[v] := C[v] + 1.$$

In other applications, we need to preserve the identity of the intervals allocated to a node v. Then we append to each node v of T a secondary structure, linked list $\mathscr{L}[v]$, whose records are the identifiers of the intervals.

Perfectly symmetrical to INSERT is the DELETE operation, expressed by the following primitive (here we assume we are just interested in maintaining the parameter $C[v]$):

> **procedure** DELETE$(b, e; v)$
> **begin if** $(b \le B[v]$ **and** $(E[v] \le e)$ **then** $C[v] := C[v] - 1$
>   **else begin if** $(b < \lfloor (B[v] + E[v])/2 \rfloor)$ **then** DELETE$(b, e; \text{LSON}[v])$;
>   **if** $(\lfloor (B[v] + E[v])/2 \rfloor < e)$ **then** DELETE$(b, e; \text{RSON}[v])$
>   **end**
> **end**.

(Note that only deletions of previously inserted intervals guarantee correctness.)

The segment tree is an extremely versatile data structure, as we shall see in connection with numerous applications (Chapters 2 and 8). We only note that if we wish to know the number of intervals containing a given point x, a simple binary search in T (i.e., the traversal of a path from the root to a leaf) readily solves the problem.

1.2.3.2 The doubly-connected-edge-list (DCEL)

The doubly-connected-edge-list (DCEL) is particularly suited to represent a planar graph embedded in the plane [Muller-Preparata (1978)]. A planar embedding of a planar graph $G = (V, E)$ is the mapping of each vertex in V to a point in the plane and each edge in E to a simple curve between the two images of extreme vertices of the edge, so that no two images of edges intersect except at their endpoints. It is well-known that any planar graph admits of a planar embedding where all edges are mapped to straight line segments [Fary (1948)].

Let $V = \{v_1, \ldots, v_N,\}$ and $E = \{e_1, \ldots, e_M\}$. The main component of the DCEL of a planar graph (V, E) is the *edge node*. There is a one-to-one correspondence between edges and edge nodes, i.e., each edge is represented

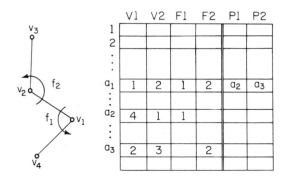

Figure 1.3   Illustration of the DCEL.

exactly once. An edge node consists of four information fields $V1$, $V2$ $F1$, and $F2$, and two pointer fields $P1$ and $P2$: therefore the corresponding data structure is easily implemented with six arrays with the same names, each consisting of M cells. The meanings of these fields are as follows. The field $V1$ contains the origin of the edge and the field $V2$ contains its terminus; in this manner the edge receives a conventional orientation. The fields $F1$ and $F2$ contain the names of the faces which lie respectively on the left and on the right of the edge oriented from $V1$ to $V2$. The pointer $P1$ (resp. $P2$) points to the edge node containing the first edge encountered after edge $(V1, V2)$ when one proceeds counterclockwise around $V1$ (resp. $V2$). Names of faces and vertices may be taken as integers. As an example, a fragment of a graph and the corresponding fragment of the DCEL are shown in Figure 1.3.

It is now easy to see how the edges incident on a given vertex or the edges enclosing a given face can be obtained from the DCEL. If the graph has N vertices and F faces, we can assume we have two arrays $HV[1:N]$ and $HF[1:F]$ of headers of the vertex and face lists: these arrays can be filled by a scan of arrays $V1$ and $F1$ in time $O(N)$. The following straightforward procedure, VERTEX(j), obtains the sequence of edges incident on v_j as a sequence of addresses stored in an array A.

```
        procedure VERTEX(j)
begin a := HV[j];
      a_0 := a;
      A[1] := a;
      i := 2;
      if (V1[a] = j) then a := P1[a] else a := P2[a];
      while (a ≠ a_0) do
          begin A[i] := a;
                if (V1[a] = j) then a := P1[a] else a := P2[a];
                i := i + 1
      end
end.
```

Clearly VERTEX(j) runs in time proportional to the number of edges incident on v_j. Analogously, we can develop a procedure, FACE(j), which obtains the sequence of edges enclosing f_j, by replacing HV and $V1$ with HF and $F1$, respectively, in the above procedure VERTEX(j). Notice that the procedure VERTEX traces the edges counterclockwise about a vertex while FACE traces them clockwise about a face.

Frequently, a planar graph $G = (V, E)$ is represented in the edge-list form, which for each vertex $v_j \in V$ contains the list of its incident edges, arranged in the order in which they appear as one proceeds counterclockwise around v_j. It is easily shown that the edge-list representation of G can be transformed to the DCEL representation in time $O(|V|)$.

1.3 Geometric Preliminaries

1.3.1 General definitions and notations

The objects considered in Computational Geometry are normally sets of points in Euclidean space.[3] A coordinate system of reference is assumed, so that each point is represented as a vector of cartesian coordinates of the appropriate dimension. The geometric objects do not necessarily consist of finite sets of points, but must comply with the convention to be *finitely specifiable* (typically, as finite strings of parameters). So we shall consider, besides individual points, the straight line containing two given points, the straight line segment defined by its two extreme points, the plane containing three given points, the polygon defined by an (ordered) sequence or points, etc.

This section has no pretence of providing formal definitions of the geometric concepts used in this book; it has just the objectives of refreshing notions that are certainly known to the reader and of introducing the adopted notation.

By E^d we denote the *d-dimensional Euclidean space*, i.e., the space of the d-tuples $(x_1, \ldots, x_d)$ of real numbers x_i, $i = 1, \ldots, d$ with metric $(\sum_{i=1}^{d} x_i^2)^{1/2}$. We shall now review the definition of the principal objects considered by Computational Geometry.

Point. A d-tuple $(x_1, \ldots, x_d)$ denotes a point p of E^d; this point may be also

---

[3] The restriction to Euclidean Geometry (a special, but extremely important case of metric geometry) enables us to resort to our immediate experience, but it is also suggested by the fact that the vast majority of applications are formulated in Euclidean space. However, such a restriction is inessential for many of the applications to be considered in the next chapters. We shall later return to this important item (Section 1.3.2).

interpreted as a *d*-component vector applied to the origin of E^d, whose free terminus is the point p.

Line, plane, linear variety. Given two distinct points q_1 and q_2 in E^d, the linear combination

$$\alpha q_1 + (1 - \alpha)q_2 \qquad (\alpha \in \mathbb{R})$$

is a *line* in E^d. More generally, given k linearly independent points $q_1, \ldots, q_k$ in E^d ($k \leq d$), the linear combination

$$\alpha_1 q_1 + \alpha_2 q_2 + \cdots + \alpha_{k-1}q_{k-1} + (1 - \alpha_1 - \cdots - \alpha_{k-1})q_k$$

$$(\alpha_j \in \mathbb{R}, \quad j = 1, \ldots, k - 1)$$

is a *linear variety* of dimension $(k - 1)$ in E^d.

Line segment. Given two distinct points q_1 and q_2 in E^d, if in the expression $\alpha q_1 + (1 - \alpha)q_2$ we add the condition $0 \leq \alpha \leq 1$, we obtain the *convex combination* of q_1 and q_2, i.e.,

$$\alpha q_1 + (1 - \alpha)q_2 \qquad (\alpha \in \mathbb{R}, \quad 0 \leq \alpha \leq 1).$$

This convex combination describes the *straight line segment* joining the two points q_1 and q_2. Normally this segment is denoted as $\overline{q_1 q_2}$ (unordered pair).

Convex set. A domain D in E^d is *convex* if, for any two points q_1 and q_2 in D, the segment $\overline{q_1 q_2}$ is entirely contained in D.

It can be shown that the intersection of convex domains is a convex domain.

Convex hull. The *convex hull* of a set of points S in E^d is the boundary of the smallest convex domain in E^d containing S.

Polygon. In E^2 a *polygon* is defined by a finite set of segments such that every segment extreme is shared by exactly two edges and no subset of edges has the same property. The segments are the *edges* and their extremes are the *vertices* of the polygon. (Note that the number of vertices and edges are identical.) An *n*-vertex polygon is called an *n-gon*.

A polygon is *simple* if there is no pair of nonconsecutive edges sharing a point. A simple polygon partitions the plane into two disjoint regions, the *interior* (bounded) and the *exterior* (unbounded) that are separated by the polygon (*Jordan curve theorem*). (Note: in common parlance, the term polygon is frequently used to denote the union of the boundary and of the interior.)

A simple polygon P is *convex* if its interior is a convex set.

A simple polygon P is *star-shaped* if there exists a point z not external to P such that for all points p of P the line segment $\overline{zp}$ lies entirely within P. (Thus, each convex polygon is also star-shaped.) The locus of the points z having the above property is the *kernel* of P. (Thus, a convex polygon coincides with its own kernel.)

Planar graph. A graph $G = (V, E)$ (vertex set V, edge set E) is *planar* if it can be embedded in the plane without crossings (see Section 1.2.3.2). A straight line planar embedding of a planar graph determines a partition of the plane called *planar subdivision* or *map*. Let v, e, and f denote respectively the numbers of vertices, edges, and regions (including the single unbounded region) of the subdivision. These three parameters are related by the classical *Euler's formula* [Bollobás (1979)]

$$v - e + f = 2. \tag{1.1}$$

If we have the additional property that each vertex has degree ≥ 3, then it is a simple exercise to prove the following inequalities

$$\begin{cases} v \leq \frac{2}{3}e, & e \leq 3v - 6 \\ e \leq 3f - 6, & f \leq \frac{2}{3}e \\ v \leq 2f - 4, & f \leq 2v - 4 \end{cases} \tag{1.2}$$

which show that v, e and f are pairwise proportional. (Note that the three rightmost inequalities are unconditionally valid.)

Triangulation. A planar subdivision is a *triangulation* if all its bounded regions are triangles. A *triangulation of a finite set S* of points is a planar graph on S with the maximum number of edges (this is equivalent to saying that the triangulation of S is obtained by joining the points of S by nonintersecting straight line segments so that every region internal to the convex hull if S is a triangle.)

Polyhedron. In E^3 a *polyhedron* is defined by a finite set of plane polygons such that every edge of a polygon is shared by exactly one other polygon (adjacent polygons) and no subset of polygons has the same porperty. The vertices and the edges of the polygons are the *vertices* and the *edges* of the polyhedron; the polygons are the *facets* of the polyhedron.

A polyhedron is *simple* if there is no pair of nonadjacent facets sharing a point. A simple polyhedron partitions the space into two disjoint domains, the *interior* (bounded) and the *exterior* (unbounded). (Again, in common parlance the term polyhedron is frequently used to denote the union of the boundary and of the interior.)

The surface of a polyhedron is isomorphic to a planar subdivision. Thus the numbers v, e, and f of its vertices, edges, and facets obey Euler's formula (1.1).

A simple polyhedron is *convex* if its interior is a convex set.

1.3.2 Invariants under groups of linear transformations

Geometry could be approached in a purely axiomatic way, as a system consisting of sets of objects—such as points, lines, planes, etc.—and a collection of relations between them. Such objects need not bear an intuitive connection to our experience; indeed the correct handling of the axioms would lead to the discovery of the properties of the system. Such an approach, taken primarily by Hilbert at the end of the nineteenth century [Hilbert (1899)] has had an extraordinary influence on the development of the discipline. To contrast appropriately the axiomatic approach with the more traditional intuitive view, we just need to recall that the sets of points and lines could even be assumed to be finite.

From a more utilitarian viewpoint, however, we prefer to mitigate this position by seeing geometry as a rigorous abstraction from ordinary experience. In this view its basic constituents are models, founded upon intuition, of their physical counterparts—for example, the straight line is the abstraction of the path of a ray of light—and its results (theorems) are directly interpretable in experiential terms. This approach is not in contrast to the preceding one: it only qualifies the nature of the axioms by tying them to geometric intuition. It should be noted, however, that this restriction still leaves a wide scope: indeed, in some sense, the adoption of non-Euclidean geometry in the theory of relativity is still dictated by geometric intuition.

In our study of geometric algorithms, it is only natural to assume as our environment the Euclidean space. After all, the Euclidean space is a perfectly comfortable and adequate model for many domains of experience in which geometric applications arise. In particular, it is not by chance that the (Euclidean) plane and space are the realms of some of the most important uses of geometric methods (computer graphics, computer-aided-designs, etc.). We shall also assume that in the Euclidean space we have an orthonormal[4] cartesian system of coordinates.

It is interesting, however, to explore the extent of validity of individual algorithmic results beyond the confines of the Euclidean space. Specifically, we are interested in characterizing the classes of transformations of the space (and of a given instance of a problem in this space) that preserve the validity of given algorithms.

Less informally, a point in E^d, can be interpreted as the d-component vector $(x_1, x_2, \ldots, x_d)$ of its coordinate (also compactly denoted here by $\mathbf{x}$). We now consider mappings $T: E^d \to E^d$ that transform E^d into itself. In our discussion, such mappings are always to be interpreted as a movement of the points with respect to a fixed reference (the *alibi* interpretation), rather than as a change of coordinate system while the points remain fixed (the *alias* interpretation). (See, for example [Birkhoff–MacLane (1965)].) In particular, we are presently interested in *linear* mappings, i.e., mappings in which the coordinates of the *new* position of a point are described by linear expressions of the coordinates of the point's *old* position. Thus, if $(x_1, x_2, \ldots, x_d)$ are the coordinates of a point p, the image p' of p in the mapping T has coordinates

---

[4] That is, the reference axes are pairwise orthogonal and their respective unitary segments have identical length.

$$x_i' = \sum_{j=1}^{d} a_{ji} x_j + c_i \qquad (i = 1, 2, \ldots, d) \tag{1.3}$$

or, more compactly,

$$\mathbf{x}' = \mathbf{x} A + \mathbf{c} \tag{1.4}$$

where $A = \|a_{ij}\|$ is a $d \times d$ matrix, $\mathbf{c}$ is a fixed d-component vector, and all vectors are *row* vectors.

Equation (1.4) is the general form of an *affine mapping*. We can develop some preliminary intuition on affine mappings by separately considering the two cases $A = I$ (the identity matrix) and $\mathbf{c} = 0$. When $A = I$, Equation (1.4) becomes

$$\mathbf{x}' = \mathbf{x} + \mathbf{c}, \tag{1.5}$$

which describes a transformation where each point is subject to a fixed displacement $\mathbf{c}$: such transformations are naturally called *translations*. On the other hand, when $\mathbf{c} = 0$ we have

$$\mathbf{x}' = \mathbf{x} A$$

i.e., a linear transformation of the space which maps the origin to itself (i.e., the origin is a *fixed point*). Note that, in general, a d-dimensional affine mapping can be viewed as a $(d + 1)$-dimensional linear mapping in homogeneous coordinates by simply extending the vector of a point $(x_1, \ldots, x_d)$ by means of an additional component $x_{d+1} = 1$. Thus, Equation (1.4) can be rewritten as

$$(\mathbf{x}', 1) = (\mathbf{x}, 1) \begin{bmatrix} A & 0 \\ \mathbf{c} & 1 \end{bmatrix}. \tag{1.6}$$

A significant classification of affine mappings is based on the properties of the matrix A. In particular, different branches of geometry can be viewed as the study of the properties that are *invariant* under a set of transformations. This very significant approach was proposed by Klein about a century ago [Klein (1872)] and has now become a fundamental part of geometric teaching.

We begin by considering the case in which A is an arbitrary nonsingular matrix. Clearly, the set of all such transformations form a group (under composition), as can be easily verified.[5] This group is called the *affine group*, and *affine geometry* is concerned with the properties which are preserved (invariant) under transformations in this group. The basic invariant of affine geometry is *incidence*, that is, the membership of a point p on a line l.

Next we consider the specialization obtained by requiring

$$AA^T = \lambda^2 I, \tag{1.7}$$

---

[5] Closure, associativity, existence of the identity and of inverses are immediate consequences of analogous properties for the group of nonsingular $d \times d$ matrices.

where λ is a real constant (a superscript "T" denotes "transpose"). This property characterizes an important subgroup of the affine group, known as the *similarity group*. If (1.7) holds, then it is easy to verify that the ratio of distances between points are preserved (and so are angles and perpendicularity). Indeed, without loss of generality, we consider the case where $\mathbf{c} = 0$ in (1.4). The norm (square of the length) of a vector $\mathbf{x}$ (a segment with an extreme at the origin) is given by the inner product $\mathbf{x}\mathbf{x}^T$; thus the norm of the image $\mathbf{x}'$ of $\mathbf{x}$ under a similarity mapping is

$$\mathbf{x}'\mathbf{x}'^T = \mathbf{x}A(\mathbf{x}A)^T = \mathbf{x}AA^T\mathbf{x}^T = \mathbf{x}\lambda^2 I\mathbf{x}^T = \lambda^2 \mathbf{x}\mathbf{x}^T,$$

which shows that the vector $\mathbf{x}$ has been subjected to a dilatation by a factor $\pm\lambda$. Next, given two vectors $\mathbf{x}$ and $\mathbf{y}$, we have for the inner product of their images $\mathbf{x}'$ and $\mathbf{y}'$

$$\mathbf{x}'\mathbf{y}'^T = \lambda^2 \mathbf{x}\mathbf{y}^T.$$

Since $\mathbf{x}'\mathbf{y}'^T = |\mathbf{x}'|\cdot|\mathbf{y}'|\cos(\mathbf{x}',\mathbf{y}')$ and $\mathbf{x}\mathbf{y}^T = |\mathbf{x}|\cdot|\mathbf{y}|\cos(\mathbf{x},\mathbf{y})$, we have

$$\cos(\mathbf{x}',\mathbf{y}') = \cos(\mathbf{x},\mathbf{y}),$$

which substantiates the earlier claim.

A further specialization (of the similarity group) is obtained by requiring

$$|A| = \pm 1,$$

where $|A|$ denotes the determinant of A. The subgroup of transformations, characterized by this property, is called *orthogonal*. The invariant of the orthogonal group is *distance*. Indeed, the determinant of $A \cdot A^T$ is

$$|A \cdot A^T| = |A|\cdot|A^T| = |A|^2 = 1;$$

but, by (1.7), $|AA^T| = \lambda^2$, whence $\lambda = \pm 1$. The distance $d(\mathbf{x},\mathbf{y})$ between two points $\mathbf{x}$ and $\mathbf{y}$ is given by the absolute value of the square root of the norm of their difference, that is,

$$d(\mathbf{x},\mathbf{y}) = \sqrt{(\mathbf{x} - \mathbf{y})\cdot(\mathbf{x} - \mathbf{y})^T}.$$

Therefore we have

$$d(\mathbf{x}',\mathbf{y}') = \sqrt{(\mathbf{x}' - \mathbf{y}')\cdot(\mathbf{x}' - \mathbf{y}')^T} = \sqrt{(\mathbf{x} - \mathbf{y})AA^T(\mathbf{x} - \mathbf{y})^T}$$
$$= \sqrt{(\mathbf{x} - \mathbf{y})\cdot(\mathbf{x} - \mathbf{y})^T} = d(\mathbf{x},\mathbf{y}).$$

The affine transformations that preserve distance (and therefore, also preserve area, angles, and perpendicularity) are the *rigid motions*, and form the foundations of Euclidean geometry.

We now return to Equation (1.6) describing a particular transformation of a $(d + 1)$-dimensional vector space and remove any restriction on the form of the last column of the transformation matrix. Thus we obtain the relation

$$\xi' = \xi B \tag{1.8}$$

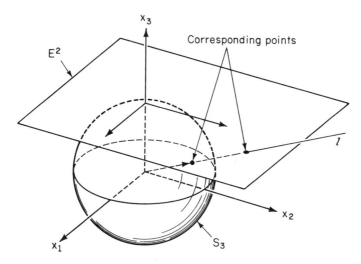

Figure 1.4 Illustration of the correspondence between homogeneous and inhomo-genous (conventional) coordinates.

where ξ' and ξ are Euclidean vectors of $(d + 1)$-dimensional space V_{d+1} and B is a $(d + 1) \times (d + 1)$ matrix, assumed to be nonsingular.

We restrict our attention to the direction of a vector, rather than to its length: this means that two collinear vectors ξ and $c\xi$ ($c \neq 0$) are considered equivalent. Thus, as representative of this equivalence class we may choose the point where the line $l = \{c\xi : c \in \mathbb{R}\}$ pierces the unit sphere S^{d+1} of E^{d+1} (note that S^{d+1} is a d-dimensional variety, i.e., its points are identified by d parameters). If we now choose the value of c for which the last component of $c\xi$ is equal to 1, we obtain the point where the line l pierces the hyperplane $x_{d+1} = 1$ (see Figure 1.4 for $d = 2$). So we have a one-to-one correspondence between points of the plane $x_{d+1} = 1$ and points of the hemisphere of S^{d+1} corresponding to $x_{d+1} > 0$; this correspondence is called a *central projection* (from the origin of E^{d+1}). Notice that the hyperplane $x_{d+1} = 1$ is itself a space E^d of coordinates $x_1, \ldots, x_d$.

We thus have accomplished the interpretation of a vector $(\xi_1, \ldots, \xi_d, \xi_{d+1})$ ($\xi_{d+1} \neq 0$) applied to the origin of E^{d+1} as a point $(x_1, \ldots, x_d)$ of E^d, so that $x_j = \xi_j/\xi_{d+1}$. If we represent the point by means of the $(d + 1)$ components of the vector we obtain the classical representation of points in *homogeneous coordinates*, which enables us to represent the points at infinity by letting $\xi_{d+1} = 0$.

Returning to the consideration of the action performed by the group of transformations, if in (1.8) we let B be of the form $B = \begin{bmatrix} A & 0 \\ c & 1 \end{bmatrix}$, we obtain a new interpretation of the group of affinities of E^d. For concreteness and ease of reference, we consider the case $d = 2$. We note that the central projection of S^3 to the plane $x_3 = 1$ establishes an invertible transformation between great circles of S^3 and lines of E^2, except for the "equatorial" circle (on $x_3 = 0$), which maps to the line at infinity of E^2. However, any transformation (1.8)

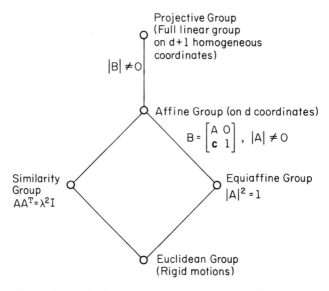

Figure 1.5   Illustration of the inclusion relation on groups of linear transformations (Hasse diagram).

maps a plane by the origin to a plane by the origin, in particular the equatorial plane can be mapped to any other plane by the origin. If we interpret this in the central projection to $x_3 = 1$, the line at infinity can be mapped to any line at finite. This freedom enables us to map any conic curve (ellipse, parabola, hyperbola) to a circle, so that all conics are equivalent under the group transformation (1.8). This group is referred to as the *projective group*.

The preceding discussion is conveniently summarized by means of a Hasse diagram [Birkhoff–MacLane (1965)] of the relation "subgroup of," illustrated in Figure 1.5.

1.3.3 Geometric duality. Polarity

We now consider an alternative interpretation of the transformation (1.8), again referring for concreteness to the case $d = 2$. So far, we have seen that (1.8) maps points to points and lines to lines in E^2, i.e., it preserves the dimension of the objects being mapped. Assuming $|B| \neq 0$, we rewrite (1.8) as

$$\boldsymbol{\eta} = \boldsymbol{\xi} B, \tag{1.9}$$

and hereafter interpret vectors denoted by the letter "$\boldsymbol{\xi}$" as representing a direction (i.e., a line by the origin) and vectors denoted by the letter "$\boldsymbol{\eta}$" as representing the normal to a plane passing by the origin. Thus relation (1.9) is interpreted as the transformation in E^3 of a line (represented by $\boldsymbol{\xi}$) to a plane (represented by $\boldsymbol{\eta}$) and is called a correlation. This is a special instance for $d = 2$ of the general property that in E^{d+1} relation (1.9) maps a linear variety of dimension $s \leq d + 1$ to its *dual* linear variety of dimension $d + 1 - s$: in this interpretation (1.9) is referred to as a *duality transformation*, and indeed, the same matrix B can be used to describe a correlation that maps planes to lines. Given a line-to-line correlation described by a matrix B, we seek a plane-to-line correlation described by a matrix D so that the pair (B, D) preserves *incidence*, that is, if line $\boldsymbol{\xi}$ belongs to plane $\boldsymbol{\eta}$, then line $\boldsymbol{\eta}D$ belongs to plane $\boldsymbol{\xi}B$. *Clearly line $\boldsymbol{\xi}$ belongs to plane $\boldsymbol{\eta}$ if and only if*

$$\boldsymbol{\xi} \cdot \boldsymbol{\eta}^T = 0.$$

Correspondingly we require

$$\boldsymbol{\xi}B(\boldsymbol{\eta}D)^T = \boldsymbol{\xi}BD^T\boldsymbol{\eta}^T = 0.$$

Since the latter holds for any choice of $\boldsymbol{\xi}$ and $\boldsymbol{\eta}$, we obtain $BD^T = kI$ or

$$D = k(B^{-1})^T,$$

for some constant k.

Note that the product BD maps lines to lines and planes to planes. Given the incidence-preserving pair $(B, (B^{-1})^T)$, we now further require that the product $B \cdot (B^{-1})^T$ map each line to itself and each plane to itself, i.e., we require $B(B^{-1})^T = kI$ or, equivalently, $B = kB^T$. In the latter identity we must have $k = \pm 1$; of particular interest is the case where

$$B = B^T, \tag{1.10}$$

i.e., B is a nonsingular 3×3 symmetric matrix.

We now consider the effect of a correlation such as (1.9) in the central projection to the plane $x_3 = 1$. We recognize that points are mapped to lines and lines to points; in the remainder of this section we shall further analyze this mapping in E^2, and refer to $\boldsymbol{\xi}$ as a point and to $\boldsymbol{\eta}$ as a line. If the 3×3 matrix B satisfies the condition $B = B^T$, letting $\mathbf{x} = (x_1, x_2, x_3)$ it is well known [Birkhoff–Maclane (1965)] that

$$\mathbf{x}B\mathbf{x}^T = 0$$

is the equation in homogeneous coordinates of a conic in the plane (referred to as the *conic defined by B*). Consider now a fixed $\boldsymbol{\xi}$ (interpreted as the homogeneous coordinate representation of a point in the plane) such that $\boldsymbol{\xi}B\boldsymbol{\xi}^T = 0$. By definition, point $\boldsymbol{\xi}$ lies on the conic defined by B; if we call point $\boldsymbol{\xi}$ the *pole* and line $\boldsymbol{\xi}B\mathbf{x}^T = 0$ the *polar* of ξ, we see that a pole on the conic is incident to its own polar. This mapping of points to lines, and *vice versa*, is called a *polarity*, and is useful in the development of geometric algorithms. Indeed, our intuition is more attuned to dealing with points than with lines (in E^2) or planes (in E^3), and we shall capitalize on this ability in later chapters.

In particular, let us choose B as the matrix defining the unit circle in the plane E^2 (refer to Figure 1.6). In this case

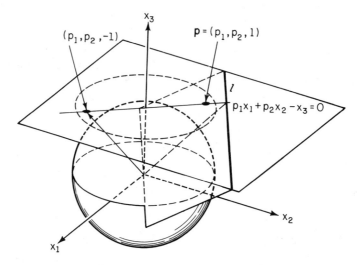

Figure 1.6   Illustration of the correspondence between pole and polar in the polarity with respect to the unit circle.

$$B = \begin{bmatrix} 1 & 0 & 0 \\ 0 & 1 & 0 \\ 0 & 0 & -1 \end{bmatrix},$$

that is, a point $\mathbf{p} = (p_1, p_2, 1)$ is mapped to a line l whose equation is $p_1 x_1 + p_2 x_2 - x_3 = 0$ (in homogeneous coordinates). The distance of $\mathbf{p}$ from the origin of the plane is $\sqrt{p_1^2 + p_2^2}$, while the distance of l from the origin is $1/\sqrt{p_1^2 + p_2^2}$. This shows that in this special polarity

$$\text{distance}(\mathbf{p}, \mathbf{0}) \times \text{distance}(\text{polar}(\mathbf{p}), \mathbf{0}) = 1.$$

Notice also that in this polarity the transformation involves no computation but simply a different interpretation of a triplet (p_1, p_2, p_3), either as the coordinates $(p_1/p_3, p_2/p_3)$ of a point or as the coefficients of the straight line whose equation is $p_1 x_1 + p_2 x_2 - p_3 x_3 = 0$ in homogeneous coordinates.

In a later chapter we shall examine interesting applications of this geometric duality.

1.4 Models of Computation

A basic methodological question, which must be confronted as a premise to any study of algorithms, is the careful specification of the adopted *computational model*. Indeed, an algorithm proposed to solve a given problem must be evaluated in terms of its *cost* as a function of the size of the instance of the problem. The fundamental importance of the model is established by the following observation:

> *A model of computation specifies the primitive operations that may be executed and their respective costs.*

The *primitive operations* are those for each of which we charge a *fixed* cost, although this cost may vary from operation to operation. For example, if the primitive operations involve individual digits of numbers (such as the boolean functions of two binary variables, as holds in the Turing machine model) the cost of an addition of two integers grows with the operand length, whereas this cost is constant in a model where the operands have a fixed length (as in any model that aims at representing actual digital computers). In choosing a model we normally must make compromises between realism and mathematical tractability, selecting a scheme that captures the essential features of the computational devices to be used, and yet it is sufficiently simple to permit a thorough analysis.

But what sort of model is appropriate for geometric applications? To develop an answer to this central question, we must examine carefully the nature of the problems to be considered.

As noted in Section 1.3, the points in E^d are fundamental objects considered by Computational Geometry. Although a point is viewed as a vector of cartesian coordinates, it is desirable that the choice of the coordinate system do not significantly affect the running time of any geometric algorithm. This implies that the model of computation must allow the necessary transformation (of cartesian reference) at a cost per point that depends possibly upon the number of dimensions but *not* upon the number of points involved.

Stated another way, a set of N points in d dimensions can be expressed with respect to a chosen cartesian reference system in time proportional to N (recall that of the same order is the time used to "read" the points as inputs), so that we may assume at the onset that the points are already given in the chosen cartesian reference.

The problems encountered in computational geometry are of several types, which—for our current purposes—may be conveniently grouped into the three following categories:

1. *Subset selection.* In this kind of problems we are given a collection of objects and asked to choose a subset that satisfies a certain property. Examples are finding the two closest of a set of N points or finding the vertices of its convex hull. The essential feature of a subset selection problem is that no new objects need be created; the solution consists entirely of elements that are given as input.

2. *Computation.* Given a set of objects, we need to compute the value of some geometric parameter of this set. The primitives allowed in the model must be powerful enough to permit this calculation. Suppose, for example, that

we are working with a set of points having integer coordinates. In order to find the distance between a pair of points we not only need to be able to represent irrational numbers, but to take square roots as well. In other problems we may even require trigonometric functions.

3. *Decision*. Naturally associated with any "Subset Selection" or "Computation" problem there is a "Decision" problem. Specifically:

(i) If computation problem $\mathscr{A}$ requests the value of a parameter A, the associated decision problem $D(\mathscr{A})$, requests a YES/NO answer to a question of the type: "Is $A \geq A_0$?," where A_0 is a constant.

(ii) If subset selection problem $\mathscr{A}$ requests the subset of a given set S satisfying a certain property P, $D(\mathscr{A})$ requests a YES/NO answer to a question of the type: "Does set S' satisfy P?," where S' is a subset of S.

We readily observe that we must be able to deal with *real* numbers (not just integers), and—in a practical computer—with suitable approximations thereof. In constructing the model, however, we are allowed the conceptual abstraction to dispense with the round-off problem in the approximate representation of real numbers; specifically, we will adopt a *random-access machine* (RAM) similar to that described in [Aho–Hopcroft–Ullman (74)], but in which *each storage location is capable of holding a single real number. The following operations are primitive and are available at unit cost (unit time)*:

1. The arithmetic operations $(+, -, \times, /)$.
2. Comparisons between two real numbers $(<, \leq, =, \neq, \geq, >)$.
3. Indirect addressing of memory (integer addresses only).

Optional (for applications requiring them):

4. k-th root, trigonometric functions, EXP, and LOG (in general, analytic functions).

This model will be referred to as the *real RAM*. It closely reflects the kinds of programs that are typically written in high-level algebraic languages such as FORTRAN and ALGOL, in which it is common to treat variables of type REAL as having unlimited precision. At this level of abstraction we may ignore such questions as how a real number can be read or written in finite time.

The establishment of lower bounds to performance measures for a given problem is one of the fundamental objectives in algorithmic analysis, because it provides a gauge for evaluating the efficiency of algorithms. This is, in general, a very difficult task. It is sometimes possible to relate the difficulty of a problem to that of another problem of known difficulty by use of the technique of *transformation of problems*.[6] Specifically, suppose we have two problems, problem $\mathscr{A}$ and problem $\mathscr{B}$, which are related so that problem $\mathscr{A}$

---

[6] This technique is very frequently referred to as *reduction*. Since "reduction" seems to imply the transformation to a *simpler* problem (which is not the case) we prefer to avoid the connotation.

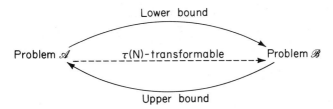

Figure 1.7   Transfer of upper and lower bounds between transformable problems.

can be solved as follows:
1. The input to problem $\mathscr{A}$ is converted into a suitable input to problem $\mathscr{B}$.
2. Problem $\mathscr{B}$ is solved.
3. The output of problem $\mathscr{B}$ is transformed into a correct solution to problem $\mathscr{A}$.

We then say that problem $\mathscr{A}$ *has been transformed* to problem $\mathscr{B}$.[7] If the above transformation Steps 1 and 3 together can be performed in $O(\tau(N))$ time—where N is as usual the "size" of problem $\mathscr{A}$—then we say that $\mathscr{A}$ is $\tau((N)$- *transformable to* $\mathscr{B}$ and express it concisely as

$$\mathscr{A} \propto_{\tau(N)} \mathscr{B}.$$

In general, transformability is not a symmetric relation; in the special case when $\mathscr{A}$ and $\mathscr{B}$ are mutually transformable, we say that they are *equivalent*.

The following two propositions characterize the power of the transformation technique, under the assumption that the transformation preserves the order of the problem size, and are self-evident.

Proposition 1 (Lower Bounds via Transformability). *If problem $\mathscr{A}$ is known to require $T(N)$ time and $\mathscr{A}$ is $\tau(N)$-transformable to $\mathscr{B}$ ($\mathscr{A} \propto_{\tau(N)} \mathscr{B}$), then $\mathscr{B}$ requires* at least $T(N) - O(\tau(N))$ *time.*

Proposition 2 (Upper Bounds via Transformability). *If problem $\mathscr{B}$ can be solved in $T(N)$ time and problem $\mathscr{A}$ is $\tau(N)$-transformable to $\mathscr{B}$ ($\mathscr{A} \propto_{\tau(N)} \mathscr{B}$), then $\mathscr{A}$ can be solved in* at most $T(N) + O(\tau(N))$ *time.*

The situation is pictorially illustrated in Figure 1.7, which shows how lower and upper bounds are transferred from one problem to the other. This transfer holds whenever $\tau(N) = O(T(N))$, i.e., when the transformation time does not dominate the computation time.

Referring to our previous classification of problems, let us consider some

---

[7] Transformability (or reducibility) is usually defined to be a relation on languages, in which case no output transformation is necessary because the output of a string acceptor is either zero or one. For geometry problems we need the greater flexibility afforded by the more general definition.

problem $\mathscr{A}$ (either a computation or a subset selection problem) and its associated decision problem $D(\mathscr{A})$. It is immediate to recognize that $D(\mathscr{A}) \propto_{O(N)} \mathscr{A}$, because:

1. If $\mathscr{A}$ is a computation problem, no input transformation is needed (so Step 1 of the transformation procedure is void) and the solution to $\mathscr{A}$ must be just compared, in constant time $O(1)$, to the fixed value supplied by $D(\mathscr{A})$.
2. If $\mathscr{A}$ is a subset selection problem, again the input S' of $D(\mathscr{A})$ is supplied as input to $\mathscr{A}$ (so Step 1 is void) and the solution to $\mathscr{A}$ is checked in time $O(N)$ to verify that its cardinality coincides with that of S'.

This is a crucial observation, because it indicates that when aiming at lower bounds, we may restrict our attention to "decision problems".

When the "real RAM" executes an algorithm for a decision problem, its behavior is described as a sequence of operations of two types: arithmetic and comparisons. In this sequence, comparisons play a fundamental role, because, depending upon the outcome, the algorithm has a branching option at each comparison. The situation is pictorially illustrated in Figure 1.8, where each round node represents an arithmetic operation, and each triangular node represents a comparison. In other words the computation executed by the RAM may be viewed as a path in a rooted tree. This rooted tree embodies the description of an extremely important computation model, called the "algebraic decision tree," which we now formalize.

An *algebraic decision tree* [Reingold (1972); Rabin (1972); Dobkin–Lipton (1979)] on a set of variables $\{x_1, \ldots, x_n\}$ is a program with statements $L_1, L_2, \ldots, L_p$ of the form:

1. L_r: Compute $f(x_1, \ldots, x_n)$; if f: 0 then go to L_i else go to L_j (: denotes any comparison relation);
2. L_s: Halt and output YES (accepted input in decision problem);
3. L_t: Halt and output NO (rejected input in decision problem).

In 1, f is *an algebraic function* (a polynomial of degree $degree(f)$). The program is further assumed to be loop-free, i.e., it has the structure of a *tree T*, such that each nonleaf node v is described by

$$f_v(x_1, \ldots, x_n): 0,$$

where f_v is a polynomial in $x_1, \ldots, x_n$ and : denotes a comparison relation. (Note that, in this description, each "computation path" in Figure 1.8 has been compressed into the next comparison node.) The root of T represents the initial step of the computation and its leaves represent the possible terminations and contain the possible answers. Without loss of generality, we assume that the tree T is *binary*.[8]

---

[8] Note that the degree of the nodes of T is the multiplicity of alternative outcomes of the comparisons. The binary tree hypothesis is based on the fact that a k-way branching can be resolved into $(k-1)$ 2-way branchings.

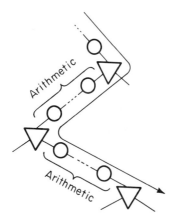

Figure 1.8   A computation as a path in a decision tree.

Although an algebraic decision tree program may be extremely less compact than the corresponding RAM program, both programs behave *identically* on the classes of problems we are considering. In particular, the worst-
case running time of the RAM program is at least proportional to the length
of the longest path from the root in the decision tree. This substantiates the
importance of the decision-tree model, because the tree structure is quite
amenable to the derivation of bounds on its depth.

An algebraic decision tree is said to be of the *d*-th *order* if *d* is the maximum
degree of the polynomials $f_v(x_1, \ldots, x_n)$ for each node v of T. The 1-st order, or
linear, decision tree model has been a very powerful tool in establishing lower
bounds to a variety of problems; we shall discuss the very clever arguments
that have been developed in the subsequent chapters, in the appropriate
context (see Sections 2.2, 4.1.3, 5.2, 8.4). However, there are two reasons for
dissatisfaction with linear decision trees. The first, and more fundamental, is
that it may happen that any known algorithm for the problem under consideration uses functions of degree ≥ 2, so that a lower-bound based on the linear
decision tree model is not significant; barring this situation—and this is the
second reason—a bound based on the linear decision tree model does not
apply to yet unknown algorithms using higher degree functions.

Extremely important contributions to settle this matter for $d \geq 2$ have been
recently made by Steele–Yao (1982) and Ben-Or (1983), using classical concepts of real algebraic geometry. Their approach is based on the following
very terse idea: Let $x_1, x_2, \ldots, x_n$ be the parameters of the decision problem,
each instance of which may be viewed as a point of the n-dimensional Euclidean space E^n. The decision problem identifies a set of points $W \subseteq E^n$, that
is, it provides a YES-answer if and only if $(x_1, \ldots, x_n) \in W$ (we say then that
the *decision tree T solves the membership problem for W*). Suppose that we
independently know, by the nature of the problem, the number $\#(W)$ of
disjoint connected components of W. Each computation corresponds to a

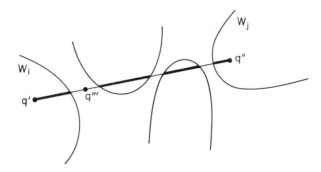

Figure 1.9   Illustration for the proof that $W_i = W_j$.

unique path $v_1, v_2, \ldots, v_{l-1}, v_l$ in T, where v_1 is the root and v_l is a leaf; associated with each vertex v_j of this path there is a function $f_{v_j}(x_1, \ldots, x_n)$, $(j = 1, \ldots, l - 1)$, so that $(x_1, \ldots, x_n)$ satisfies a constraint of the type

$$f_{v_j} = 0 \quad \text{or} \quad f_{v_j} \geq 0 \quad \text{or} \quad f_{v_j} > 0. \tag{1.11}$$

To gain some intuition, we first consider the specialized argument for $d = 1$ (linear decision or computation tree model) due to Dobkin and Lipton (1979). The argument goes as follows. Let $W \subseteq E^n$ be the membership set for a given decision problem, and let $\#(W)$ denote the number of its disjoint connected components. Let T be the (binary) linear decision tree embodying the al- gorithm $\mathscr{A}$ that tests membership in W. Associated with each leaf of T is a domain of E^n, and each leaf is either "accepting" or "rejecting." Specifically, let $\{W_1, \ldots, W_p\}$ be the components of W, $\{l_1, \ldots, l_r\}$ be the set of leaves, and D_j the domain associated with l_j. Leaf l_j is classified as

$$\begin{cases} \text{accepting,} & \text{if } D_j \subseteq W \\ \text{rejecting,} & \text{otherwise.} \end{cases}$$

A lower bound to r is obtained by showing $r \geq \#(W)$. Indeed, we construct a function $Y: \{W_1, W_2, \ldots, W_p\} \to \{1, 2, \ldots, r\}$ defined by $Y(W_i) = \min \{j: j \in \{1, 2, \ldots, r\}$ and $D_j \cap W_i \neq \varnothing\}$. Suppose now, for a contradiction, that there are two distinct subsets, W_i and W_j, such that $Y(W_i) = Y(W_j) = h$. Since algorithm $\mathscr{A}$ solves the membership of a point q in W_i, leaf l_h is accepting. On the other hand, by the definition of Y, $Y(W_i) = h$ implies that $W_i \cap D_h$ is nonempty. Therefore, let q' be a point in $W_i \cap D_h$. Similarly, q'' is a point in $W_j \cap D_h$. Since T is a *linear* decision tree, the region of E^n corresponding to D_h is the intersection of half-spaces, that is, it is a *convex* region. This implies that any convex combination of points of D_h belongs to D_h (Section 1.3.1). Consider now the segment $\overline{q'q''}$ (refer to Figure 1.9). This segment intersects at least two components of W (it certainly intersects W_i and W_j), and, since these components are disjoint, it contains at least one point $q''' \notin W$. But, by con- vexity, the entire $\overline{q'q''}$ belongs to D_h, and so does q''', which is therefore declared internal to W, a contradiction. In conclusion, T must have at least

p leaves. Since the shallowest binary tree with a given number of leaves is balanced, the depth of T is at least $\log_2 p = \log_2 \# (W)$. This is summarized in the following theorem.

Theorem 1.1 (Dobkin–Lipton). *Any linear decision tree algorithm that solves the membership problem in $W \subseteq E^n$ must have depth at least $\log_2 \# (W)$, where $\# (W)$ is the number of disjoint connected components of W.*

Unfortunately, the above technique is restricted to linear computation tree algorithms, since it rests crucially on the property that the domain of E^n associated with a leaf of the tree is convex. When the maximum degree of the polynomials f_v is ≥ 2 this useful property disappears. Therefore more sophisticate concepts are needed.

Intuitively, when higher degree polynomials are used the domain associated with a specific leaf of the decision tree may consist of *several* disjoint components of W. If we succeed in bounding the number of components assigned to a leaf in terms of the leaf's depth, we shall shortly see that we succeed in bounding the depth of T.

The key to solve this difficult problem is fortunately provided by a clever adaptation [Steele–Yao (1982); Ben-Or (1983)] of a classical result in algebraic geometry independently proved by Milnor (1964) and Thom (1965). Milnor and Thom's result goes as follows.

Let V be the set of points (technically called an *algebraic variety*) in the m-dimensional cartesian space E^m defined by polynomial equations

$$g_1(x_1, \ldots, x_m) = 0, \ldots, g_p(x_1, \ldots, x_m) = 0. \tag{1.12}$$

Then if the degree of each polynomial g_i ($i = 1, \ldots, p$) is $\leq d$, the number $\# (V)$ of disjoint connected components[9] of V is bounded as

$$\# (V) \leq d(2d - 1)^{m-1}.$$

Unfortunately, V is defined in terms of equations (see (1.12)) whereas the constraints associated with a path in T consists of both equations and inequalities. This difficulty has been adeptly circumvented by Ben-Or by transforming our situation to meet the hypothesis of the Milnor–Thom Theorem, as we now illustrate.

Define $U \subseteq E^n$ as the set of points satisfying the following constraints (here $\mathbf{x} = (x_1, \ldots, x_n)$):

$$\begin{cases} q_1(\mathbf{x}) = 0, \ldots, q_r(\mathbf{x}) = 0 \\ p_1(\mathbf{x}) > 0, \ldots, p_s(\mathbf{x}) > 0 \\ p_{s+1}(\mathbf{x}) \geq 0, \ldots, p_h(\mathbf{x}) \geq 0 \end{cases} \tag{1.13}$$

---

[9] In reality, Milnor and Thom prove the stronger result that $d(2d - 1)^{m-1}$ is an upper bound to the sum of the Betti numbers of V, whereas the number $\# (V)$ of connected components is only one such number (the zero-th Betti number).

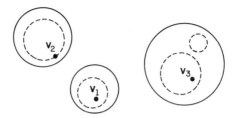

Figure 1.10   Construction of U_ε from U; U is shown with solid lines, U_ε with dotted lines.

where the q_i's and p_j's are polynomials and $d = \max\{2, \text{degree}(q_i), \text{degree}(p_j)\}$. Note that we have three types of constraints: equations, open inequalities, and closed inequalities. Let $\#(U)$ denote the number of connected components of set U.

The first step of the transformation consists of replacing the open inequalities with closed inequalities. Since $\#(U)$ is finite (see [Milnor (1968)]), we let $\#(U) \triangleq t$ and pick a point in each component of U. Denoting these points $\mathbf{v}_1, \ldots, \mathbf{v}_t$, we let

$$\varepsilon = \min\{p_i(\mathbf{v}_j): i = 1, \ldots, s; j = 1, \ldots, t\}.$$

Since each $\mathbf{v}_j \in U$, then $p_i(\mathbf{v}_j) > 0$ (for $i = 1, \ldots, s$) and $\varepsilon > 0$. Clearly the closed point set U_ε defined by

$$\begin{cases} q_1(\mathbf{x}) = 0, \ldots, q_r(\mathbf{x}) = 0 \\ p_1(\mathbf{x}) \geq \varepsilon, \ldots, p_s(\mathbf{x}) \geq \varepsilon \\ p_{s+1}(\mathbf{x}) \geq 0, \ldots, p_h(\mathbf{x}) \geq 0 \end{cases} \qquad (1.14)$$

is contained in U, and $\#(U_\varepsilon) \geq \#(U)$, since each $\mathbf{v}_j$ is in a distinct component of U_ε (see, for an intuitive illustration, Figure 1.10).

The second step of the transformation consists of introducing a slack variable y_j for each $j = 1, \ldots, h$ (i.e., for each one of the closed inequalities), so that E^n is viewed as a subspace of E^{n+h} as

$$\begin{cases} q_1(\mathbf{x}) = 0, \ldots, q_r(\mathbf{x}) = 0 \\ p_1(\mathbf{x}) - \varepsilon - y_1^2 = 0, \ldots, p_s(\mathbf{x}) - \varepsilon - y_s^2 = 0 \\ p_{s+1}(\mathbf{x}) - y_{s+1}^2 = 0, \ldots, p_h(\mathbf{x}) - y_h^2 = 0. \end{cases} \qquad (1.15)$$

Clearly, the set U^* of all solutions to (1.15) is an algebraic variety in E^{n+h} defined by polynomials of degree at most d and satisfies the conditions of the Milnor–Thom Theorem; therefore, the number $\#(U^*)$ of connected components of U^* is bounded as

$$\#(U^*) \leq d(2d - 1)^{n+h-1}.$$

Notice that U_ε is the projection of U^* into E^n; since the projection is a continuous function, $\#(U_\varepsilon) \leq \#(U^*)$. Recalling that $\#(U) \leq \#(U_\varepsilon)$, we conclude that

$$\#(U) \le \#(U_\varepsilon) \le \#(U^*) \le d(2d-1)^{n+h-1}$$

which is the sought result. Indeed, if (1.13) is the set of constraints obtained in traversing a root-to-leaf path in T, h—the number of inequalities—is at most as large as the path length. It follows that the leaf of this path has associated with itself at most $d(2d-1)^{n+h-1}$ connected components of the solution set W of the problem. If h^* is the depth of T (the length of the longest root-to-leaf path), then T has at most 2^{h^*} leaves; each leaf accounts for at most $d(2d-1)^{n+h^*-1}$ components of W, whence

$$\#(W) \le 2^{h^*} d(2d-1)^{n+h^*-1}$$

or, equivalently

$$h^* \ge \frac{\log_2 \#(W)}{1 + \log_2(2d-1)} - \frac{n \log_2(2d-1)}{1 + \log_2(2d-1)} - \frac{\log_2 d + \log_2(2d-1)}{1 + \log_2(2d-1)}.$$

$$(1.16)$$

We summarize this important result as a theorem.

Theorem 1.2.[10] *Let W be a set in the cartesian space E^n and let T be an algebraic decision tree of fixed order d ($d \ge 2$) that solves the membership problem in W. If h^* is the depth of T and $\#(W)$ is the number of disjoint connected components of W, then $h^* = \Omega(\log \#(W) - n)$.*

Theorem 1.2 subsumes the case $d = 1$, since any fixed $d \ge 2$ implies the bound for polynomials of lower degree (by just setting the higher order coefficients to zero). In essence, the use of superlinear polynomial decision functions does not radically change the nature of the problem; it merely shortens the minimum depth of the computation trees by a multiplicative constant dependent upon the maximum degree d. The last theorem is the cornerstone of lower-bound arguments to be presented in later chapters.

---

[10] In reality, Ben-Or proves a somewhat stronger result, independent of d.

CHAPTER 2
Geometric Searching

To describe searching in its simplest abstract setting, suppose we have some accumulated data (called the "file") and some new data item (called the "sample"). Searching consists of relating the sample to the file. Accessory operations—conceptually not a part of searching—may involve absorbing the sample into the file, deleting the sample from the file if already present, and so on. As Knuth (1973) points out, searching means to locate the appropriate record (or records) in a given collection of records.

Although all searching problems share some basic common feature, the specific geometric setting gives rise to questions of a somewhat unique flavor. First of all, in geometric applications, files are primarily not just "collections," as in several other information processing activities. Rather, they are frequently representations of more complex structures, such as polygons, polyhedra, and the like. Even the case, say, of a collection of segments in the plane appears deceptively unstructured: indeed, each segment is associated with its end-point coordinates, which are implicitly related by the metric structure of the plane.

The fact that coordinates of points are thought of as real numbers entails that a search may frequently not find a file item matching the sample, but rather it will *locate* the latter in relation to the file. This provides an additional point of contrast between geometric searching and conventional searching.

2.1 Introduction to Geometric Searching

This chapter develops the basic tools of geometric searching that will be used in the succeeding chapters to solve rather formidable problems. The search message, performing the interrogation of the file, is normally called *query*.

The types of the file and of the admissible queries will greatly determine the organization of the former and the algorithms for handling the latter. A concrete example will provide the necessary motivation into this important facet of the problem.

Suppose we have a collection of geometric data, and we want to know if it possesses a certain property (say, convexity). In the simplest case the question will only be asked once, in which event it would be wasteful to do any preconditioning in the hope of speeding up future queries. A one-time query of this type will be referred to as *single-shot*. Many times, however, queries will be performed repeatedly on the same file. Such queries will be referred to as *repetitive-mode queries*.

In this case it may be worthwhile to arrange the information into an organized structure to facilitate searching. This can be accomplished only at some expense, though, and our analysis must focus on four separate cost measures:

1. *Query time.* How much time is required, in both the average and worst cases, to respond to a single query?
2. *Storage.* How much memory is required for the data structure?
3. *Preprocessing time.* How much time is needed to arrange the data for searching?
4. *Update time.* Given a specific item, how long will it take to add it to or to delete it from the data structure?

The various trade-offs among query time, preprocessing time, and storage are well-illustrated by the following instance of the problem of *range searching* [Knuth (1973), p. 550],[1] which arises frequently in geographic applications and database management:

PROBLEM S.1 (RANGE SEARCHING—COUNT). Given N points in the plane, how many lie in a given rectangle with sides parallel to the coordinate axes?[2] That is, how many points (x, y) satisfy $a \leq x \leq b, c \leq y \leq d$ for given $a, b, c,$ and d? (See Figure 2.1.)

It is clear that a single-shot range query can be performed in (optimal) linear time, since we need examine only each of the N points to see whether it satisfies the inequalities defining the rectangle. Likewise, linear space suffices because only the $2N$ coordinates need to be saved. There is no preprocessing time and the update time for a new point is constant. What kind of data structure can be used to speed the processing of repetitive-mode queries? It seems too difficult to organize the points so that an arbitrary new rectangle can be accommodated easily. We also cannot solve the problem in advance for *all* possible rectangles because of their infinite number. The following solution

---

[1] This problem will be studied extensively in Section 2.3.

[2] The rectangle is assumed to have its sides parallel to the coordinate axes.

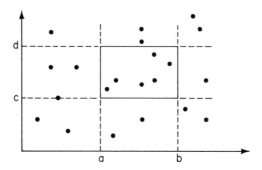

Figure 2.1   A Range Query. How many points lie in the rectangle?

is an example of the *locus method* of attacking geometry problems: the query is mapped to a point in a suitable search space and the search space is subdivided into regions (*loci*) within which the answer does not vary. In other words, if we call equivalent two queries producing the same answer, each region of the partition of the search space corresponds to an equivalence class of queries.

A rectangle itself is an unwieldy object; we would prefer to deal with points. This suggests, for example, that we might replace the rectangle query by four subproblems, one for each vertex, and combine their solutions to obtain the final answer. In this case the subproblem associated with a point p is to determine the number of points $Q(p)$ of the set that satisfy both $x \leq x(p)$, and $y \leq y(p)$, that is, the number of points in the southwest quadrant determined by p. (See Figure 2.2.)

The concept we are dealing with here is that of *vector dominance*. (See also Section 4.1.3.) We say that a point (vector) $\mathbf{v}$ *dominates* $\mathbf{w}$ if and only if and for all indices i, $v_i \geq w_i$. In the plane , $\mathbf{w}$ is dominated by $\mathbf{v}$ if and only if it lies in $\mathbf{v}$'s southwest quadrant. $Q(p)$ is thus the number of points dominated by p. The connection between dominance and range queries is apparent in Figure 2.3.

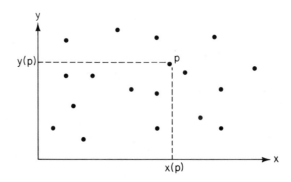

Figure 2.2   How many points lie to the southwest of p?

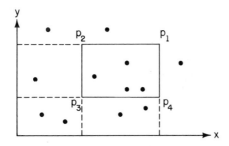

Figure 2.3   A range search as four dominance queries.

The number $N(p_1p_2p_3p_4)$ of points contained in rectangle $p_1p_2p_3p_4$ is given by

$$N(p_1p_2p_3p_4) = Q(p_1) - Q(p_2) - Q(p_4) + Q(p_3). \qquad (2.1)$$

This follows from the combinatorial principle of inclusion–exclusion, [Liu (1968)]: all points in the rectangle are certainly dominated by p_1. We must remove those dominated by p_2 and also those dominated by p_4, but this will cause some points to be eliminated twice—specifically, the ones dominated by both p_2 and p_4 and these are just the points lying in p_3's southwest quadrant.

We have thus reduced the problem of range searching to one of performing four point dominance queries. The property that makes these queries easy is that there are nicely shaped regions of the plane within which the dominance number Q is constant.

Suppose we drop perpendiculars from the points to the x- and y-axes, and extend the resulting lines indefinitely. This produces a mesh of $(N + 1)^2$ rectangles, as shown in Figure 2.4.

For all points p in any given rectangle, $Q(p)$ is a constant. This means that dominance searching is just a matter of determining which region of a rectilinear mesh a given point lies in. This question is particularly easy to answer. Having sorted the points on both coordinates, we need only perform two binary searches, one on each axis, to find which rectangle contains the point. Thus the query time is only $O(\log N)$. Unfortunately, there are $O(N^2)$ rectangles, so quadratic storage is required. We must compute, of course, the dominance number for each rectangle. This can readily be done for any single rectangle in $O(N)$ time, which would lead to an overall $O(N^3)$ time for preprocessing; however, by a less naive approach,[3] the preprocessing time can be reduced to $O(N^2)$.

If the storage and preprocessing requirements of the above approach are found to be excessive, then one may wish to investigate alternative approaches. Typically, these investigations expose a general feature of algorithm

---

[3] This work on range searching is reported in [Bentley–Shamos (1977)].

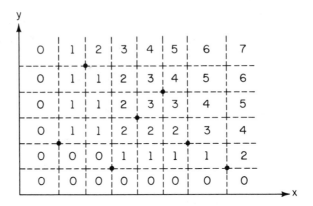

Figure 2.4   Mesh of rectangles for dominance searching.

Table I

Query	Storage	Preprocessing	Comments
$O(\log N)$	$O(N^2)$	$O(N^2)$	Above method
$O(\log^2 N)$	$O(N \log N)$	$O(N \log N)$	[4]
$O(N)$	$O(N)$	$O(N)$	No preprocessing

design, that is, a trade-off among different measures of performance. Our range searching problem is no exception; the trade-off materializes in the combinations of resources illustrated in Table I.

From the preceding discussion, two major paradigms of geometric searching problems emerge:

1. *Location problems*, where the file represents a partition of the geometric space into regions, and the query is a point. Location consists of identifying the region the query point lies in.
2. *Range-search problems*, where the file represents a collection of points in space, and the query is some standard geometric shape arbitrarily translatable in space (typically, the query in 3-space is a ball or a box). Range-search consists either of retrieving (*report* problems) or of counting (*census* or *count* problems) all points contained within the query domain.

Although the techniques used for solving the two types of problems are not independent—we have just seen that range searching can be transformed into a point-location problem—it is convenient to illustrate them separately. The rest of this chapter is devoted to this task.

---

[4] This result, also derived in [Bentley–Shamos (1977)], will be illustrated in Section 2.3.4.

2.2 Point-Location Problems

2.2.1 General considerations. Simple cases

Point-location problems could be appropriately renamed *point-inclusion problems*. Indeed, "point p lies in region R" is synonymous with "point p is contained in region R." Of course, the difficulty of the task will essentially depend upon the nature of the space and of its partition.

As is typical of so much of computational geometry at this time, the planar problem is well understood, while very little is known for E^3 and even less for higher number of dimensions.

Even in the more restricted context of the plane, the nature of the partition to be searched determines the complexity of the task, when evaluated under the three basic measures outlined in the preceding section. We shall begin by considering plane partitions—or *planar subdivisions*, as they are normally called—constructed with straight-line segments; subsequently, we shall relax this seemingly restrictive assumption. Before proceeding, however, we must examine a technical detail: The planar subdivisions to be considered are such that, after removing from the plane the region boundaries, the resulting open sets are connected. This is to ensure that two points of the same region be connected by a curve which does not contain any point of the boundary.

The first problem that comes to mind is when the plane is subdivided into just two regions, one of which is unbounded and the other is a polygon P. Clearly, P is a *simple* polygon, as a consequence of the two-region partition and of the Jordan Curve Theorem for polygons (see Section 1.3.1). Thus we can formulate:

PROBLEM S.2 (POLYGON INCLUSION). Given a simple polygon P and a point z, determine whether or not z is internal to P.

Here again, the difficulty of the problem depends on whether P, besides simplicity, possesses some other useful attribute. Intuitively, a *convex* polygon appears as a somewhat simpler object. Therefore we consider

PROBLEM S.3 (CONVEX POLYGON INCLUSION). Given a convex polygon P and a point z, is z internal to P?

We can dispose of the single-shot approach immediately, and the result holds for nonconvex polygons as well.

Theorem 2.1. *Whether a point z is internal to a simple N-gon P can be determined in $O(N)$ time, without preprocessing.*

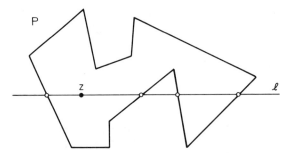

Figure 2.5   Single-shot inclusion in a simple polygon. There is one intersection of l with P to the left of z, so z is inside the polygon.

PROOF. Consider the horizontal line l passing through z (see Figure 2.5). By the Jordan Curve theorem, the interior and exterior of P are well defined. If l does not intersect P, then z is external. So, assume that l intersects P and consider first the case that l does not pass by any vertex of P. Let L be the number of intersections of l with the boundary of P to the left of z. Since P is bounded, the left extremity of l lies in the exterior of P. Consider moving right on l from $-\infty$, toward z. At the leftmost crossing with the boundary of P, we move into the interior; at the next crossing, we return outside again, and so on. Thus z is internal if and only if L is odd. Next consider the degenerate situation where l passes through vertices of P. An infinitesimal counterclockwise rotation of l around z does not change the classification of z (internal/external) but removes the degeneracy. Thus, if we imagine to perform this infinitesimal rotation, we recognize: if both vertices of an edge belong to l, this edge must be ignored; if just one vertex of an edge belongs to l, then it must be counted if it is the vertex with larger ordinate, and ignored otherwise. In summary, we have the following algorithm.

begin $L := 0$;
        for $i := 1$ **until** N **do if** edge (i) is not horizontal **then**
                if (lower extreme of edge (i) intersects l to the left of z)
                    then $L := L + 1$;
        if (L is odd) **then** z is internal **else** z is external
end.

It is now straightforward to recognize that this simple algorithm runs in $O(N)$ time.   □

For queries in the repetitive mode, we first consider the case in which P is convex. The method relies on the convexity of P, specifically, on the property that the vertices of a convex polygon occur in angular order about any internal point. Any such point q can be easily found, for example, as the centroid of the triangle determined by any three vertices of P. We then consider the N rays from q which pass through the vertices of P (Figure 2.6).

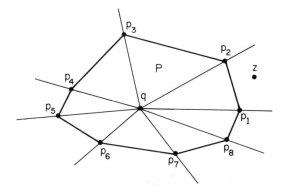

Figure 2.6   Division into wedges for the convex inclusion problem. 1. By binary search we learn that z lies in wedge $p_1 q p_2$. 2. By comparing z against edge $\overline{p_1 p_2}$ we find that it is external.

These rays partition the plane into N wedges. Each wedge is divided into two pieces by a single edge of P. One of these pieces is wholly internal to P, the other wholly external. Treating q as the origin of polar coordinates, we may find the wedge in which z lies by a single binary search, since the rays occur in angular order. Given the wedge, we need only compare z to the unique edge of P that cuts it, and we will learn whether z is internal to P.

The preprocessing for the outlined approach consists of finding q as the centroid of three vertices of P and of arranging the vertices $p_1, p_2, \ldots, p_N$ in a data structure suitable for binary search (a vector, for example). This is obviously done in $O(N)$ time, since the sequence $(p_1, \ldots, p_N)$ is given. Next we turn our attention to the search procedure:

Procedure CONVEX INCLUSION
1. Given a query point z, determine by binary search the wedge in which it lies. Point z lies between the rays defined by p_i and p_{i+1} if and only if angle (zqp_{i+1}) is a right turn and angle (zqp_i) is a left turn.[5]
2. Once p_i and p_{i+1} are found, then z is internal if and only if $(p_i p_{i+1} z)$ is a left turn.

Theorem 2.2. *The inclusion question for a convex N-gon can be answered in $O(\log N)$ time, given $O(N)$ space and $O(N)$ preprocessing time.*

---

[5] Note that to decide whether angle $(p_1 p_2 p_3)$ is right or left turn corresponds to evaluating a 3×3 determinant in the points' coordinates. Specifically, letting $P_i = (x_i, y_i)$, the determinant

$$\begin{vmatrix} x_1 & y_1 & 1 \\ x_2 & y_2 & 1 \\ x_3 & y_3 & 1 \end{vmatrix}$$

gives twice the *signed area* of the triangle $(p_1 p_2 p_3)$, where the sign is $+$ if and only if $(p_1 p_2 p_3)$ form a counterclockwise cycle. So, $p_1 p_2 p_3$ is a left turn if and only if the determinant is positive.

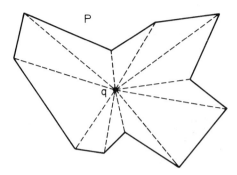

Figure 2.7   A star-shaped polygon.

What property do convex polygons possess that enables them to be searched quickly? In order to be able to apply binary search the vertices must occur in sequence about some point. One immediately realizes that convexity is only a sufficient condition for this property to hold. Indeed there is a larger class of simple polygons, containing the convex polygons, which exhibits this property: this is the class of *star-shaped* polygons (see Section 1.3.1). Indeed, a star-shaped polygon P (Figure 2.7) contains at least one point q such that $\overline{qp_i}$ lies entirely within P for any vertex P_i of P, $i = 1, \ldots, N$.

To determine whether or not a point is internal to a star-shaped polygon, we may use the preceding algorithm directly, if an appropriate origin q from which to base the search can be found. The set of feasible origins within P has been defined in Section 1.3.1 as the *kernel* of P. The construction of the kernel of a simple N-gon is beyong the scope of this chapter: we shall return to this problem in Section 7.2.6, where we will show that—surprisingly—the kernel can be found in $O(N)$ time. At present, we assume that the kernel of P is known (if nonempty), so that we can select a reference point about which to create wedges for inclusion searching. We have

Theorem 2.3. *The inclusion question for an N-vertex star-shaped polygon can be answered in $O(\log N)$ time and $O(N)$ storage, after $O(N)$ preprocessing time.*

We can now turn our attention to unrestricted simple polygons, which we shall call "general polygons." There is a hierarchy that is strictly ordered by the subset relation

$$\text{CONVEX} \subset \text{STAR-SHAPED} \subset \text{GENERAL}. \qquad (2.2)$$

We have just seen that star-shaped inclusion is asymptotically no more difficult than convex inclusion. What can we say for the general case? One approach to the problem is motivated by the fact that every simple polygon is the union of some number of special polygons—such as star-shaped or convex polygons, or, for that matter, triangles. Unfortunately, when decomposing a simple N-vertex polygon into polygons of some special type, the minimum cardinality k of the decomposition may itself be $O(N)$.[6] Therefore, this approach ends up transforming an N-vertex simple polygon into an N-vertex planar graph embedded in the plane. It appears therefore that inclusion in a simple polygon is no easier problem than the apparently more general location of a point in a planar subdivision, although no proof of equivalence is known. Thus we shall now concentrate on the latter problem.

2.2.2 Location of a point in a planar subdivision

We recalled in Chapter 1 that a planar graph can always be embedded in the plane so that its edges are mapped to straight-line segments. Such embedded graphs will be referred to as *planar straight-line graphs*, or PSLG. A PSLG determines in general a subdivision of the plane; if the PSLG contains no vertex of degree < 2, then it is straightforward to realize that all the bounded regions of the subdivision are simple polygons. Without loss of generality, here and hereafter, G is assumed to be connected.

To be able to search such a structure, one may attempt to decompose each region into polygons for which the search operation is relatively simple. Neglecting for a moment the complexity of achieving the desired decomposition—which may indeed be nontrivial—the success of point location depends on the ability to narrow down quickly the set of components to be searched. On the other hand, the fastest known search methods are based on *bisection*, or binary search. One easily realizes, after a little reflection, that from the viewpoint of query time the ability to use binary search is more important than the minimization of the size of the set to be searched, due to the logarithmic dependence of the search time upon the latter. From this qualitative discussion one learns that a potentially successful point-location method should exhibit the following features: the planar subdivision is efficiently transformed into a new one, each of whose constituent polygons intersects a small and fixed number (possibly just one) of original regions, and can be organized so that binary search is applicable. In other words, the fundamental idea is to *create new geometric objects* to *permit binary searching* and is to be credited in this context to Dobkin and Lipton (1976). All of the known methods to be described next are essentially inspired by this principle.

2.2.2.1 The slab method

Given a PSLG G, consider drawing a horizontal line through each of its vertices, as in Figure 2.8. This divides the plane into $N + 1$ horizontal strips, referred to as "slabs." If we sort these slabs by y-coordinate as part of the preprocessing, we will be able to find in $O(\log N)$ time the slab in which a query point z lies.

---

[6] This is trivially so if the special polygons are triangles, since $k = N - 3$. But even for star-shaped components, Chvatal has shown that k may be as large as $\lceil N/3 \rceil$ [Chvatal (1975)].

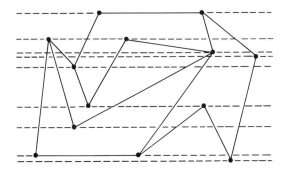

Figure 2.8   The vertices of the PSLG define the horizontal slabs.

Now consider the intersection of a slab with G, which consists of segments of the edges of G. These segments define trapezoids.[7] Since G is a planar embedding of a planar graph its edges intersect only at vertices, and since each vertex defines a slab boundary, no segments intersect within a slab. (See Figure 2.9.)

The segments can thus be totally ordered from left to right and we may use binary search to determine in $O(\log N)$ time the trapezoid in which z falls. This will give a worst-case query time of $O(\log N)$.

It only remains to analyze how much work is done in preconditioning the PSLG and storing it. Naively, it seems that we must sort all of the line segments in every slab. Furthermore, each slab may have $O(N)$ segments, so it appears that $O(N^2 \log N)$ time and $O(N^2)$ storage will be required. We will next show how to reduce the preprocessing time to $O(N^2)$. Nothing can be done (in this algorithm) to reduce the storage used since there exist PSLGs that need quadratic space (Figure 2.10).

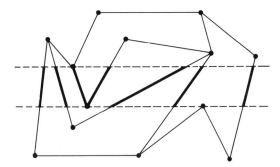

Figure 2.9   Within a slab, segments do not intersect.

---

[7] The trapezoids can obviously degenerate into triangles, as shown in Figure 2.9.

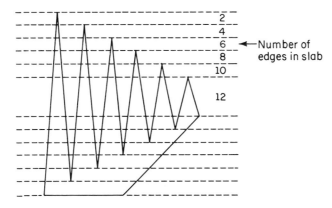

Figure 2.10   The slabs may contain a total of $O(N^2)$ segments.

Notice that if an edge of the PSLG passes through several slabs, these slabs are consecutive. This observation is the key that allows us to reduce the preprocessing time, since we may use the plane-sweep technique (see Section 1.2.2). We recall that a plane-sweep algorithm is characterized by two basic structures: (i) the *event-point schedule*, i.e., the sequence of positions to be assumed by the sweep-line, and (ii) the *sweep-line status*, i.e., the description of the intersection of the sweep-line with the geometric structure being swept. In our case, if the plane sweep proceeds, say, from bottom to top, the instantaneous sweep-line status is the left-to-right sequence of edges that traverse the slab containing the sweep line. This sequence, i.e., the left-to-right order of these edges, remains unaltered within the slab, but changes at the next slab boundary, when a new vertex v of the PSLG is reached; indeed, at this point, the edges terminating at v are deleted and are replaced by those issuing from v. The sweep-line status can be maintained as a height-balanced tree (e.g., a 2-3-tree), which, as is well-known, supports INSERT and DELETE operations in time logarithmic in its size. In addition, the plane sweep scans the slabs in ascending order, i.e., the event-point schedule is just the bottom-to-top sequence of the vertices of the PSLG. At each event point, the sweep-line status is updated and read out: indeed this read-out is precisely the segment sequence inside the next slab. From the complexity viewpoint, the work consists of the insertion and deletion of each edge—at a cost $O(\log N)$ per operation—and of the generation of the output; the former is $O(N \log N)$, since by Euler's theorem there are $O(N)$ edges in an N-vertex planar graph, while the latter has size $O(N^2)$, as noted earlier. In the preprocessing algorithm, intuition is helped if edges are thought of as directed from bottom to top. Vertices are stored in an array VERTEX, ordered by increasing y; $B[i]$ is the set of edges incident on VERTEX[i] from below (incoming edges), ordered counterclockwise; $A[i]$ is the set of edges issuing from VERTEX[i], ordered clockwise. Edges are maintained in a height-balanced tree L. Here is the algorithm:

    procedure PREPROCESSING-FOR-PLANAR-POINT-
           LOCATION
begin VERTEX$[1:2N]$:= Sort the vertices of G by increasing y;
    $L := \emptyset$;
    for $i := 1$ **until** N **do**
        begin DELETE$(B[i])$;
              INSERT$(A[i])$;
              Output L
    end
end.

Thus we have

Theorem 2.4. *Point-location in an N-vertex planar subdivision can be effected in* $O(\log N)$ *time using* $O(N^2)$ *storage, given* $O(N^2)$ *preprocessing time.*

Although this method shows optimal query time behavior, its preprocessing time and—even more—its storage requirement are rather objectionable and, for many applications, unacceptable. It was conjectured [Shamos (1975a)] that the storage requirement could be reduced to $O(N)$ while possibly increasing the query time to only $O(\log^2 N)$. This indeed was verified shortly thereafter, and will be illustrated next. This suboptimal method has been the basis of significant later developments [Edelsbrunner–Guibas–Stolfi (1985)].

2.2.2.2 The chain method

As the slab method achieves efficient searching by decomposing the original subdivision into trapezoids, the chain method of Lee and Preparata (1978) attains the same goal by using as its basic component a new type of polygon, called *monotone*, to be defined below.

The key to this method is the notion of "chain" as given by the following.

Definition 2.1. A *chain* $C = (u_1, \ldots, u_p)$ is a PSLG with vertex set $\{u_1, \ldots, u_p\}$ and edge set $\{(u_i, u_{i+1}): i = 1, \ldots, p - 1\}$.

In other words, a chain, as given by the above definition, is the planar embedding of a graph-theoretic chain and is also appropriately called a *polygonal line* [see Figure 2.11 (a)].

Consider the planar subdivision induced by a PSLG G. Suppose that in G we find a chain C (a subgraph of G), of one of the following types: either (i) C is a cycle, or (ii) both extremes u_1 and u_p of C belong to the boundary of the unbounded region. In the latter case, we extend C on both extremes with semi-infinite parallel edges. In both instances C effectively partitions the subdivision into two portions. Moreover, we shall see soon that the operation of deciding on which side of C a query point z lies—called *discrimination of z*

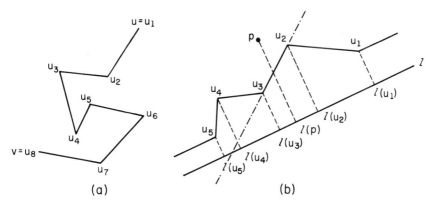

Figure 2.11 Examples of chains: (a) general, (b) monotone with respect to line l.

against C—is relatively simple. Then, if C can be chosen so that the two parts are of comparable complexity, by recursive splits we obtain a point location method which performs a logarithmic number of discriminations. This immediately poses two questions: (i) How easy is the discrimination of a point against an arbitrary chain, and are there chains for which the discrimination is easy? (ii) What is the difficulty of finding a suitable splitting chain?

With respect to the first question, one readily realizes that to discriminate against an arbitrary chain is basically the same problem as to test for inclusion in a general polygon. Thus, one should look for a restricted class of chains. One such class is provided by the following definition.

Definition 2.2. A chain $C = (u_1, \ldots, u_p)$ is said to be *monotone* with respect to a straight line l if a line orthogonal to l intersects C in exactly one point.

In other words, a monotone chain C is of type (ii) as defined above, and the orthogonal projections $\{l(u_1), \ldots, l(u_p)\}$ of the vertices of C on l are ordered as $l(u_1), \ldots, l(u_p)$.

One such chain is illustrated in Figure 2.11(b). It is clear that, given a query point z, the projection $l(z)$ of z on l can be located with a binary search in a unique interval $(l(u_i), l(u_{i+1}))$. Next, a single linear test determines on which side of the line containing $\overline{u_i u_{i+1}}$ the query point z lies. Thus, the discrimination of a point against a p-vertex chain is effected in time $O(\log p)$.[8]

This efficiency is an enticement toward the use of monotone chains in point location. Suppose that there is a set $\mathscr{C} = \{C_1, \ldots, C_r\}$ of chains montone with respect to the same line l and with the following additional properties:

Property 1. *The union of the members of $\mathscr{C}$ contains the given* PSLG[9];

---

[8] Observe that monotone chains can be viewed as limiting cases of star-shaped polygons.

[9] Notice that a given edge of G could belong to more than one chain in $\mathscr{C}$.

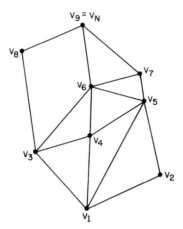

Figure 2.12   An example of a regular PSLG.

Property 2. *For any two chains C_i and C_j of $\mathscr{C}$, the vertices of C_i which are not vertices of C_j lie on the same side of C_j.*

Such set $\mathscr{C}$ is referred to as a *monotone complete set of chains of G.* Note that Property 2 means that the chains of a complete set are ordered. Therefore, one can apply the bisection principle (i.e., binary search) to $\mathscr{C}$, where the primitive, rather than a simple comparison, is now a point-chain discrimination. Thus, if there are r chains in $\mathscr{C}$ and the longest chain has p vertices, then the search uses worst-case time $O(\log p \cdot \log r)$.

But a crucial question is: given PSLG G, does it admit a monotone complete set of chains? The answer is readily found to be negative. However, we shall constructively show that a PSLG admits a monotone complete set of chains, provided it satisfies a rather weak requirement. In addition, we shall show that an arbitrary PSLG can be easily transformed into one to which the chain construction procedure is applicable. This transformation creates some new "artificial" regions, with no harm, however, to the effective solution of the point location problem.

The weak requirement is expressed by the following definition.

Definition 2.3. Let G be a PSLG with vertex set $\{v_1, \ldots, v_N\}$, where the vertices are indexed so that $i < j$ if and only if either $y(v_i) < y(v_j)$ or, if $y(v_i) = y(v_j)$, then $x(v_i) > x(v_j)$. A vertex v_j is said to be *regular* if there are integers $i < j < k$ such that (v_i, v_j) and (v_j, v_k) are edges of G. Graph G is said to be regular if each y_j is regular for $1 < j < N$ (i.e., with the exception of the two extreme vertices v_1 and v_N).

Figure 2.12 shows an example of a regular graph. We now show that a regular graph admits a complete set of chains monotone with respect to the y-axis. (Hereafter line l is assumed to be the y-axis of the plane.)

To aid the intuition, imagine edge (v_i, v_j) to be directed from v_i to v_j if $i < j$. Thus, we can speak of the sets $\text{IN}(v_j)$ and $\text{OUT}(v_j)$ of incoming and outgoing edges, respectively, of vertex v_j. The edges in $\text{IN}(v_j)$ are assumed to be ordered counterclockwise, while those in $\text{OUT}(v_j)$ are ordered clockwise. Due to the hypothesis of regularity, both these sets are nonempty for each nonextreme vertex. This fact enables us to show that, for any $v_j (j \neq 1)$, we can construct a y-monotone chain from v_1 to v_j: this is trivial for $j = 2$. Assume the statement is true for $k < j$. Since v_j is regular, by Definition 2.3, there is some $i < j$ such that (v_i, v_j) is an edge of G. But, by the inductive hypothesis, there is a chain C from v_1 to v_i monotone with respect to the y-axis; clearly, the concatenation of C and (v_i, v_j) is a y-monotone chain. To complete the proof we must show that Properties 1 and 2 above hold. Let $W(e)$, the *weight* of edge e, be the number of chains to which e belongs. In addition, we let

$$W_{\text{IN}}(v) = \sum_{e \in \text{IN}(v)} W(e)$$

$$W_{\text{OUT}}(v) = \sum_{e \in \text{OUT}(v)} W(e).$$

Then all that is needed is to show that the edge weights can be chosen so that

(1) each edge has positive weight;
(2) for each $v_j (j \neq 1, N)$, $W_{\text{IN}}(v_j) = W_{\text{OUT}}(v_j)$.

Condition (1) ensures that each edge belongs to at least one chain (Property 1), while Condition (2) guarantees that $W_{\text{IN}}(v_j)$ chains pass through v_j and can be chosen so that they do not cross (Property 2). The realization of $W_{\text{IN}} = W_{\text{OUT}}$ is a rather classical flow problem [Even (1979)], and can be achieved by two passes over G. After setting $W(e) = 1$ for each edge e, in the first pass— from v_1 to v_N—we achieve $W_{\text{IN}}(v_i) \leq W_{\text{OUT}}(v_i)$ for each nonextreme v_i. In the second pass—from v_N to v_1—we achieve $W_{\text{IN}}(v_i) \geq W_{\text{OUT}}(v_i)$, i.e., the desired balancing. Letting $v_{\text{IN}}(v) = |\text{IN}(v)|$ and $v_{\text{OUT}}(v) = |\text{OUT}(v)|$, here is the algorithm:

procedure WEIGHT-BALANCING IN REGULAR PSLG
begin for each edge e **do** $W(e) := 1$; (∗initialization∗)
    for $i := 2$ **until** $N - 1$ **do**
      begin $W_{\text{IN}}(v_i) :=$ sum of weights of incoming edges of v_i;
        $d_1 :=$ leftmost outgoing edge of v_i;
        if $(W_{\text{IN}}(v_i) > v_{\text{OUT}}(v_i))$ **then** $W(d_1) := W_{\text{IN}}(v_i) - v_{\text{OUT}}(v_i) + 1$
      end (∗of first pass∗);
    for $i := N - 1$ **until** 2 **do**
      begin $W_{\text{OUT}}(v_i) :=$ sum of weights of outgoing edges of v_i;
        $d_2 :=$ leftmost incoming edge of v_i;
        if $(W_{\text{OUT}}(v_i) > W_{\text{IN}}(v_i))$ **then** $W(d_1) := W_{\text{OUT}}(v_i) -$
          $W_{\text{IN}}(v_i) + W(d_2)$
      end (∗of second pass∗)
end.

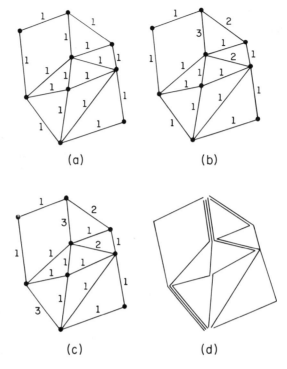

Figure 2.13 Construction of $\mathscr{C}$ for the PSLG of Figure 2.12. Edge labels are the edge weights after initialization (a), first pass (b), and second pass (c). The set of chains is displayed in (d).

This algorithm clearly runs in time linear in the number of edges and vertices of G. The procedure can be appropriately modified so that the construction of $\mathscr{C}$, i.e., the assignment of edges to chains, is done during the second pass. For brevity we omit these details and show in Figure 2.13(a)–(d) the edge weight configurations after the initialization, the first pass, and the second pass, together with the display of $\mathscr{C}$ for the PSLG of Figure 2.12. This completes the proof that a regular PSLG admits of a monotone complete set of chains.

Before analyzing the performance of the method, it is important to illustrate how to transform an arbitrary PSLG into a regular one. Consider a nonregular vertex v of G that, say, has no outgoing edge (see Figure 2.14). A horizontal line through v intercepts, in general, two edges e_1 and e_2 of G adjacent to v on either side. (Note that since v is nonextreme, at least one of these intercepts must exist.) Let v_i be the upper endpoint of edge e_i $(i = 1, 2)$, and let v^* be the one with smaller ordinate (v_2 in our example). Then the segment $\overline{vv^*}$ does not cross any edge of G and therefore can be added to the PSLG, thereby "regularizing" vertex v. This observation is crucial to the technique, which is a vertical plane sweep (see Section 1.2.2). Specifically, we

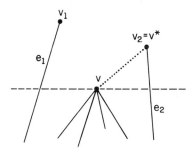

Figure 2.14   Illustrations of a typical nonregular vertex v.

sweep from top to bottom to regularize vertices with no outgoing edge and from bottom to top to regularize the other type. Considering just the first case, the event-point schedule is just the sequence $(v_N, v_{N-1}, \dots, v_1)$ of vertices. The sweep-line status structure (to be realized as a height-balanced tree) gives the left-to-right order of the intersections of the sweep-line with the PSLG and, in addition, it keeps for each interval in this partition a vertex (with lowest ordinate in that interval). In the sweep, for each vertex v reached we perform the following operations: (1) locate v in an interval in the status structure (by abscissa); (2) update the status structure; (3) if v is not regular, add an edge from v to the vertex associated with the interval determined in (1). We leave it as an exercise to construct a more formal algorithm. Notice that *regularization* of an N-vertex PSLG is accomplished in time $O(N \log N)$, due to the possible initial sorting of the vertex ordinates and to the location of N vertices in the status structure, each at $O(\log N)$ cost. We state this fact as a theorem for future reference.

Theorem 2.5. *An N-vertex PSLG can be regularized in time $O(N \log N)$ and space $O(N)$.*

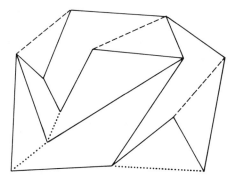

Figure 2.15   A regularized PSLG. Broken-line edges are added in the top-to-bottom sweep; dotted edges in the bottom-to-top sweep.

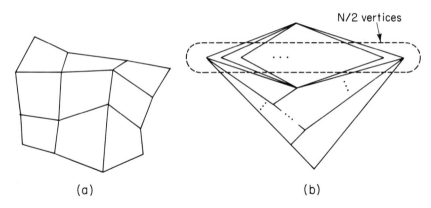

Figure 2.16   Examples of worst-case PSLGs.

In Figure 2.15 we show the result of regularization on the PSLG of Figure 2.8.

After establishing that the chain method is applicable—possibly after regularization—to any PSLG, we now analyze its performance. We have already established an $O(\log p \cdot \log r) = O(\log^2 N)$ worst-case upper bound to the search time. This bound is attainable, since there are N-vertex PSLGs with $O(\sqrt{N})$ chains, each with $O(\sqrt{N})$ edges (see Figure 2.16(a) for $N = 16$).

Next consider the PSLG of Figure 2.16(b). This graph contains, in its complete chain set, $N/2$ chains each of which has $N/2$ edges. At first sight, this example is worrisome as regards the storage requirements of the method, since it seems to indicate a discouraging $O(N^2)$ behavior! Things are not nearly as bad, however, due to the particular use the algorithm makes of chains. Indeed, chains are used in a binary search scheme. A binary search algorithm on a totally ordered set S induces a natural hierarchy on S, represented by a rooted binary tree; an actual search corresponds to tracing a root-to-leaf path in this tree. If we naturally number the chains, say, from left to right, then if an edge e belongs to more than one chain it belongs to all members of a set (an interval) of consecutive chains. Now suppose we assign the chains to the nodes a binary search tree. If an edge e is shared by the chains of an interval, then there is a unique member C^* in this interval which, in the search tree, is a common ascendant[10] of all other members of the interval. Let C be any such other member; then the discrimination of a point z against C is preceded—in the binary search scheme—by the discrimination of z against C^*. It follows that edge e may be assigned to C^* alone, and precisely it will be assigned to the hierarchically highest chain to which it belongs. As an example, this assignment is illustrated in Figure 2.17. Notice the presence of bypass pointers in Figure 2.17(c); however there are no more bypass pointers than there

---

[10] According to standard terminology, a vertex v_i of a rooted tree T is an *ascendant* or *ancestor* of a vertex v_j if there is a path from the root of T containing both v_i and v_j and v_i is closer to the root.

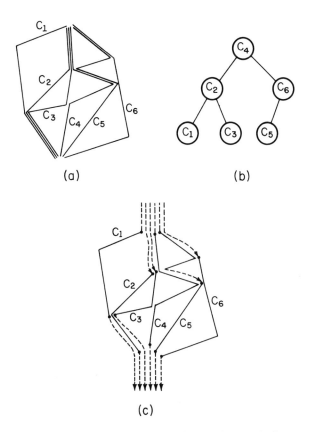

Figure 2.17 A complete set of monotone chains (a) is searched according to the hierarchy expressed by a binary tree (b). Bypass pointers are shown as broken lines (c).

are edges, whence we conclude that the entire search structure can be stored in $O(N)$ space.

As to preprocessing, we conservatively assume that the PSLG G is given as a DCEL data structure, as described in Section 1.2.3.2. Since in time linear in its size we can obtain the edge cycle around any vertex v of G, the construction of $IN(v)$ and $OUT(v)$, for each v, is completed in time $O(N)$.

The weight-balancing procedure also runs, as we observed, in time $O(N)$. Considering that a possible preliminary regularization pass also uses time $O(N \log N)$, we conclude that the preprocessing task is completed in time $O(N \log N)$. In summary

Theorem 2.6. *Point-location in an N-vertex planar subdivision can be effected in* $O(\log^2 N)$ *time using* $O(N)$ *storage, given* $O(N \log N)$ *preprocessing time.*

At the beginning of this section, we indicated that the chain method makes use of a new simple type of simple polygon. Such polygons are characterized by the following definition.

Definition 2.4. A simple polygon is said to be *monotone* if its boundary can be decomposed into two chains monotone with respect to the same straight line.

Monotone polygons are very interesting geometric objects. It follows trivially from the present discussion that inclusion in an N-vertex monotone polygon can be decided in time $O(\log N)$; we slall see later that both to triangulate a monotone polygon and to decide if a simple polygon is monotone are tasks that can be carried out in optimal time $O(N)$.

2.2.2.3 Optimal techniques: the planar-separator method and the triangulation refinement method

The existence of a method exhibiting $O(\log N)$ search time and using less than quadratic storage has been for quite some time an outstanding problem. This question was settled affirmatively by the remarkable construction of Lipton and Tarjan (1977a, 1977b, 1980). Emphasis must be placed on the phrase "existence of a method." Indeed, the technique is so formidable and its performance estimates are so generously favored by the "big oh" notation that Lipton and Tarjan themselves expressed the following caveat (1977b): "We do not advocate this algorithm as a practical one, but its existence suggests that there may be a practical algorithm with $O(\log N)$ time bound and $O(N)$ space bound." The description of this considerably elaborate technique is beyond the scope of this book. Instead we shall consider two recently proposed optimal techniques, which represent answers to the expectations of Lipton and Tarjan. The first one—the triangulation refinement method [Kirkpatrick (1983)] promises to develop into a practical algorithm, and is described in this section. The second one [Edelsbrunner–Guibas–Stolfi (1984)] is a clever refinement of the chain method and will be sketched in the "Notes" at the end of the chapter.

The "triangulation refinement technique" is due to Kirkpatrick. The N-vertex PSLG is assumed to be a triangulation (see Section 1.3.1); in case it is not a triangulation to begin with, it can be transformed into one in time $O(N \log N)$ by a simple algorithm to be described in Section 6.2.2. Recall that a triangulation on a vertex set V, embedded in the plane, is a PSLG with the maximum number of edges, which is at most $3|V| - 6$, by Euler formula. In addition—for reasons to be soon apparent—it is convenient to inscribe the triangulation within a triangular boundary, by creating an inscribing triangle and triangulating the region inbetween (see Figure 2.18). With this provision, all triangulations to be considered have a 3-vertex boundary and exactly $3|V| - 6$ edges.

Let an N-vertex triangulation G be given and suppose we construct a *sequence of triangulations* $S_1, S_2, \ldots, S_{h(N)}$, where $S_1 = G$ and S_i is obtained from S_{i-1} as follows:

Step (i). Remove a set of independent (i.e., nonadjacent) nonboundary vertices of S_{i-1} and their incident edges (The choice of this set, to be

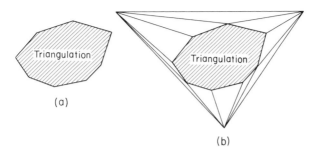

Figure 2.18   A given triangulation (a) and the same triangulation within an inscribing triangle (b).

specified later, is crucial to the performance of the algorithm.);

Step (ii). Retriangulate the polygons arising from the removal of vertices and edges.

Thus, $S_{h(N)}$ has no internal vertices (i.e., it consists of just one triangle).

Notice that all triangulations $S_1, S_2, \ldots, S_{h(N)}$ have the same boundary, since in Step (i) we are removing only internal vertices. Triangles, which are the basic constituents of the technique, are referred to with subscripted R's. A triangle R_j may appear in many triangulations; however, R_j is conventionally said to *belong to triangulation* S_i (denoted $R_j \in {}^*S_i$) if R_j is created in Step (ii) while constructing S_i.

We now build a search data structure T, whose nodes represent triangles. (For brevity we shall frequently refer to a node as "a triangle R_j" rather than to "the node representing triangle R_j".) Structure T, whose topology is that of a directed acyclic graph, is defined as follows: There is an arc from triangle R_k to triangle R_j if, when constructing S_i from S_{i-1}, we have:

1. R_j is eliminated from S_{i-1} in Step (i);
2. R_k is created in S_i in Step (ii);
3. $R_j \cap R_k \neq \emptyset$.

Obviously the triangles in S_1 have no outgoing arcs and are the only ones with this property.

It is convenient, for explanatory purposes, to display T in a stratified manner, i.e., by displaying the nodes in horizontal rows, each row corresponding to a triangulation. A sequence of triangulations is shown in Figure 2.19(a); the encircled vertices are those which are removed at each step. The corresponding structure T is illustrated in Figure 2.19(b).

Once T is available, it should be clear how point location proceeds. The primitive operation is "inclusion in a triangle," which is obviously carried out in $O(1)$ time. The initial step consists in locating the query point z in $S_{h(N)}$. Next, we trace a downward path in T and eventually stop at a triangle belonging to S_1. The construction of this path proceeds as follows: once we

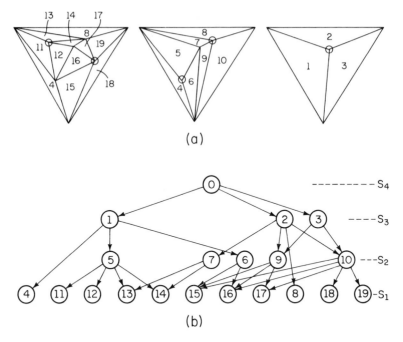

(a)

(b)

Figure 2.19 A sequence of triangulations (a) and the corresponding search-directed acyclic graph (b).

have reached a node in the path (i.e., we have located z in the corresponding triangle) we test z for inclusion in all of its descendants; since z is included in exactly one descendant, this advances the path one more arc. We may also view the search as the successive location of z in triangulations $S_{h(N)}, S_{h(N)-1}, \ldots$; since S_{i-1} is a refinement of S_i, this justifies the name of the technique.

Less informally, we assume that all descendants of a node v of T be arranged in a list $\Gamma(v)$, and let TRIANGLE(v) denote the triangle corresponding to node v. We then have the following search algorithm:

procedure POINT-LOCATION
begin if ($z \notin$ TRIANGLE(root)) **then** print "z belongs to unbounded region"
    else begin $v :=$ root;
           while ($\Gamma(v) \neq \varnothing$) **do**
               for each $u \in \Gamma(v)$ **do if** ($z \in$ TRIANGLE(u)) **then** $v := u$;
           print v
    end
end.

As mentioned earlier, the choice of the set of triangulation vertices to be removed in constructing S_i from S_{i-1} is crucial to the performance of the technique. Suppose we are able to choose this set so that, denoting by N_i the number of vertices of S_i, the following properties hold:

Property 1. $N_i = \alpha_i N_{i-1}$, with $\alpha_i \leq \alpha < 1$ for $i = 2, \ldots, h(N)$.

Property 2. Each triangle $R_j \in S_i$ intersects at most H triangles in S_{i-1}, and vice versa.

Property 1 would have the immediate consequence that $h(N) \leq \lceil \log_{1/\alpha} N \rceil = O(\log N)$, since in going from S_{i-1} to S_i at least a fixed fraction of the vertices is removed. Properties 1 and 2 would jointly imply $O(N)$ storage for T. Indeed, observe that this storage is used for nodes and for pointers to their descendants. As a corollary of Euler's Theorem on planar graphs, S_i contains $F_i < 2N_i$ triangles. The number of nodes in T representing triangles of S_i is at most F_i (only the triangles actually "belonging" to S_i appear on the corresponding tier), whence fewer than

$$2(N_1 + N_2 + \cdots + N_{h(N)}) \leq 2N_1(1 + \alpha + \alpha^2 + \cdots + \alpha^{h(N)-1}) < \frac{2N}{1 - \alpha}$$

nodes appear in T (note that $N_1 = N$, by definition). As regards the storage used by pointers, by Property 2 each node has at most H pointers, whence less than $2HN/(1 - \alpha)$ pointers appear in T. This justifies the claim.

We must now show that there is a criterion for selecting the set of vertices to be removed which achieves Properties 1 and 2. The criterion is (here K is an integer, to be carefully chosen): "Remove a set of nonadjacent vertices of degree less than K." The order in which the vertices are inspected and possibly removed is irrelevant: one starts arbitrarily with one, and marks its neighbors (they cannot be removed); then one continues until no more vertices are unmarked.

This criterion does the trick. Indeed, Property 2 is trivially enforced. Since removal of a vertex of degree less than K gives rise to a polygon with less than K edges, each of the replaced triangles intersects at most $K - 2 \triangleq H$ new triangles. To verify Property 1 it is necessary to use some properties of planar graphs. Although it is possible to develop a more detailed analysis leading to better results, the following arguments are adequate to prove the point.

i. From Euler's formula on planar graphs, specialized for a triangulation with a 3-edge boundary, the numbers N of vertices and e of edges are related by

$$e = 3N - 6.$$

ii. Unless there are no internal vertices (in which case the problem is trivial), each of the three vertices on the boundary has degree at least three. Since there are $3N - 6$ edges, and each edge contributes 2 to the total vertex degree, the total vertex degree is $< 6N$. This immediately yields that there are at least $N/2$ vertices of degree less than 12. Therefore let $K = 12$. Let v be the number of selected vertices: since each selected vertex may eliminate at most $K - 1 = 11$ adjacent vertices and the three boundary vertices are nonselectable, we have

$$v \geq \left\lfloor \frac{1}{11}\left(\frac{N}{2} - 3\right) \right\rfloor$$

If follows that $\alpha \cong 1 - 1/22 < 0.955 < 1$, thus proving the previous claim. Of course, this coarse analysis yields an estimate of α unrealistically close to 1, and the performance of the technique, as indicated by the analysis, appears worse than experiments seem to indicate.

Theorem 2.7. *Point location in an N-vertex planar subdivision can be effected in* $O(\log N)$ *time using* $O(N)$ *storage, given* $O(N \log N)$ *preprocessing time.*

An alternative proof of this theorem is afforded by the refinement of the chain method outlined in the Notes and Comments (Section 2.4).

2.2.2.4. The trapezoid method

Although the question of an optimal point-location method had been settled on the theoretical level, and practical optimal algorithms have been presented in the literature, as described in the preceding section, still a more attractive technique may not be one whose worst-case behavior is asymptotically optimal, as noted in Section 1.2.1. In other words, a compromise approach having a straightforward search procedure possibly at the expense of a slightly suboptimal storage use, may be a desirable objective. The "trapezoid method," to be illustrated now, is likely to fit these specifications [Preparata (1981); Bilardi–Preparata (1981)].[11]

There is more than one way to capture the informal idea of the method. Perhaps the most natural one is to view it as an evolution of the slab-method of Dobkin–Lipton, described in Section 2.2.2.1. As noted earlier, this technique has an extremely simple $O(\log N)$ search procedure, but is plagued with an $O(N^2)$ worst-case storage use. The latter is due to the possible presence of "long" PSLG edges, where by long edge we mean one that crosses several slabs and is therefore segmented into as many fragments. An interesting fact is that long edges—a bane for the slab method—may turn into a boon for the trapezoid method. Indeed a long edge e may conveniently partition the PSLG so that the horizontal slabs on either side of e form two totally independent structures (whereas in the Dobkin–Lipton method a slab extends on either side of e). As a result, the plane will be partitioned into trapezoids defined as follows.

Definition 2.5. A *trapezoid* has two horizontal sides and may be bounded or unilaterally unbounded or bilaterally unbounded; the other two sides—if they exist—are (portions of) PSLG edges, and there is no PSLG edge crossing the interiors of both horizontal sides.

---

[11] Experimental observations reported in [Edahiro *et al.* (1983)], confirm this expectation.

The search proceeds by locating the query point through a sequence of nested trapezoids, until it terminates in a trapezoid void of PSLG edges or portions thereof. We shall show that with this approach an edge is segmented in the worst case into fewer than $2\log_2 N$ fragments, thereby substantially improving the storage behavior while maintaining the $O(\log N)$ search time of the Dobkin–Lipton method (with an insignificant deterioration of the constant).

The trapezoid technique does not require that the graph be triangulated; indeed, as we shall see later, it does not even require that the edges be rectilinear. However, for the time being, we assume we have a PSLG G. The vertex set V of G is ordered according to increasing ordinate, while the edge set E is ordered consistently with the following (partial ordering) relation "$\prec$":

Definition 2.6. Given two edges e_1 and e_2 in E, $e_1 \prec e_2$ (read "e_1 to the left of e_2") denotes that there is a horizontal line intersecting both and the intersection with e_1 is to the left of the one with e_2. (We shall discuss later how this ordering can be algorithmically obtained (Remark 1).)

The basic mechanism, which constructs the search data structure, processes one trapezoid R at a time, and aims at partitioning R into as many trapezoids as possible. This is done by initially cutting R into "lower" and "upper" slices R_1 and R_2 by means of a horizontal line through the PSLG vertex whose ordinate is the median in the set of vertices internal to R. Each of these slices, R_1 and R_2, although geometrically a trapezoid, is not one in the technical sense of Definition 2.5, since it may have PSLG edges crossing both horizontal sides, called *spanning edges*. Next, each edge that spans either slice[12] determines a further cut of it.

The action develops as follows. After determining the median line $y = y_{\mathrm{med}}$ of the trapezoid R (see Figure 2.20), the string of PSLG edges that intersect R is scanned from left to right and split into two strings, pertaining to slices R_1 and R_2, respectively. In this process, as soon as a spanning edge of either slice is encountered, this edge is the right boundary of a new trapezoid, which can in turn be processed. Processing a trapezoid is trivial when its interior is empty (no PSLG edge crossing it).

The search data structure corresponding to the trapezoid of Figure 2.20 is illustrated in Figure 2.21. To each trapezoid R there corresponds a binary search tree $T(R)$, each node of which is associated with a linear test. It is convenient to distinguish the nodes into two types: V-nodes, if the associated test corresponds to a horizontal line, and O-nodes, if the associated test corresponds to the straight line containing a PSLG edge. Clearly the root of $T(R)$ is always a V-node. Notice also that a V-node corresponds uniquely to a vertex of the PSLG, specifically the vertex whose ordinate is the median of

---

[12] No edge can span both R_1 and R_2, otherwise it would span the trapezoid R, contrary to the definition.

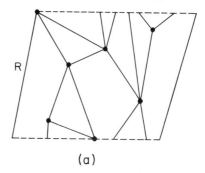

(a)

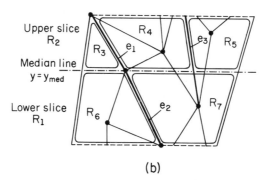

(b)

Figure 2.20   The trapezoid R of (a) is partitioned into trapezoids R_3, R_4, R_5, R_6, and R_7, as shown in (b).

vertex ordinates in the trapezoid. Since the two extreme vertices of the PSLG are not involved in trapezoid partitioning, there will be exactly $(N - 2)$ ∇-nodes in the search tree.

Less informally, the function TRAPEZOID that constructs the search tree accepts as its input a string E of PSLG edges, the sequence of vertices internal to R ordered by increasing ordinate, and the y-interval I of the trapezoid R. The procedure makes use of some auxiliary actions—such as "find median" and "balance"—which will be described later. E_1, E_2, U_1, and U_2 are lists internal to the algorithm.

```
1    function TRAPEZOID (E, V, I)
2    begin if (V = ∅) then return Λ (*a leaf of the search tree*)
3         else begin E₁ := E₂ := V₁ := V₂ := U := ∅;
4              y_med := median ordinate of V;
5              I₁ := [min(I), y_med]; I₂ := [y_med, max(I)];
6              repeat e ⇐ E;
7                   for i = 1, 2, do
8                        begin if (e has endpoint p in interior of Rᵢ)
                         then
9                             begin Eᵢ ⇐ e;
```

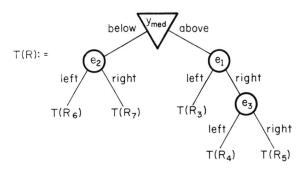

Figure 2.21   The search data structure corresponding to the trapezoid of Figure 2.20.

10	$$V_i := V_i \cup \{p\}$$
	end;
11	**if** $(e$ spans $R_i)$ or $(e = \Lambda)$ **then**
12	**begin** $U_i \Leftarrow \text{TRAPEZOID}[E_i, V_i, I_i]$;
13	**if** $(e \neq \Lambda)$ **then** $U_i \Leftarrow e$;
14	$E_i := V_i := \varnothing$
	end
	end
15	**until** $e = \Lambda$;
16	new (w); (*create a new ∇-node w, the root of $T(R)$*)
17	$Y[w] := y_{\text{med}}$; (*discriminant for node w*)
18	LTREE$[w] :=$ BALANCE(U_1);
19	RTREE$[w] :=$ BALANCE(U_2) (*function BALANCE takes an alternating sequence of trees and edges and arranges these items into a balanced tree*)
	return Tree$[w]$
	end
	end.

In this algorithm we distinguish three major activities:

i. the determination of the median ordinate of the vertices internal to R;
ii. the partitioning of the upper and lower slices into trapezoids and the production for each slice R_i $(i = 1, 2)$ of a string U_i, whose terms are edges and trees (in our example, $U_2 = T(R_3)e_1 T(R_4)e_3 T(R_5)$ and $U_1 = T(R_6)e_2 T(R_7)$). This is the bulk of the work and is done by the *repeat*-loop in lines 6–15.
iii. the balancing of the two strings U_1 and U_2 (lines 18 and 19).

The first activity could be carried out, in principle, in time $O(|V|)$ by the classical median-finding algorithm. However, there is a simpler and more straightforward way. The vertices in the trapezoid are arranged in an array by increasing ordinate. (The median of this array can be found by a single access.) The modification to TRAPEZOID goes as follows. The edge sequence is

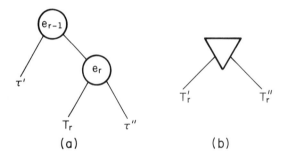

Figure 2.22   Illustration of the balancing operation.

scanned *twice*. In the first pass each vertex is simply marked with the name of the trapezoid it is assigned to; using these markings in a straightforward way we construct the vertex array for each generated trapezoid. In the second pass, *repeat*-loop 6–15 is executed (with line 10 suppressed).

   The auxiliary function BALANCE arranges an alternating string of trees and edges $U = T_1 e_1 T_2 e_2 \ldots e_{h-1} T_h$ in a balanced tree. Each tree T_j has a weight $W(T_j) \geq 0$, equal to the number of PSLG vertices contained in the corresponding trapezoid. The weight $W(U)$ of U is equal to $\Sigma_{j=1}^r W(T_j)$. BALANCE operates as follows:

1. If $W(U) = 0$ (i.e., $U = e_1 e_2 \ldots e_{h-1}$), arrange the e_i's in a balanced tree.
else 2.1. Determine an integer r so that $\Sigma_{j=1}^{r-1} W(T_j) < W(U)/2$ and $\Sigma_{j=1}^{r} W(T_j) \geq W(U)/2$
    2.2. Construct a tree as in Figure 2.22(a), where $U' = T_1 e_1 \ldots T_{r-1}$ and $U'' = T_{r+1} e_{r+1} \ldots T_h$ and $\tau' = \text{BALANCE}(U')$, $\tau'' = \text{BALANCE}(U'')$. (Note that $W(U') < W(U)/2$ and $W(U'') \leq W(U)/2$.)

BALANCE clearly runs in time $O(h \log h)$. Indeed, the split (U', T_r, U'') is obtained in time $O(h)$ and is followed by two recursive calls of the procedure on a dichotomy of the original string.

   But even more interesting is the analysis of the depth of the balanced trees. Observe at first that h, the number of tree-terms in string U, is less than N, where N is, as usual, the number of vertices of the PSLG. Indeed, referring to Figure 2.23, the number h of new trapezoids is one more than the number of edges spanning the original trapezoid. The latter is bounded above by the number of edges that can be cut by a single straight line. We now observe that these edges could be viewed as "diagonals" of a simple polygon (shown in dotted outline in Figure 2.23). If this polygon has s vertices, there are at most $s - 3$ diagonals (this happens when the diagonals determine a triangulation of the polygon). Since $s \leq N$ (the number of vertices of the PSLG), we obtain $h \leq s - 2 \leq N - 2$. Next we can prove the following theorem.

Theorem. *Given the string* $U = T_1 e_1 \ldots e_{n-1} T_h$, *the depth* $\delta(U)$ *of the tree obtained by* BALANCE(U) *is at most* $3 \log_2 W(U) + \lceil \log_2 N \rceil + 3$.

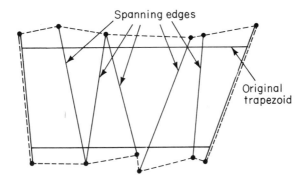

Figure 2.23   Illustration for the proof that $h < N$ (h = number of new trapezoids).

PROOF. By induction on $W(U)$. We start the induction with $W(U) = 1$. In this case, BALANCE yields a tree as in Figure 2.22(a), where τ' and τ'' are possibly empty and T_r is as in Figure 2.22(b). Here T_r' and T_r'' consist each of at most N O-nodes. Thus the depth of the tree is at most $\lceil \log_2 N \rceil + 3$, and the theorem holds. Letting $W(U) = K$ (a positive integer), we assume the theorem holds for all weights less than K. We distinguish two cases:

Case 1. $W(T_r) \leq K/2$. Since also $W(U')$, $W(U'') \leq K/2$ by the inductive hypothesis $\delta(U')$, $\delta(U'')$, $\delta(T_r) \leq 3 \log_2 K/2 + \lceil \log_2 h \rceil + 3 \leq 3 \log_2 K - 3 + \lceil \log_2 N \rceil + 3$, so that the arrangement of Figure 2.22(a) extends the inductive hypothesis.

Case 2. $W(T_r) > K/2$. Since $W(U')$, $W(U'') \leq K/2$, we argue for τ' and τ'' as in Case 1. As to T_r, we observe that it corresponds to a trapezoid. Thus, by the action of procedure TRAPEZOID, T_r has the structure illustrated in Figure 2.22(b), where T_r' and T_r'' are themselves balanced trees, and, because of the median partitioning, $W(T_r')$, $W(T_r'') \leq W(T_r)/2 \leq K/2$. Therefore for some $h'' < N$, $\delta(T_r')$, $\delta(T_r'') \leq 3 \log W(T_r)/2 + \lceil \log_2 h'' \rceil + 3 \leq 3 \log K - 3 + \lceil \log_2 N \rceil + 3$, which completes the proof. $\square$

Since $W(U) \leq N$ for the PSLG, we conclude that the depth of the search tree for the PSLG is at most $4\lceil \log_2 N \rceil + 3$.

Finally, we consider the *repeat*-loop of procedure TRAPEZOID. For each edge scanned (i.e., extracted from list E), the procedure performs a bounded amount of work expressed by lines 6 and, alternatively, 8–10 or 12–14. Therefore the global work done by the loop is proportional to the total number of edge fragments produced by the algorithm. We shall now estimate an upper bound for this number. Referring to Figure 2.24, an edge is cut for the first time when its endpoints lie in distinct slices determined by a median line of a trapezoid R. After this cut, the upper fragment falls in a trapezoid R'. If the median line of R' cuts e (worst-case) then e spans the lower slice of R' and further cuts can occur only in the upper slice of R'. Analogously we argue for the fragment of e in the lower slice of R. Thus, expressing the number of cuts of

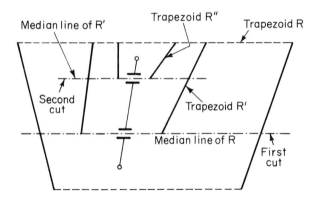

Figure 2.24   Illustration of the edge-segmenting mechanism.

the upper fragment of edge e as a function $c(s')$ of the number s' of vertices of R', we have the crude bound

$$c(s') \leq 1 + c(s''),$$

where s'' is the number of vertices of trapezoid R'' as shown in Figure 2.24. Since $s'' < s'/2$, we obtain that $c(s') \leq \log_2 s'$, and, because $s' < N/2$, the total number of cuts of e is at most $2 \log_2 N - 2$. This estimate proves two results: first, the total work done by the *repeat*-loop is $O(N \log N)$, since there are $O(N)$ edges and each edge is segmented into $O(\log N)$ fragments. Since both the median-finding and the balancing tasks run globally in time $O(N \log N)$, the entire construction of the search data structure is completed in time $O(N \log N)$. Second, the search data structure has an O-node for each edge fragment and a ∇-node for each vertex; thus the storage used is also $O(N \log N)$.[13] In summary

Theorem 2.8. *A point can be located in an N-vertex planar subdivision with fewer than* $4 \log_2 N$ *tests, using* $O(N \log N)$ *storage and preprocessing time.*

Remark 1. Earlier we deferred discussion of how the edge set E can be ordered consistently with the partial ordering "$\prec$" given in Definition 2.6. The technique that can be used to achieve this objective is a simple modification of the plane-sweep described in Section 2.2.2.2 to "regularize" a planar graph.

In a top-to-bottom sweep, the vertices of G are scanned and the sweep-line status is updated as described earlier. The byproduct of the sweep is the recording of all edge adjacencies obtained either as a result of deletions or of insertion of edges. An edge adjacency is expressed by an ordered edge pair, (left, right) (see Figure 2.25), which turns out to be a pair of the transitive reduction "$\ll$" of the partial ordering in question. Obviously, if G has N

---

[13] In the worst case, the storage utilization is unequivocally $O(N \log N)$. However, in a specialized instance of PSLG [Bilardi–Preparata (1981)] the average-case storage is $O(N)$ and it is conjectured that the same average-case behavior applies to general PSLGs.

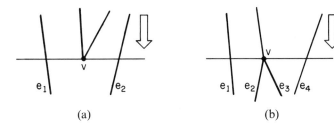

Figure 2.25   Illustration of the computation of relation "$\ll$," as a result of processing vertex v. In case (a), pair (e_1, e_2) is generated; in case (b), pairs (e_1, e_2), (e_2, e_3), and
(e_3, e_4) are produced.

vertices, relation "$\ll$" can be computed in time $O(N \log N)$. Once "$\ll$" is
available, the constant orderings can be obtained in time $O(N)$ by a standard
topological sorting technique [Knuth (1973), p. 262].

Remark 2. The described method is not restricted to PSLGs. Indeed the
straight-line segments may be replaced by other curves if the following two
properties hold: (i) the curves are single valued in one selected coordinate
(say, y) and (ii) the discrimination of a point with respect to any such curve
can be done in constant time. For example, these conditions are clearly met by
arcs of circles or other conics, if they have no horizontal tangent except at
their extremes. To adapt Kirkpatrick's method (Section 2.2.2.3) to this situation apparently presents substantial difficulties, because of the need to retriangulate a polygon with curvilinear edges.

2.3 Range-Searching Problems

2.3.1 General considerations

As we indicated in Section 2.1, range-searching problems may be viewed as
dual, in some sense, of the point-location problems we have discussed so far.
   Here the file is a collection of records, each of which is identified by an
ordered d-tuple of "keys" $(x_1, x_2, \ldots, x_d)$. We naturally view a d-tuple of keys
as a point in a d-dimensional Cartesian space. Several user actions—or
queries—are conceivable on such a file. However, in the present context we
shall deal exclusively with *range queries*. The query specifies a domain in the d-
dimensional space and the outcome of the search is to *report* the subset of the
file points contained in the query domain. An alternative, more restrictive,
search objective is simply to *count* the number of points in the query domain.
(A unifying framework for these two types of problems is one where each of
the records is associated with an element of a commutative semigroup with an
operation "$*$", and the query executes "$*$" over all records in the query range.

Thus, in the report mode each record is associated with its name and "$*$" is set union; in the count mode each record is associated with the integer 1 and "$*$" is ordinary addition [Fredman (1981)].)

The Cartesian space formulation is the abstraction of a variety of very important applications, often referred to as "multikey searching" [Knuth (1973), Vol. 3]. For example, the personnel office of a company may wish to know how many employees whose age is between 30 and 40 years earn a salary between $27,000 and $34,500. Aside from this somewhat contrived example, applications of this nature arise in geography, statistics [Loftsgaarden and Queensberry (1965)], and design automation [Lauther (1978)].

Returning to our abstract problem, the single-shot mode of operation is scarcely of interest, since it invariably reduces to some exhaustive search of the space. Thus our interest is confined to repetitive-mode applications, and here again there is an initial investment in preprocessing and storage to obtain a long term gain in searching time. In this context, the ultimate objective is to solve the problem in its widest generality; that is, to be able to handle a space of arbitrarily many dimensions and a variety of query domains, as concerns both shape and size. Whereas the question of dimensionality can be adequately tackled (at some computational cost, of course), unfortunately with regard to the nature of the query most that is known to date concerns hyperrectangular domains. (A hyperrectangular domain is the Cartesian product of intervals on distinct coordinate axes.[14]) Although the case of the hyperrectangular domain is very important, it by no means exhausts the set of desirable choices. Indeed, the search for all points at bounded distance d from a query point z, referred to as *bounded distance search*, involves a spherical query domain (i.e., a hypersphere of radius d centered in z). As will be apparent below, the techniques that have been successfully developed for the hyperrectangular case are not applicable to the spherical one. If comparable performance is ever going to be achieved for the latter case, it is likely to be based on entirely different approaches: Indeed, some recent progress [Chazelle–Cole–Preparata–Yap (1984)] in bounded distance search for $d = 2$ (called *circular range search*) is based on notions of "proximity" to be presented in Chapters 5 and 6. These considerations strictly apply, however, to a mode of operation where the searching technique does not access any other file item but those contained in the query domain. If this condition is relaxed, then a spherical domain can be adequately approximated by means of a family of hyperrectangular domains; while such an approach could have a disastrous worst-case behavior, in "real," practical situations it may perform very well (for example, the k-D tree method of Section 2.3.2).

---

[14] This choice of search domain is frequently referred to as *orthogonal query*, possibly because the hyperplanes bounding the hyperrectangle are each orthogonal to a coordinate axis. Sometimes the hyperrectangle is referred to as *rectilinear*—a qualifier possibly suggested by the shape of paths in the L_1-metric in the plane (see Chapter 8). However, this use of "rectilinear" appears as a misnomer (for further discussion of this point, see Section 8.2).

A common feature of all methods to be considered in this chapter is that the search data structures employed are *static*, i.e., they are not modified once they are built. This obviously requires that their constituent items—in our case a set of points of a d-dimensional space—be all given beforehand. The corresponding *dynamic* data structures, i.e., structures that also support insertions and deletions of items, appears to be considerably more complex; their investigation is still in a very active stage, and will not be discussed in detail in this text, except for some brief remarks and bibliographic references in Section 2.4.

It is now appropriate to seek a yardstick for a comparative evaluation of different methods. Normally, this is in the form of a lower bound to some performance measures. Before proceeding further, however, it is appropriate to analyze at an informal level the interplay of the various performance measures.

As usual, we assume that the file consists of a fixed collection S of N records. Each query will produce as a response a subset S' of S. A possible approach consists in precomputing the responses to all possible range-search queries. Since the possible responses are a finite set (the power set of the finite set S), the infinite set of queries is partitioned into a finite set of equivalence classes (two queries are equivalent if they produce the same response). Thus, processing a query would reduce to mapping the query to a standard representative of its equivalence class, which in turn would provide immediate access to the response. Such direct access method (to be discussed in Section 2.3.3) would exhibit a rather simple query processing, at the expenses of enormous requirements in storage and in the corresponding preprocessing time. Of the latter two measures, however, storage is a more important concern than preprocessing time; indeed, preprocessing time represents a one-time expenditure, which may be tolerated, whereas storage represents a long-term investment, not reusable for the lifetime of the query-answering system. Therefore, it has become customary, if not to neglect, at least to de-emphasize preprocessing time and to concentrate on query time and file storage. (In many applications, however, preprocessing time and storage are of the same order, so that the latter covers both measures.) The method sketched above exhibits low query time and high storage; conceivably, reductions in storage are obtained at the expense of an increase in query time, so we are facing a trade-off between the two measures. According to current practice, we shall characterize a range-search method by the pair (file storage, query time).

In turn, the notion of "query time" deserves some further discussion. The work expended in processing a query naturally depends upon the file size N and the subset S' of S contained within the query range, but also upon the type of query (i.e., upon the nature of the semigroup operation "$*$" alluded to earlier). Indeed, a count-mode query will in all cases report a single integer (the cardinality k of S'), whereas a report-mode query will report the names of each of the k members of S'. In principle, we may distinguish two types of activities in processing a query:

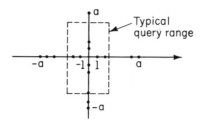

Figure 2.26   Illustration of set S for $d = 2$.

1. *Search*, i.e., the overhead activity leading to the items of the accessed set (usually, a sequence of comparisons);
2. *Retrieval*, i.e., the activity of assembling the query response (for report-mode queries, the actual retrieval of the accessed subset).

In count-mode queries, the whole computational work is conveniently absorbed into the "search"; in this case, the worst-case query time is expected to be a function $f(N, d)$ of the file size and the number of dimensions. More subtle is the case of report-mode queries, since the computational work is now a mixture of search and retrieval activites, and the fraction pertaining to the latter is bounded below by the size k of S'. If we insist that *exactly* the members of S' be accessed by the query, then conceivably the searches for the two types of queries have identical costs. If we allow instead that the accessed set be a slightly larger superset of the target S', then there is the possibility of reducing the storage requirement and of "charging" to the retrieval task part of the cost of the search activity; this is the basic idea of *filtering search* [Chazelle (1983c)] to be further discussed at the end of this chapter.[15] Thus, in general,an upper bound to query time for a report-mode query will be expressed in the form $O(f(N, d) + k \cdot g(N, d))$; the two terms are to be interpreted respectively either as "search time" and "retrieval time" if just the retrieval set is to be accessed, or as query times for "small" ($k = O(f(N, d))$) and "large" ($f(N, d) = o(k)$) retrieval sets in the more general case.

   With respect to lower bounds, we note immediately that $\Omega(k)$ is a trivial lower bound to retrieval time, since the length of the output is proportional to k. As regards search time, we resort to the binary decision-tree model, which—as, is well-known—counts the number of comparisons performed to access the elements S' of S within the query range (note that we assume that just the elements to be retrieved be accessed). This number of comparisons is bounded below by the logarithm in base 2 of the number $Q(S)$ of the distinct subsets of S obtainable as query responses. Referring to Figure 2.26, consider

---

[15] There is a basic difference between filtering search and a search technique that approximates the range with ranges of a different type (e.g., a spherical range with hyperrectangular ranges). While both methods are not of the "exact access" type, in filtering search the size of the accessed set is at most a *fixed* multiple of that of the retrieved set.

the following set S [Bentley–Maurer (1980)]: Assuming for simplicity that $N = 2ad$, S is a set of all points with a single nonzero integer coordinate in the interval $[-a, a]$. Then the selection of two arbitrary integers in $[-a, -1]$ and $[1, a]$, respectively, for each coordinate axis, determines a query range whose associated subset of points is nonempty and distinct from all other subsets. Since the above selection can be effected in $a^{2d} = (N/2d)^{2d}$ ways, the lower bound to the number of binary decisions is $\Omega(\log(N/2d)^{2d}) = \Omega(d\log N)$. Although a more subtle combinatorial analysis [Saxe (1979)] could yield a more accurate estimate of the number of orthogonal ranges with distinct retrieved sets, the above simple argument gives a lower bound $\Omega(d\log N)$ to the search time. Unfortunately—as noted by Lueker (1978)—the decision-tree model is apparently inadequate, since for large d all known methods exhibit a search time exponential in d rather than linear (i.e., an upper bound of the form $O(\log N)^d$). This question was addressed by Fredman (1981), who proved a lower bound $\Omega((\log N)^d)$ to the search time for *dynamic* query systems, by bounding the complexity of the semigroup operations "$*$," introduced earlier in this section; it is not entirely clear, however, how Fredman's result applies to the situation being considered. Finally, we note that $\Omega(Nd)$ is a trivial lower bound to the storage space used by the search data structure.

In the rest of this chapter, we shall be mainly interested in report-mode algorithms rather than in count-mode algorithms (for the latter, see Exercise 4). Note that an algorithm of the first type can be trivially used to solve a problem of the second type; however, the inverse transformation is in general not possible, and specific efficient techniques for the count-mode can be developed. For all the described algorithms the search time is of the form $O(f(N, d) + k)$.

The next sections illustrate some quite interesting and clever methods that have been proposed for the range-search problem. They should shed some light on the difficulties which have so far prevented the design of an optimal algorithm and have just resulted in configuring a trade-off between the two important resources of query time and storage space.

To set the appropriate stage for the techniques to be described later we shall start from the simplest instance of range searching, which, although apparently trivial, contains the essential traits of the problem: one-dimensional range-searching.

A set of N points on the x-axis represents the file, and the query range is an interval $[x', x'']$ (referred to as the x-range). The device which enables an efficient (optimal) range searching is *binary search*, that is, the *bisection* of the ordered set to be searched. Indeed, by binary search the left extreme x' of the x-range is located on the x axis; this completes the search, for retrieval is obtained by traversing the x-axis in the direction of increasing x, until the right extreme x'' is reached. The data structure which supports the outlined action is a *threaded binary tree*, i.e., a balanced binary tree whose leaves are additionally connected as a list reflecting the ordering of abscissae; the tree and the list are visited in the search and retrieval phases, respectively. Note that the outlined method is optimal both in query time—$\theta(\log N + k)$—and in storage—$\theta(N)$.

In the following discussion of the different approaches to d-dimensional range searching, we shall first consider the simplest case (i.e., $d = 2$) to bring out the essential features of each method, uncluttered by the burden of dimensionality. Once this objective is achieved, we shall pursue, whenever convenient, the generalization to an arbitrary number of dimensions.

2.3.2 The method of the multidimensional binary tree (k-D tree)

We have just recognized that the notion of bisection is crucial to the development of an optimal technique for one-dimensional range searching. It is therefore natural to try to generalize bisection to the two-dimensional case when the file is a collection of N points in the (x, y) plane. What bisection does is to successively split an interval—bounded or unbounded—into two parts. In two dimensions, one could view the entire plane as an unbounded rectangle to be initially split into two half-planes by means of a line parallel to one of the axes, say y. Next, each of these half-planes could be further split by a line parallel to the x-axis, and so on, letting at each step the direction of the cutting line alternate, for example, between x and y. How do we choose the cutting lines? By exactly the same principle used in standard bisection, i.e., the principle of obtaining approximately equal numbers of elements (points) on either side of the cut. We realize that a legitimate generalization of bisection has been obtained, and, indeed the idea of the *multidimensional binary tree* is born [Bentley (1975)].

Less informally, we call a (generalized) *rectangle* the region of the plane defined by the cartesian product $[x_1, x_2] \times [y_1, y_2]$ of an x-interval $[x_1, x_2]$ and a y-interval $[y_1, y_2]$, including the limiting cases where any combination of the choices $x_1 = -\infty$, $x_2 = \infty$, $y_1 = -\infty$, $y_2 = \infty$ is allowed. So we shall treat as rectangles also an unbounded strip (either on one or two sides), a quadrant, or even the entire plane.

The process of partitioning S by partitioning the plane is best illustrated in conjunction with the building of the two-dimensional binary tree T. With each node v of T we implicitly associate a rectangle $\mathcal{R}(v)$ (as defined earlier) and the set $S(v) \subseteq S$ of the points contained in $\mathcal{R}(v)$; explicitly—i.e., as actual parameters of the data structure—we associate with v a selected point $P(v)$ of $S(v)$ and a "cutting line" $l(v)$ passing by $P(v)$ and parallel to one of the coordinate axes.

The process begins by defining the root of T and by setting $\mathcal{R}(\text{root})$ to be the entire plane and $S(\text{root}) = S$; next we identify the point $p \in S$ so that $x(p)$ is the median of the abscissae of the points of $S(\text{root})$, and set $P(\text{root}) = p$ and $l(\text{root})$ the line whose equation is $x = x(p)$. Point p subdivides S into two sets of approximately identical size, which are assigned to the offsprings of the root. This splitting process stops when we reach a rectangle not containing any point: the corresponding node is a leaf of the tree T.

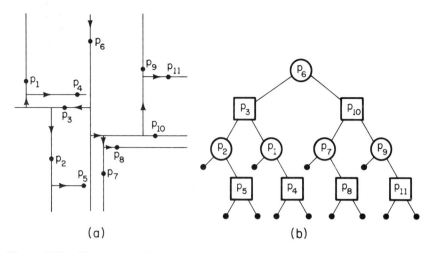

(a)   (b)

Figure 2.27   Illustration of the two-dimensional binary-tree search method. The partition of the plane in (a) is modeled by the tree in (b). Search begins with the vertical line through P_6. In (b) we have the following graphic convention: circular nodes denote vertical cuts; square nodes denote horizontal cuts; solid nodes are leaves.

The technique is illustrated with an example in Figure 2.27, for a set of $N = 11$ points. We have also indicated with different graphical symbols nodes of three different types: circular, for nonleaf nodes with a vertical cutting line; square, for nonleaf nodes with a horizontal cutting line; solid, for leaves. The structure just obtained is frequently referred to as a *2-D tree*, abbreviation for "two-dimensional binary search tree."

We now investigate the use of the 2-*D* tree in range searching. The algorithmic paradigm is a pure instance of divide-and-conquer. Indeed, consider the interaction of a rectangular region $\mathscr{R}(v)$, associated with node v of T, with a rectangular range D, such that $\mathscr{R}(v)$ and D have a nonempty intersection. Region $\mathscr{R}(v)$ is cut into two rectangles R_1 and R_2 by line $l(v)$ through $P(v)$. If $D \cap \mathscr{R}(v)$ is entirely contained in R_i ($i = 1, 2$), then the search continues with the single (region, range) pair (R_i, D). If, on the other hand, $D \cap \mathscr{R}(v)$ is split by $l(v)$, this means that $l(v)$ has a nonempty intersection with D and therefore D may contain $P(v)$. Thus we first test if $P(v)$ is in the interior of D, and, if so, place it in the retrieved set; next, we continue the search with the two (region, range) pairs (R_1, D) and (R_2, D). At any leaf reached in this search, the process terminates.

More formally, a generic node v of T is characterized by three items $(P(v), t(v), M(v))$. Point $P(v)$ had already been defined. The other two parameters jointly specify the line $l(v)$, that is, $t(v)$ indicates whether $l(v)$ is horizontal or vertical, and in the first case, $l(v)$ is the line $y = M(v)$, in the second case it is $x = M(v)$. The algorithm accumulates the retrieved points in a list U external to the procedure, initialized as empty. Denoting by $D = [x_1, x_2] \times [y_1, y_2]$ the query range, the search of the tree T is effected by a call SEARCH(root(T), D) of the following procedure:

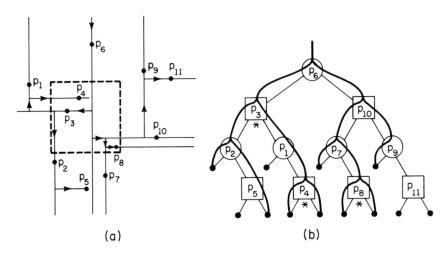

(a)   (b)

Figure 2.28   Illustration of a range search for the file previously given. Part (b) shows the nodes, actually visited by the search.

procedure SEARCH(v, D)
begin if ($t(v)$ = vertical) **then** $[l, r] := [x_1, x_2]$ **else** $[l, r] := [y_1, y_2]$;
    if ($l \leq M(v) \leq r$) **then if** ($P(v) \in D$) **then** $U \Leftarrow P(v)$;
    if ($v \neq$ leaf) **then**
        begin if ($l < M(v)$) **then** SEARCH(LSON[v], D);
                if ($M(v) < r$) **then** SEARCH(RSON[v], D)
        end
end.

In Figure 2.28(a) we illustrate an example of range search for the file previously given in Figure 2.27(a). Figure 2.28(b) illustrates in particular the visit of nodes of T performed by the search algorithm outlined above. Note that only at nodes where the search visit bifurcates (such as $p_6, p_3, p_2, p_4, p_{10}, p_7, p_8$) does the test for inclusion of a point in the range take place. Nodes marked with an asterisk ($*$) are those where such a test is successful and a point is picked up. Indeed the retrieved set is $\{p_3, p_4, p_8\}$.

From the performance viewpoint, the 2-D tree uses $\theta(N)$ optimal storage (one node per point of S). It can also be constructed in optimal time $\theta(N \log N)$ in the following manner. A vertical cut of a set S of points is done by computing the median of the x-coordinates of the points of S in time $O(|S|)$ (by using the algorithm of [Blum et al. (1973)]), and by forming the partition of S within the same time bound; analogously, for a horizontal cut. Thus in time $O(N)$ the initial set is split and the two resulting half-planes are each equipped with its set of $N/2$ points; the recurrence for the running time $T(N)$ is trivially

$$T(N) \leq 2T(N/2) + O(N)$$

which yields the claimed processing time. A more direct implementation

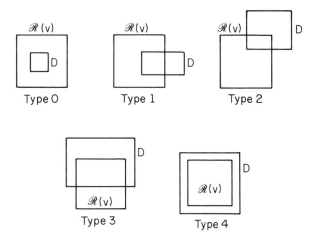

Figure 2.29   Illustration of the different types of intersections of D with $\mathcal{R}(v)$, when $D \cap \mathcal{R}(v) \neq \varnothing$.

avoids recourse to the complicated median-finding algorithm. Since the method just outlined effectively sorts the sets of x- and y-coordinates by recursive median finding, we may resort to a preliminary $O(N \log N)$-time sorting to form the ordered arrays of abscissae and ordinates of the points of S, called respectively the x- and y-array. The initial cut is done by accessing, in constant time, the median of the x-array and by marking, say, the points with smaller abscissae (in time $O(N)$). Using these markings, the two sorted subarrays are readily formed and the process is recursively repeated on them.

The worst-case analysis of the query time is considerably more complex. Not surprisingly this analysis was developed [Lee–Wong (1977)] well after the original proposal of multidimensional binary trees. The query time is clearly proportional to the *total number* of nodes of T visited by the search algorithm, since at each node the search algorithm spends a constant amount of time. This time is obviously well spent at node v if $P(v)$ is retrieved (*productive* node); otherwise, the node is *unproductive*. Lee and Wong's analysis aims at constructing a worst-case situation, that is, one corresponding to a thickest subtree of visited nodes, all unproductive. As noted earlier, each node v of T corresponds to a generalized rectangle $\mathcal{R}(v)$. The intersection of a query-range D and one such generalized rectangle $\mathcal{R}(v)$ may be of different "types" depending upon the number of sides of $\mathcal{R}(v)$ which have nonempty intersection with D. Specifically, if this number is i, the intersections is said to be of type i, for $i = 0, 1, 2, 3, 4$. (See Figure 2.29.) The only kind of intersection which is always productive is type 4; all others may be unproductive. In particular, we restrict ourselves to intersections of type 2 and 3. (Note that types 2 and 3 may start appearing at two and three levels from the root, respectively.) Referring to Figure 2.30(a) we can easily construct the situation where a type-2 intersection at height m is unproductive and gives rise to one type-2

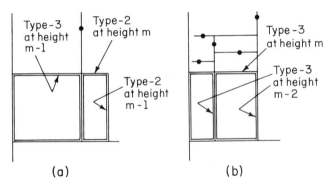

Figure 2.30   Self-replicating situations corresponding to unproductive nodes in T.

and one type-3 intersection at height $(m - 1)$ (both of which can be constructed as unproductive). Similarly, as shown in Figure 2.30(b) with the same restriction to unproductive nodes alone, we may construct the situation where a type-3 intersection at height m in T gives rise to two type-3 intersections at height $(m - 2)$. Thus denoting by $U_i(m)$ the attainable number of unproductive nodes in a subtree of T of height m, whose root is of type i $(i = 2, 3)$, we obtain the recurrences

$$\begin{cases} U_2(m) = U_2(m - 1) + U_3(m - 1) + 1 \\ U_3(m) = 2U_3(m - 2) + 3. \end{cases} \tag{2.3}$$

The solution is that both $U_2(m)$ and $U_3(m)$ are $O(\sqrt{m})$.

We conclude that the worst case behavior of the query time is $O(\sqrt{N})$ for an N-term file, even when the retrieved set is empty. This negative worst-case result should be contrasted with good performance obtained in simulations [Bentley (1975)] and justified by a heuristic argument [Bentley–Stanat (1975)].

The generalization of the described method to d dimensions is straightforward. Here we shall deal with cutting hyperplanes orthogonal to the coordinate axes, and all that needs to be specified is the criterion by which the orientations of the cutting hyperplanes are to be chosen. If the coordinates are $x_1, x_2, \ldots, x_d$, a possible criterion is to *rotate* on the cyclic sequence $(1, 2, \ldots, d)$ of coordinate indices. The ensuing partition of the d-space is modelled by an N-node binary tree, known as *multidimensional binary tree*, or *k-D tree*.[16] The performance analysis can also be judiciously extended to the d-dimensional case, but we do not discuss it here. The results are summarized in the following theorem.

---

[16] The expression "*k-D* tree" is D. E. Knuth's abbreviation for "*k*-dimensional binary-search tree." It is not very descriptive, but it has entered the current jargon by now.

Theorem 2.9. *With the k-D tree method, for $d \geq 2$, range searching of an N-point d-dimensional set can be effected in time $O(dN^{1-1/d} + k)$, using $\theta(dN)$ storage, given $\theta(dN \log N)$ preprocessing time. Thus, the k-D tree method is a $(dN, dN^{1-1/d})$-algorithm.*

The optimal space and preprocessing time behavior unfortunately does not offset the discouraging worst-case search-time performance. It is therefore not surprising that other methods have been proposed as alternatives. These methods will now be described.

2.3.3 A direct access method and its variants

The worst-case inefficiency of the k-D tree method is a sufficient motivation to look for methods that perform satisfactorily with respect to search time, possibly at the expense of the other two relevant performance measures. With this objective, we shall start from the least sophisticated approach and develop successive refinements of it.

As alluded to in Section 2.3.1, the most brutal approach to minimizing search time consists in precomputing the answers to all possible range search queries. One is then tempted to say that search time could be reduced to $O(1)$, since a single memory access would complete the search. This, of course, raises a puzzling question for its apparent violation of the lower bound discussed in Section 2.3.1. We shall soon see that there is a catch, and every item will fit in place.

Following a line of thought introduced in Section 2.1, suppose that, given a set S of N points in the plane, we trace a horizontal and a vertical line through each of these points. These lines, of course, partition the plane into $(N + 1)^2$ rectangular cells. Given a range $D = [x_1, x_2] \times [y_1, y_2]$, point (x_1, y_1) belongs to a cell C_1 and point (x_2, y_2) belongs to a cell C_2. If we now move at will either point, within its respective cell, we recognize that the accessed set (i.e., the subset of S contained in D) is invariant. In other words, pairs of cells form equivalence classes with respect to range searching. Therefore, the number of distinct ranges we may have to consider is bounded above by

$$\binom{N + 1}{2} \times \binom{N + 1}{2} = O(N^4).$$

If we precompute the retrieved set for each of these pairs of cells (thereby attaining $O(N^5)$ storage), we have a scheme whereby a single access completes the search. We realize, however, that to achieve this objective an arbitrary range D must be mapped to a pair of cells, or, equivalently, an arbitrary point must be mapped to a cell. It is clear that this can be done by binary searches on the sets of abscissae and ordinates of the points of S. This "normalization," however, costs $O(\log N)$ time, which, added to the $O(1)$ time of the single access, resolves the apparent conflict we stumbled on earlier. In conclusion, in the plane this approach yields a $(N^5, \log N)$-algorithm, too onerous in the use of storage.

To find ways to reduce the storage, we note that—in spite of our unsuccessful efforts to generalize it—one-dimensional searching is optimal in both measures (search time and storage). This remark suggests an improvement of the above "cell method." Indeed, we may combine one-dimensional searching with direct access range location [Bentley–Maurer (1980)]. Specifically, given range $D = [x_1, x_2] \times [y_1, y_2]$, we may perform a direct access on coordinate x followed by one-dimensional searching on coordinate y. This requires that all the distinct ordered pairs of abscissae of points of S (x-ranges) be directly accessible; each such pair (x', x'') in turn points to the binary search tree of the ordinates of the points contained in the strip $x' \leq x \leq x''$. It is convenient at this point to transform the original problem by replacing each real-valued coordinate with its rank in its set of coordinate. This transformation, called *normalization*, is accomplished by a sorting operation on each set of coordinates. After normalization of the data, an arbitrary x-range is transformed in time $O(\log N)$ into a pair of integers (i, j) in $[1, N + 1]$ $(i < j)$. Such a pair corresponds to an address in a direct-search array,[17] which contains a pointer to a binary tree with $(j - i)$ leaves. Thus, the total amount of storage is proportional to

$$\sum_{i-1}^{N} \sum_{j=i+1}^{N+1} (j - i) = \frac{[(N + 1)^3 - (N + 1)]}{6} = O(N^3).$$

Note that while we have succeeded in reducing the storage from $O(N^5)$ to $O(N^3)$, the search time has remained $O(\log N)$, since both range normalization and one-dimensional range searching use $O(\log N)$ time. The situation for our current example is illustrated in Figure 2.31 (note that the points are indexed in order of increasing abscissa). The range to be accessed in the x-array is $[2, 8]$, in normalized abscissae (see Figure 2.31(b)); from here a pointer directs the search toward a binary tree. Here the search *per se* is completed by locating P_8, after which retrieval of the sequence (P_8, P_3, P_4) is effected.

We still have little reason to be satisfied with the above proposal, not only because $O(N^3)$ storage is still a very demanding requirement, but also because in the multidimensional case the trick to introduce one-dimensional searching can be done just once, thereby knocking just a factor of N^2 from the storage use. Indeed, a straightforward analysis shows that $O(N^{2d-1})$ is the storage used when this approach is extended to d dimensions.

It is evident that the direct-access part of this two-phase search scheme is the

---

[17] For example, if this array starts at address A, the pair (i, j) could be uniquely mapped to address

$$A - 2 + \frac{2N(i - 1) + i - i^2}{2} + j.$$

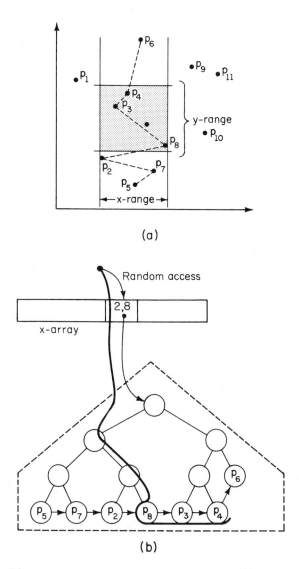

(a)

(b)

Figure 2.31   Direct-access range searching. The address of x-range is located by random access after normalization. This address contains a pointer to a binary search tree.

one responsible for the high storage use. Thus, improvement should be sought in this area. Bentley and Maurer proposed a *multistage* approach [Bentley–Maurer (1980)],[18] which we illustrate now in the 2-stage case. The idea is to use successively a coarse gauge and a fine gauge (see Figure 2.32). All coordinates are normalized, and the coarse gauge is subdivided into intervals k units

---

[18] The structures to be described were originally called "multilevel k-ranges."

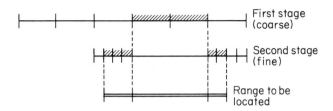

Figure 2.32   Illustration of the 2-stage direct-access scheme.

long, while there is a fine gauge for each such interval, whose divisions have
unit spacing. It is therefore clear that an arbitrary range is subdivided into at
most *three* intervals, one in the coarse gauge and two in the fine gauge (Figure
2.32). Each gauge corresponds to a direct-search array as described above.
Specifically, if the coarse gauge contains N^α divisions ($0 < \alpha \le 1$), there are
$O(N^{2\alpha})$ coarse-gauge intervals, and the storage of the corresponding data
structure is $O(N^{2\alpha}) \times O(N^{1+2\alpha})$. Similarly, each fine gauge contains $N^{1-\alpha}$
divisions, and there are N^α such structures, for a total storage of order N^α
$\times N^{(1-\alpha)2} \times N^{(1-\alpha)} = N^{3-2\alpha}$. The minimum value for the storage is attained
for $\alpha = \frac{1}{2}$ and is $O(N^2)$ without sacrificing the logarithmic behavior of the
range search time. Indeed, the search time has been increased approximately
by a factor of 3, and we have an $O(N^2, \log N)$-algorithm.

At this point, one may extend the approach by using a sequence of l gauges
of increasing fineness, with the objective to further reduce the storage at the
cost, of course, of an $O(l \log n)$ search time. This relatively simple analysis
leads to the following theorem.

Theorem 2.10. *For any $1 > \varepsilon > 0$, range searching of an N-point two-dimensional file can be effected with an $(N^{1+\varepsilon}, \log N)$-algorithm based on the multistage direct access technique.*

The most interesting conclusion one can draw from the preceding discussion of direct-access techniques is the following. One-dimensional range
searching can be done optimally. Therefore the objective to pursue is to
transform the multidimensional problem into a collection of one-dimensional
problems. The direct-access scheme is a way to effect this transformation.
Basically, in the two-dimensional case we partition an arbitrary interval into
at most *three* standard intervals (and in the d-dimensional case we still
ultimately obtain a *fixed* number of standard intervals). This approach contains the conceptual seed of the range-tree method, to be described next.[19]

---

[19] Chronologically, however, the range-tree method was developed sometime before the direct-
access methods [Bentley (1979)]. However, on a purely conceptual basis, it may be viewed as an
evolution of the latter. Embryonic ideas related to the notion of range-tree can be found in
[Bentley–Shamos (1977)].

2.3.4 The range-tree method and its variants

The search time efficiency of the direct-access method is due to the construction of a collection of "standard" intervals. The gain in storage efficiency in going from a single-stage to a double-stage scheme is due to the drastic reduction of the number of such standard intervals. One may want to further pursue this line, that is, to develop a scheme that minimizes the number of standard intervals. This is the idea underlying the *range-tree* method.

Let us consider a set of N abscissae on the x-axis, normalized to the integers in $[1, N]$ by their rank. These N abscissae determine $N - 1$ elementary intervals $[i, i + 1]$, for $i = 1, 2, \ldots, N - 1$. The device we shall use to effect the partition of an arbitrary interval $[i, j]$ is the *segment tree* we introduced in Section 1.2.3.1 of Chapter 1. We simply recall, for the reader's convenience, that an arbitrary interval whose extremes belong to the set of N given abscissae can be partitioned by the segment tree $T(1, N)$ into at most $2\lceil \log_2 N \rceil - 2$ standard intervals. Each standard interval is associated with a node of $T(1, N)$ and the nodes identifying the partition of $[i, j]$ are called the *allocation nodes* of $[i, j]$.

With this background, the use of the segment tree in range searching is straightforward. We begin, as usual, with the two-dimensional case. The segment tree $T(1, N)$ is employed for searching the x-coordinate. This search identifies a unique set of nodes (the allocation nodes). Each such node v corresponds to a set of $(E[v] - B[v])$ abscissae (see Section 1.2.3.1 for definitions), i.e., to a set of $(E[v] - B[v])$ points in the plane. The y-coordinates of these points are arranged in a standard threaded binary tree for range search in the y-direction. In summary, we construct a new data structure, called *range tree*, whose primary structure is the segment tree pertaining to the abscissae of the given point set S. Each node of this tree has a pointer to a threaded binary tree (secondary structures). For our running example, this is illustrated in Figure 2.33.

The generalization to d dimensions can be done very naturally. A set S of N points is given in d-dimensional space, with coordinate axes $x_1, x_2, \ldots, x_d$. These coordinates are processed in a specified order: first x_1, then x_2, and so on. We also assume that all coordinate values are normalized. The range tree is recursively constructed as follows:

i. A primary segment tree T^* corresponding to the set $\{x_1(p): p \in S\}$. For each node v of T^*, let $S_d(v)$ denote the set of points projecting to the interval $[B[v], E[v])$ in the x_1-coordinate. Define

$$S_{d-1}(v) \triangleq \{(x_2(p), \ldots, x_d(p)): p \in S_d(v)\},$$

a $(d - 1)$-dimensional set.
ii. Node v of T^* has a pointer to the range tree for $S_{d-1}(v)$.

Before proceeding to the performance analysis of the method, we pause a moment to observe that the segment tree can be viewed not only as a device for

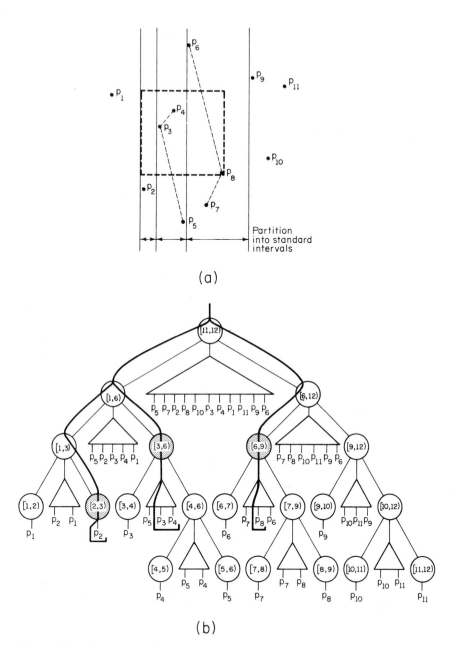

(a)

(b)

Figure 2.33   Illustration of the range-tree method applied to our running example. The given range is partitioned into three standard intervals (a). The search activity is illustrated in (b).

segmenting an interval into a logarithmic number of fragments, but also as a "recipe" for a divide-and-conquer approach to range searching.

Turning now to the analysis of performance, we observe at first that, for $d \geq 2$, each of the allocation nodes of the x_1-interval of the query range ($O(\log N)$ in number) initiates a separate $(d - 1)$-dimensional range searching problem. Specifically, node v initiates a search on $n(v) \triangleq E[v] - B[v]$ points. Denoting by $Q(N, d)$ the search time for a file of N d-dimensional points, we have the simple recurrence

$$Q(N, d) = O(\log N) + \sum_{\substack{v \in \text{allocation set} \\ \text{of } x_1\text{-interval of range}}} Q(n(v), d - 1). \qquad (2.4)$$

Here the first term in the right-side is due to searching the primary segment tree, and the second term accounts for the ensuing $(d - 1)$-dimensional subproblems. Since there are at most $2\lceil \log_2 N \rceil - 2$ allocation nodes and $n(v) \leq N$ (trivially), we obtain

$$Q(N, d) = O(\log N)Q(N, d - 1).$$

Since $Q(N, 1) = O(\log N)$—a binary search—we readily obtain for the search time

$$Q(N, d) = O((\log N)^d). \qquad (2.5)$$

Denoting by $S(N, d)$ the storage used by the range tree, we have the recurrence relation

$$S(N, d) = O(N) + \sum_{\substack{\text{all nodes } v \\ \text{of primary tree}}} S(n(v), d - 1), \qquad (2.6)$$

where the first term in the right side is due to the storage used by the primary tree and the second term is due to all the $(d - 1)$-dimensional range trees. An estimate of this term is very simple if N is a power of 2; otherwise a simple approximation will be equally effective. There are at most two nodes v with $n(v) = 2^{\lceil \log_2 N \rceil - 1}$, at most four with $n(v) = 2^{\lceil \log_2 N \rceil - 2}$, and so on. Thus we can upper-bound the sum as

$$\sum_{\substack{\text{all nodes } v \\ \text{of primary tree}}} S(n(v), d - 1) \leq \sum_{i=0}^{\lceil \log_2 N \rceil} 2^{\lceil \log_2 N \rceil - i} \cdot S(2^i, d - 1).$$

With this approximation and the observation that $S(N, 1) = O(N)$—the storage of the threaded binary tree—we obtain the solution

$$S(N, d) = O(N(\log N)^{d-1}). \qquad (2.7)$$

It is a rather straightforward exercise to see that the same line of arguments holds when evaluating the preprocessing time. Indeed a recurrence relation identical to (2.6), with all its consequences, applies to this measure of performance. We can summarize the preceding discussion in the following theorem.

Theorem 2.11. *Range searching of an N-point d-dimensional file can be effected by an $(N(\log N)^{d-1}, (\log N)^d)$-algorithm, given $O(N(\log N)^{d-1})$ preprocessing time. Such algorithm is based on the range-tree technique.*

In particular, for $d = 2$ the three measures—query time, storage, preprocessing time—become $O((\log N)^2 + k)$, $O(N \log N)$ and $O(N \log N)$,[20] respectively. A very attractive storage behavior has been achieved, albeit with some increase in the search time.

Before we resign ourselves to this trade-off as perhaps inherent to the problem, it is worth re-examining if any loss of efficiency occurs in the search process. Willard (1978) and Lueker (1978) independently did exactly this and proposed a variant of the scheme that achieves optimal search time in two dimensions. We now discuss their idea.

Each node v of the primary segment tree is linked to the *list* $Y(v)$ (no longer a binary tree!) of the points projecting to the interval of the node. This list is ordered so that the ordinates of the points are nondecreasing. To effect retrieval, we need to locate the initial element of the sublist to be retrieved. In the range-tree scheme, this location is done by binary search for each node v of the primary tree. If we consider the two offsprings of v, we know that $Y(\text{LSON}[v])$ and $Y(\text{RSON}[v])$ form a partition of $Y(v)$. Suppose now we know that p is the initial (leftmost) element of $Y(v)$ to be retrieved. Without loss of generality, if $p \in Y(\text{LSON}[v])$ there is a unique element p' in $Y(\text{RSON}[v])$ such that $y(p')$ is the smallest for which $y(p) \le y(p')$. Therefore two pointers from p in $Y(v)$ uniquely identify the corresponding initial elements in the Y-lists of the offspring of v.

The conclusions are now relatively straightforward. In the range tree we can replace each threaded binary tree with its leaf-list, for all nodes but the root. Indeed, the root structure is still a threaded binary tree, because it must support search in logarithmic time. For each nonleaf node of the segment tree, there are two pointers from each item of its Y-list to items of the Y-lists of its offsprings. This technique is sometimes referred to as "layering" and the modified range tree may be called the "layered range tree." The use of the modified range tree for our running example is shown in Figure 2.34.

The storage use does not change, since a threaded binary tree with s leaves or an s-record list with two extra pointers per record use both $O(s)$ storage. However, the search time is cut down to $O(\log N)$, because a binary search at the root performs an initial location of the left bound of the y-range. The pointer structure provides constant time updating of the left bound per each node visited in the "tour" of the segment tree. The net effect of the generalization of this scheme to higher dimensions is the removal of a $\log N$ factor from the search time performance. So we have

---

[20] Note that the range-tree method can be trivially adapted to the point-counting mode, thereby achieving $O(\log^2 N)$, $O(N \log N)$, and $O(N \log N)$ for the three measures of complexity. This confirms a result anticipated in Section 2.1 (Table I).

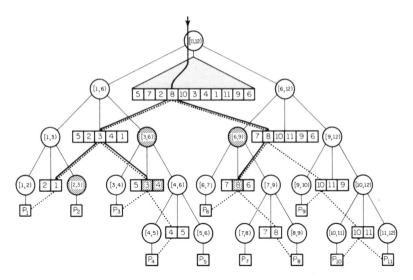

Figure 2.34 Illustration of the tour of the modified range tree for our running example. Only a fraction of the Y-list pointers is shown.

Theorem 2.12. *Range searching of an N-point d-dimensional file for $d \geq 2$ can be effected by an $(N(\log N)^{d-1}, (\log N)^{d-1})$-algorithm, given $(N(\log N)^{d-1})$ preprocessing time. Such algorithm is based on Willard–Lueker modification of the range tree, also known as layered range tree.*

2.4 Notes and Comments

Since the appearance of the paper of Dobkin–Lipton, considerable progress has been made in solving the point-location problem. As mentioned in Section 2.2, all methods presented create new geometric objects as aids in the search. So, the slab method partitions the plane into elementary empty trapezoids with two horizontal sides: two cascaded binary searches identify a unique trapezoid. The triangulation refinement method, by retriangulating portions of the PSLG, creates a suitable small set of *new* triangles that are easily searched. The trapezoid method achieves an efficient exploration of the given graph by identifying trapezoids with two horizontal sides which rapidly narrow the search domain. Finally, the chain method extracts from the planar straight-line graph a set of polygonal chains that partition the plane into montone polygons, which can be easily searched. In spite of its suboptimality, this method can be refined so as to yield an optimal and yet extremely simple technique: This amelioration is based on a clever combination of the chain method, of Willard–Lueker's pointer structure, and of "filtering search." We now briefly describe filtering search.

Filtering search [Chazelle (1983c)][21] is a general approach to report-mode searching problems, based on the computationally valid tenet that one should tolerate accessing a

---

[21] Seminal, very preliminary ideas of this philosophy can be traced to [Bentley–Maurer (1979)].

superset of the target set if there are benefits in other measures of performance (such as storage or search time), provided the superset cardinality is guaranteed not to exceed a given (small) multiple of the cardinality of the target set. In other words, the superset is the "scoop" to be retrieved and to be subsequently "filtered" in order to extract the target set; in other words, 'filtering search' could be appropriately categorized as "scoop-and-filter" search. This methodology has been successfully applied to several problems, such as orthogonal range queries, proximity searching (see Chapter 6), intersection problems, etc.

Returning to point location, we now outline how the philosophies of filtering search and of Willard–Lueker's pointer structure can be implanted on the chain method [Edelsbrunner–Guibas–Stolfi (1985)]. In the primary tree structure (where each node v is associated with a monotone chain C) there is a pointer from v to a secondary structure, whose main component is a sequence $T(v)$ containing not only the ordinates of the edges assigned to chain C (as in Lee–Preparata's version), but is augmented with *every other term* of the sequence obtained by merging the (augmented) sequences of the two offsprings of v. As in Willard–Lueker's scheme for range trees, the sequence $T(v)$ is searched by means of a binary tree only at the root of the primary structure. At all other nodes, $T(v)$ is accessed via a pointer structure which makes $O(\log N)$ search time achievable. The criterion of propagating "every other term" from a level to the next, ensures linear storage; however, twice as many regions as necessary may be accessed, and must be filtered in the spirit of filtering search.

All of the techniques presented in this chapter for planar point location are tuned to worst-case performance. In the realm of average-case techniques, a recently reported "bucket" method [Edahiro–Kokubo–Asano (1983)] appears to experimentally outperform the previously reported methods; among the latter, the trapezoid method showed the best experimental performance.

In connection with range searching, we have seen in Section 2.3 that the only method using linear storage (the k-D tree technique) has a high search time. Thus redundancy of representation seems to be the key to fast search time. Both the direct-access technique and the range-tree technique are examples of this approach. Indeed, they are instances of applications called by Bentley and Saxe "decomposable searching problems" [Saxe–Bentley (1979)], where the answer to the query over the entire space is obtained by combining (in this case, by joining) the answers to the query specialized to a suitable collection of subsets of the space. It remains an important open question whether an $(N, \log N)$ 2-dimensional range searching algorithm exists. Recently, Chazelle (1983c) using the filtering-search approach, has been able to improve the space requirement from $N \log N$ to $N \log N / \log \log N$, thus at least showing that $N \log N$ is not a lower bound.

Other versions of the range searching problem can be classified on the basis of the type of search domain. One such class is obtained by choosing the search domain as a k-gon (*polygon-range search*); assuming, as usual, an N-point file, Willard has developed an $(N, KN^{0.77})$-algorithm [Willard (1982)] later improved to an $(N, KN^{0.695})$ algorithm [Edelsbrunner–Welzl (1983)]. An interesting special case of this problem occurs when the polygon range is selected as a half-plane; for the resulting *half-planar search* an optimal $(N, \log N)$ result has been recently proposed [Chazelle–Guibas–Lee (1983)] (this technique, however, is not applicable to the corresponding counting problem). Of great interest is also the case where the search domain is a circle (*circular range search*, or *disk search*). If the disk radius is *fixed*, the problem can be solved by means of the locus approach, where the plane is partitioned into regions so that all points of a given region generate identical responses. This approach could be implemented by resorting, for example, to the trapezoid method, thereby obtaining an $(N^3 \log N, \log N)$ algorithm. This algorithm, however, is well outperformed by an $(N, \log N)$ technique recently proposed [Chazelle–Edelsbrunner (1984)]. More interesting is the *variable disk search*, where the circular domain has arbitrary radius and center; the best known technique for this problem has an $(N(\log N \log N)^2, \log N)$ behavior and will be outlined in Chapter 6 [Chazelle–Cole–Preparata–Yap (1984)].

Other searching problems refer to different geometric settings, where the (query, file item) pair is no longer of the types (point, region)–as in point location–or (region, point)–as in range searching. For example, we may have the pairs (polygon, polygon), (segment, segment), etc: in these cases the relation being searched is "intersection," so that these tasks are more appropriately studied in the context of intersection problems (Chapter 7). In particular, when the (query, file item) pair is (orthogonal range, orthogonal range), we have a host of interesting problems—either of the intersection or of the inclusion type—that are more appropriately discussed in Chapter 8 (The Geometry of Rectangles).

Finally, significant progress has been made in the study of dynamic searching structures, i.e., structures that, in addition to queries, support insertions and deletions. While one-dimensional dynamic structures (AVL trees, etc.) have been known for over twenty years, only more recently the multi-dimensional problem has been attacked. Following the pioneering work of Bentley (1979), general techniques to convert static structures, satisfying some weak constraints, into dynamic ones have been developed.

We shall mention the technique of van Leeuwen and Wood (1980) whose general principle is to organize the file as a collection of separate data structures, so that each update can be confined to one (or, possibly, a fixed small number) of them; however, to avoid shifting the burden from the updates to the query activity, one must refrain from excessive fragmentation, since queries normally involve the entire collection. Overmars' thesis (1983) is the most current and comprehensive update on the state-of-the-art of "dynamizing" and the reader is strongly encouraged to refer to it.

2.5 Exercises

1. Develop a more subtle analysis of the triangulation refinement method of Kirkpatrick and select K to attain $\alpha \cong 59/63$. (Hint: Bound separately the vertices of degree larger than K and the vertices of degree not larger than K.)

2. *Seidel.* Let S be a set of N points in the plane and A a planar subdivision with $\theta(N)$ regions.
   (a) Show that it takes $\Omega(N \log N)$ time to locate all points of S in A.
   (b) Now assume a triangulation of S is available. A triangulation contains information on how points of S are located with respect to each other. This information might help in locating all points of S in A. Show that in spite of knowing a triangulation of S it still takes $\Omega(N \log N)$ time to locate all points of S in A.

3. Prove Theorem 2.10. Assume that the search structure consists of l stages of increasingly finer gauges, and obtain a $(N^{c_2}, c_1 \log N)$-algorithm with minimal c_2. Express both c_1 and c_2 as functions of l.

4. For $d \geq 2$ can the Willard–Lueker implementation of the range tree be modified to obtain an $(N(\log N)^{d-1}, (\log N)^{d-1})$-algorithm for the count-mode range searching problem? Fully justify your answer.

5. *Edelsbrunner–Guibas–Stolfi.* Consider the chain method for planar point location on an N-vertex map and modify it as follows. Begin by assigning the chains to nodes

of a primary tree structure T as in (Section 2.2.2.2) [Lee–Preparata (1978)], so that each node v of T is assigned the sequence $Y(v)$ of the extremes of the edges of its chain; next, for each nonleaf node v form $Y^*(v)$ by augmenting $Y(v)$ with every other term of the MERGE($Y^*(\text{LSON}[v])$, $Y^*(\text{RSON}[v])$). Each $Y^*(v)$ represents a partition of the vertical line: establish a pointer from each interval of $Y^*(v)$ to each interval of either $Y^*(\text{LSON}[v])$ or $Y^*(\text{RSON}[v])$ it overlaps. This defines the search data structure T^*.

(a) Show that T^* can be stored in $O(N)$ space.

(b) Show that planar point location is effected in $O(\log N)$ time.

6. Apply the locus approach to solve the following problem (fixed-radius circular range search): Given N points in the plane and a constant $d > 0$, report (possibly, with a logarithmic-time overhead) the points that are at most at distance d from a given query point q.

CHAPTER 3

Convex Hulls: Basic Algorithms

The problem of computing a convex hull is not only central to practical applications, but is also a vehicle for the solution of a number of apparently unrelated questions arising in computational geometry. The computation of the convex hull of a finite set of points, particularly in the plane, has been studied extensively and has applications, for example, in pattern recognition [Akl–Toussaint (1978); Duda–Hart (1973)], image processing [Rosenfeld (1969)] and stock cutting and allocation [Freeman (1974); Sklansky (1972); Freeman–Shapira (1975)].

The concept of convex hull of a set of points S is natural and easy to understand. By definition, it is the smallest convex set containing S. Intuitively, if S consists of a finite set of points in the plane, imagine surrounding the set by a large, stretched rubber band; when the band is released it will assume the shape of the convex hull.

In spite of the intuitive appeal of the convex hull concept, the history of algorithms to compute convex hulls is illustrative of a general pattern in algorithmic research. Unfortunately, the simple definition of convex hull recalled above is not of a constructive nature. Thus, appropriate notions must be identified that are conducive to algorithm development.

The construction of the convex hull, in two or more dimensions, is the subject of this chapter; in the next chapter we shall consider applications, variants, and some related—but inherently different—problems. To avoid some repetition, it is convenient to develop a suitable framework for the notions pertaining to convex hulls in arbitrary dimension [Grünbaum (1967); Rockafellar (1970); McMullen–Shephard (1971); Klee (1966)]; these notions have very simple specializations in the ordinary plane and space. The next section is devoted to this task; frequent reference will be made to concepts introduced in Section 1.3.1.

3.1 Preliminaries

The points lie in the d-dimensional space E^d, and we shall use the familiar identification of points and d-dimensional vectors applied to the origin of E^d.

We begin by considering the notion of an affine set.

Definition 3.1. Given k distinct points $p_1, p_2, \ldots, p_k$ in E^d, the set of points

$$p = \alpha_1 p_1 + \alpha_2 p_2 + \cdots + \alpha_k p_k, \qquad (\alpha_j \in \mathbb{R}, \quad \alpha_1 + \alpha_2 + \cdots + \alpha_k = 1)$$

$$(3.1)$$

is the *affine set* generated by $p_1, p_2, \ldots, p_k$, and p is an *affine combination* of p_1, $p_2, \ldots, p_k$.

Note that the affine combination specializes the notion of linear combination by the added condition $\alpha_1 + \cdots + \alpha_k = 1$. If $k = 2$, the resulting affine set is the straight line through two points. Clearly, affine sets are points, lines, planes, hyperplanes, etc., that is, all structures with the intuitive nature of "flatness"; indeed, *flat* is a synonym of affine set (along with the term "affine variety"). The correspondence between vector subspaces and affine sets is expressed by the fact that each affine set is the *translation* (by a fixed vector) of a vector subspace (a *linear set*). For example, in E^3 linear sets are lines and planes through the origin (and the origin itself), while affine sets are points and lines and planes in general position.

Given k points $p_1, p_2, \ldots, p_k$ in E^d, they are said to be *affinely independent* if the $(k - 1)$ vectors $p_2 - p_1, \ldots, p_k - p_1$ are linearly independent. Intuitively, we translate the finite point set so that one point (in our case, p_1) is brought to the origin, and test the linear independence of the resulting vector set; obviously, the choice of which point is brought to the origin is immaterial. Given k affinely independent points $p_1, p_2, \ldots, p_k$, they form the affine base of a $(k - 1)$-dimensional affine set (i.e., the dimension of an affine set is the dimension of the linear set of which it is a translate).

Definition 3.2. Given a subset L of E^d, the *affine hull* aff(L) of L is the smallest affine set containing L.

In other words, for any two points in L, the entire line determined by these two points belongs to aff(L). As examples, the affine hull of a segment is a line, of a plane polygon is a plane, and so on.

Next, we turn our attention to the notion of a finitely generated convex set.

Definition 3.3. Given k distinct points $p_1, p_2, \ldots, p_k$ in E^d, the set of points

$$p = \alpha_1 p_1 + \alpha_2 p_2 + \cdots + \alpha_k p_k \qquad (\alpha_j \in \mathbb{R}, \, \alpha_j \geq 0, \, \alpha_1 + \alpha_2 + \cdots + \alpha_k = 1)$$

$$(3.2)$$

is the *convex set* generated by $p_1, p_2, \ldots, p_k$, and p is a *convex combination* of $p_1, p_2, \ldots, p_k$.

Again, note that the convex combination specializes the notion of affine combination by the added condition $\alpha_j \geq 0$, $j = 1, 2, \ldots, k$. If $k = 2$, the resulting convex set is the segment joining the two points. The *dimension* of a convex set is the dimension of its affine hull: for example, the dimension of a convex plane polygon is two, since its affine hull is a plane. We now have the following important concept.

Definition 3.4. Given an arbitrary subset L of points of E^d, the *convex hull* conv(L) of L, is the smallest convex set containing L.

In our study, we shall be exclusively concerned with the case where L is finite (or a finitely generated convex set). To characterize the structure of conv(L) for a finite point set L we need to generalize the notion of convex polygon and convex polyhedron.

Definition 3.5. A *polyhedral set* in E^d is the intersection of a finite set of closed half-spaces (a half-space is the portion of E^d lying on one side of a hyperplane).

We note that a polyhedral set is convex, since a half-space is convex and the intersection of convex sets is also convex. In particular, convex plane polygons and space polyhedra—as defined in Section 1.3.1—are 2- and 3-dimensional instances of (bounded) polyhedral sets. Generally, we shall refer to a bounded d-dimensional polyhedral set as a *convex d-polytope* (or, briefly, a d-polytope or a polytope).

The desired characterization of convex hulls is provided by the following theorem:

Theorem 3.1 [McMullen–Shepard (1971), pp. 43–47]. *The convex hull of a finite set of points in E^d is a convex polytope; conversely, a convex polytope is the convex hull of a finite set of points.*

A convex polytope is described by means of its boundary, which consists of *faces*. Each *face* of a convex polytope is a convex set (that is, a lower-dimensional convex polytope); a *k-face* denotes a k-dimensional face (that is, a face whose affine hull has dimension k). If a polytope P is d-dimensional, its $(d-1)$-faces are called *facets*, its $(d-2)$-faces are called *subfacets*, its 1-faces are *edges*, and its 0-faces are *vertices*. Clearly, edges and vertices retain their usual connotations in all dimensions. For a 3-polytope, facets are plane polygons, while subfacets and edges coincide. As we shall see later, these four classes of faces play an important role in convex hull algorithms. For uniformity, it may be useful to refer to the given d-polytope as a d-face, while the empty set becomes a (-1)-face. If P is the convex hull of a finite set S in E^d, a face of P is the convex hull of a subset T of S (i.e., it is determined by a subset of S); however, not all subsets of S determine a face.

Some types of polytopes deserve special attention. A d-polytope P is a d-*simplex* (or briefly, a *simplex*) if it is the convex hull of $(d + 1)$ affinely independent points. In this case any subset of these d vertices is itself a simplex and is a face of P. Thus every k-face contains 2^{k+1} faces[1] (of dimensions k, $k - 1, \ldots, 0, -1$). For example, for $d = 0, 1, 2$, and 3, the corresponding simplex is a vertex, an edge, a triangle, and a tetrahedron, respectively; note, for example, that a tetrahedron (a 3-face, according to the previous convention) contains one 3-face (itself), four 2-faces (triangles), six 1-faces (edges), four 0-faces (vertices) and one (-1)-face (the empty set), for a total of $16 = 2^4$ faces.

A d-polytope is called *simplicial* if each of its facets is a simplex; equivalently, each of the facets of a simplicial d-polytope contains exactly d subfacets. On the other hand (we could say "dually") a d-polytope is called *simple* if each of its vertices is incident with exactly d edges. Indeed, it is easy to realize that a simplicial polytope is the dual of a simple polytope, in the framework of the general topological duality that maps a set of dimension $s \leq d$ to a set of dimension $d - s$ (consider, also the correspondence under the more special duality embodied by polar transformations, Section 1.3.3). Simplicial and simple polytopes are quite important not only because of their structural attractiveness and—forgive the pun—simplicity, but also because they naturally arise in two typical (and dual!) situations. Indeed, referring for concreteness to the familiar 3-space, the convex hull of a finite set of points in general position is a simplicial 3-polytope (that more than 3 vertices lie on the same facet is an event of zero probability), while the intersection of a finite set of half-spaces in general position is a simple 3-polytope.

Considerable attention has been devoted to the combinatorial nature of polytopes, specifically the relationship between the numbers of faces of different dimensions. It suffices here to recall that the number $F(d, N)$ of facets of a d-polytope with N vertices could be as large as [Klee (1966)]

$$F(d, N) = \begin{cases} \dfrac{2N}{d} \cdot \dbinom{N - \dfrac{d}{2} - 1}{\dfrac{d}{2} - 1}, & \text{for } d \text{ even} \\[3em] 2 \cdot \dbinom{N - \left\lfloor \dfrac{d}{2} \right\rfloor - 1}{\left\lfloor \dfrac{d}{2} \right\rfloor}, & \text{for } d \text{ odd.} \end{cases} \tag{3.3}$$

Concisely, we can say that $F(d, N) = O(N^{\lfloor d/2 \rfloor})$. The fact that the number of

---

[1] Indeed, in this case the power set of the set of vertices is the set of faces. Its Hasse diagram is, as is well known, a $(k + 1)$-dimensional cube. Such diagram is technically called, in general, the *facial graph* of the polytope.

facets is, at worst, exponential in the number of vertices and *vice versa* (by duality), poses serious difficulties in the representation of d-polytopes for large values of d. The situation is fortunately much simpler in the important cases of $d = 2, 3$.

Specifically, for $d = 2$ the 2-polytope is a convex polygon. It is important to realize that a polygon—convex or otherwise—is an *ordered sequence* of vertices. Such sequence is adequately represented either as an array or as a bidirectional list.

Representation problems are also not too severe in three dimensions. A polyhedron may be specified completely by giving its vertices, edges, and faces. Because of Euler's formula (Section 1.3.1), the numbers of vertices, edges, and faces of a three-dimensional polyhedron are linearly related, which means that an N-vertex polyhedron can be completely represented in only $O(N)$ space. Furthermore, the skeleton of such a polyhedron (its set of edges) is a planar graph,[2] so we may represent the polyhedron by means of any data structure suitable to represent a planar graph (such as the adjacency list or the doubly-connected-edge-list described in Section 1.2.3.2).

Before proceeding to the description of algorithms, it is appropriate to formally state the problem and to address the important question of lower bounds to the complexity. Since convex hull algorithms are concerned with the *boundary* of the convex hull, for a given hull conv(L) we shall denote its boundary by CH(L). Note, however, that, according to common use, both conv(L) and CH(L) will be referred to as "convex hull."

3.2 Problem Statement and Lower Bounds

We begin by stating two fundamental versions of the convex hull problem:

PROBLEM CH1 (CONVEX HULL). Given a set S of N points in E^d, construct its convex hull (that is, the complete description of the boundary CH(S)).

PROBLEM CH2 (EXTREME POINTS). Given a set S of N points in E^d, identify those that are vertices of conv(S).

It should be clear that problem CH1 is asymptotically at least as hard as CH2 because the output of CH1 becomes a valid solution to CH2 if we merely recopy the vertex set produced by the former as an unordered list of points. That is, one problem is transformable to the other or

$$\text{EXTREME POINTS} \propto_N \text{CONVEX HULL.}$$

---

[2] This follows from Steinitz' Theorem. See [Grünbaum (1967), p. 235].

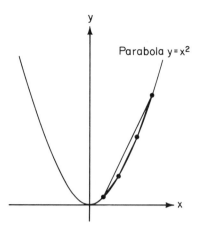

Figure 3.1   Illustration for the proof of Theorem 3.2.

It is natural to ask whether the former is asymptotically easier than the latter or if they are in fact of the same complexity. Since any collection of points in two dimensions is trivially embedded in E^d, with $d > 2$, any lower bound result obtained for $d = 2$ remains *a fortiori* valid for $d > 2$. Therefore, in the following discussion we shall consider the planar instances of Problems CH1 and CH2.

We begin by considering problem CH1, Planar convex hull. The fact that the vertices of the convex hull polygon appear in order—indeed, we may refer to the *ordered convex hull*—points to a natural connection with the problem of sorting. Indeed the following theorem formalizes the fact that any algorithm for problem CH1 must be able to sort.

Theorem 3.2. *Sorting is linear-time transformable to the convex hull problem; therefore, finding the ordered convex hull of N points in the plane requires $\Omega(N \log N)$ time.*

PROOF. We exhibit the transformability; the conclusion follows from Proposition 1 of Section 1.4. Given N real numbers $x_1, \ldots, x_N$, all positive, we must show how a convex hull algorithm can be used to sort them with only linear overhead. Corresponding to the number x_i we construct the point (x_i, x_i^2), and associate the number i with it (see Figure 3.1). All of these points lie on the parabola $y = x^2$. The convex hull of this set, in standard form, will consist of a list of the points sorted by abscissa. One pass through the list will enable us to read off the x_i in order.[3]   □

---

[3] M. I. Shamos originally proved this theorem by mapping the x_i onto the unit circle. The parabola mapping, suggested by S. Eisenstat, is superior because it requires only rational arithmetic operations.

Because the transformation involves only arithmetic operations, Theorem 3.2 holds in many computational models; namely, those in which multiplication is permitted and sorting is known to require $\Omega(N \log N)$ time. As noted earlier, Theorem 3.2 applies in all dimensions greater than one.[4]

Turning our attention to the EXTREME POINTS problem, CH2, we realize immediately that no elementary argument as the one given above is forthcoming. Indeed, this problem remained unsolved for some time; the first breakthrough was the work of A. C. Yao (1981), and a definitive answer was obtained by combining the powerful algebraic decision-tree technique of Ben-Or (see Section 1.4) with a recent result by Steele and Yao (1982).

Following a familiar pattern of thought, we consider the decision problem associated with CH2, which is formulated as follows.

PROBLEM CH3 (PLANAR EXTREME POINTS TEST). Given N points in the plane, are they vertices of their convex hull?

Before discussing the extremely important result of Steele–Yao and Ben-Or, let us consider why the linear decision tree model is inadequate for our problem. Briefly, no existing convex hull algorithm uses exclusively linear tests, so that a linear-decision-tree bound would not apply. The typical primitive operation is of type: Given three points p, p', and p'', does p lie to the left, or to the right, or on the directed segment from p' to p''? The polynomial embodying the corresponding test is the determinant

$$\Delta = \begin{vmatrix} x & y & 1 \\ x' & y' & 1 \\ x'' & y'' & 1 \end{vmatrix}, \tag{3.4}$$

where $p = (x, y)$, $p' = (x', y')$, and $p'' = (x'', y'')$, which, as observed in Section 2.2.1, gives twice the signed area of Triangle($pp'p''$). Clearly, polynomial (3.4) is *quadratic*.

This is an unfortunate circumstance, because a simple and elegant argument based on the linear-decision-tree model had been presented earlier by Avis (1979). That argument made crucial use of the fact that a linear test $f(x_1, \ldots, x_m): 0$ defines a hyperplane in the m-dimensional Euclidean space E^m, and each root-to-leaf path defines the common intersection of convex sets (half-spaces correspond to ">" or "<" test outcomes, hyperplanes correspond to "=" test outcomes), which is itself a convex set. That is, all input vectors leading to a given leaf of the decision-tree form a convex set in E^m (the decision region associated with that leaf). This property, however, disappears when the tests are of higher degree.

The clever but extremely complex argument presented by A. C. Yao was

---

[4] The convex hull of a set of points in one dimension is the smallest interval that contains them, which can be found in linear time.

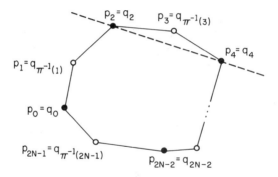

Figure 3.2   Illustration for the proof of Theorem 3.3.

restricted to the *quadratic* decision tree model. While the restriction to quadratic computation trees is adequate to cover the existing algorithms, the following question spontaneously arises: If we allow polynomials of degree higher than quadratic, can one still prove analogous results? Or, if not, can one develop faster algorithms by using such possibly more powerful tests? Yao conjectured against an affirmative answer. However, only recently has this question been definitively settled by the following argument.

Suppose we are given an algebraic-decision-tree algorithm of fixed order $d \geq 2$, which is claimed to solve CH3 for sets of $2N$ points in the plane, i.e., it accepts an input vector with $4N$ components (two coordinates per given point) and determines if their convex hull has $2N$ extreme points. Note that the $4N$-component input vector is correctly thought of as a point in E^{4N}. The decision problem we are considering is characterized by a "decision set" $W \subset E^{4N}$ (see Section 1.4). We then repeat here, for convenience, Ben-Or's result.

Theorem 1.2. *Let W be a set in E^m and let T be an algebraic decision tree of fixed order d that solves the membership problem in W. If h^* is the depth of T and $\#(W)$ is the number of disjoint connected components of W, then*

$$h^* = \Omega(\log \#(W) - m).$$

In order to use this result we must obtain a lower bound to $\#(W)$, which can be done in the following manner.

Let $(p_0, p_1, \ldots, p_{2N-1})$ be the clockwise sequence of vertices of a convex polygon (see Figure 3.2). The input to the algorithm is in the form of a string $\mathbf{z} = (z_0, z_1, \ldots, z_{4N-1})$ of real numbers, obtained as follows: given $2N$ points $q_0, q_1, \ldots, q_{2N-1}$ in the plane with $q_i \equiv (x_i, y_i)$, we set $z_{2i} = x_i$ and $z_{2i+1} = y_i$, for $i = 0, 1, \ldots, 2N - 1$. For fixed $(p_0, p_1, \ldots, p_{2N-1})$, we can construct $N!$ distinct instances of input strings for the given instance of the problem, by selecting an arbitrary permutation π of the integers $\{0, 1, \ldots, N - 1\}$ and by setting

$$q_{2s} = p_{2s}, \qquad q_{2s+1} = p_{\pi(2s+1)} \qquad (s = 0, \ldots, N - 1). \qquad (3.5)$$

Notice that all possible input sequences for the given problem have identical subsequences of even-indexed points (shown as solid circles in Figure 3.2). For given π, we denote by $\mathbf{z}(\pi)$ the sequence of the coordinates of $(q_0, \ldots, q_{2N-1})$.

For each π we now construct an array, $A(\pi)$, of N^2 entries as follows. Denoting by $\Delta(u, v, w)$ the signed area of the triangle (u, v, w), and by $A(\pi)[j]$ the j-th entry of $A(\pi)$, we have

$$A(\pi)[Ns + r] = \Delta(q_{2s}, q_{(2s+2) \bmod 2N}, q_{2r+1}), \qquad 0 \le s < N \quad \text{and} \quad 0 \le r < N. \tag{3.6}$$

It is easy to realize from (3.5) and (3.6) and from Figure 3.2 that, for any given s, there is just one value of r for which $A(\pi)[Ns + r] < 0$ (this happens when $2s + 1 = \pi^{-1}(2r + 1)$). Now let π_1 and π_2 be two *distinct* permutations of $\{0, 1, \ldots, N - 1\}$, and let us consider $A(\pi_1)$ and $A(\pi_2)$. We claim that there is at least one integer j, such that $A(\pi_1)[j]$ and $A(\pi_2)[j]$ have opposite signs. Indeed, let us consider the N entries of $A(\pi_1)$ that are negative, that is, $A(\pi_1)[Ni + r_i]$, $0 \le i < N$. If $A(\pi_1)[Ni + r_i] \cdot A(\pi_2)[Ni + r_i] > 0$, then both points $q_{\pi_1^{-1}(Ni+r_i)}$ and $q_{\pi_2^{-1}(Ni+r_i)}$ lie to the left of the directed segment $\overrightarrow{p_{2i}p_{2i+2}}$: but we know there is just one point in the set $\{p_0, \ldots, p_{2N-1}\}$ with this property, and this point is p_{2i+1}. Thus $q_{\pi_1^{-1}(Ni+r_i)} = q_{\pi_2^{-1}(Ni+r_i)} = p_{2i+1}$. If this holds for all values of i, then the two permutations coincide, contrary to the hypothesis.

So there is a j such that $A(\pi_1)[j]$ and $A(\pi_2)[j]$ have opposite signs. Observe now that $\mathbf{z}(\pi_1)$ and $\mathbf{z}(\pi_2)$ are two distinct points of E^{4N}, and consider a curve ρ in E^{4N} joining them. A point $\mathbf{z}$ travelling on ρ from $\mathbf{z}(\pi_1)$ to $\mathbf{z}(\pi_2)$ describes a continuous deformation in the plane of the configuration corresponding to π_1 to the configuration corresponding to π_2, while also $A(\pi_1)$ is transformed continuously to $A(\pi_2)$. Since $A(\pi_1)[j] \cdot A(\pi_2)[j] < 0$, by the intermediate value theorem there is a point $\mathbf{z}^*$ on ρ for which, in the deformed configuration, three points become collinear (i.e., their triangle has area 0), and so at most $(2N - 1)$ points are extreme. This proves that in passing from $\mathbf{z}(\pi_1)$ to $\mathbf{z}(\pi_2)$ we must go out to W, so that $\mathbf{z}(\pi_1)$ and $\mathbf{z}(\pi_2)$ belong to different components of W. Since this holds for any choice of π_1 and π_2 it follows that $\#(W) \ge N!$.

Combining this result with Ben-Or's theorem we have

Theorem 3.3. *In the fixed-order algebraic-computation-tree model, the determination of the extreme points of a set of N points in the plane requires $\Omega(N \log N)$ operations.*

PROOF. In our case $m = 4N$ and $\#(W) \ge N!$. Ben-Or's theorem says that $(\log \#(W) - m) \ge (N \log N - 4N) = \Omega(N \log N)$ operations are necessary. $\square$

This completes our discussion of the complexity of the problem. Before leaving the topic, we observe that the planar instances of both the ordered hull problem CH1 and the extreme point problem CH2 bear an intimate relation with sorting. Indeed, CH1 is explicitly related to it through a direct reduction, while the linkage of CH2 is mediated through the cardinality of the symmetric group. This connection is profound and we shall refer to it frequently in the rest of this chapter.

We shall now begin the description of convex hull algorithms.

3.3 Convex Hull Algorithms in the Plane

3.3.1 Early development of a convex hull algorithm

Referring to the nonconstructive nature of Definition 3.4, we must now find mathematical results that will lead to efficient algorithms. It is conventional in works of this type to present a succession of well-motivated and polished results, but to do so here would reveal very little of the algorithm design process that is central to computational geometry. Accordingly, we begin by discussing some unfruitful leads.

Definition 3.6. A point p of a convex set S is an *extreme point* if no two points $a, b \in S$ exist such that p lies on the open line segment $\overline{ab}$.

The set E of extreme points of S is the smallest subset of S having the property that $\operatorname{conv}(E) = \operatorname{conv}(S)$, and E is precisely the set of vertices of $\operatorname{conv}(S)$.

It follows that two steps are required to find the convex hull of a finite set:

1. Identify the extreme points. (This is problem CH2 in the plane.)
2. Order these points so that they form a convex polygon.

We need a theorem that will enable us to test whether a point is an extreme point.

Theorem 3.4. *A point p fails to be an extreme point of a plane convex set S only if it lies in some triangle whose vertices are in S but is not itself a vertex of the triangle.*[5] (See Figure 3.3.)

This theorem provides an algorithm for eliminating points that are not extreme: There are $\Omega(N^3)$ triangles determined by the N points of S. Whether a point lies in a given triangle can be determined in a constant number of operations so we may learn whether a specific point is extreme in $O(N^3)$ time. Repeating this procedure for all N points of S requires $O(N^4)$ time. While our algorithm is extremely inefficient, it is conceptually simple and it demon-

---

[5] This follows immediately from the proofs of Theorems 10 and 11 of [Hadwiger–Debrunner (1964)]. The generalization to d dimensions is obtained by replacing "triangle" by "simplex on $d + 1$ vertices." Note that triangles may degenerate to three collinear points.

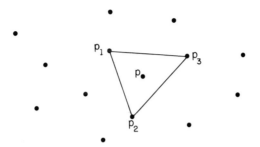

Figure 3.3 Point p is not extreme because it lies inside triangle ($p_1 p_2 p_3$).

strates that identifying extreme points can be accomplished in a finite number of steps.

We have spent $O(N^4)$ time just to obtain the extreme points, which must be ordered somehow to form the convex hull. The nature of this order is revealed by the following theorems.

Theorem 3.5. *A ray emanating from an interior point of a bounded convex figure F intersects the boundary of F in exactly one point.*[6]

Theorem 3.6. *Consecutive vertices of a convex polygon occur in sorted angular order about any interior point.*

Imagine a ray, centered at an interior point q of polygon P that makes a counterclockwise sweep over the vertices of P, starting from the positive x-axis. As it moves from vertex to vertex, the polar angle[7] subtended by the ray increases monotonically. This is what we mean by the vertices of P being "sorted" (see Figure 3.4).

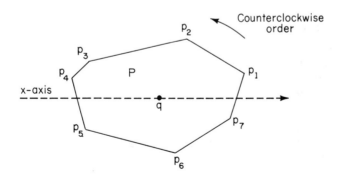

Figure 3.4 The vertices of P occur in sorted order about q.

---

[6] This is a consequence of [Valentine (1964)], Theorem 1.10, and the Jordan Curve Theorem.

[7] Polar angles are measured in the usual way, counterclockwise from the x-axis.

Given the extreme points of a set, we may find its convex hull by constructing a point q that is known to be internal to the hull and then sorting the extreme points by polar angle about q. As point q we may select the centroid[8] of the extreme points: indeed it is well known that the centroid of a set of points is internal to its convex hull.[9] The centroid of a set of N points in k dimensions can be computed trivially in $O(Nk)$ arithmetic operations.

A different method of finding an internal point is due to Graham, who observes that the centroid of any three noncollinear points will suffice [Graham (1972)]. He begins with two arbitrary points and examines the remaining $N - 2$ points in turn, looking for one that is not on the line determined by the first two. This process uses $O(N)$ time at worst, but almost always takes only constant time—if the points are drawn from an absolutely continuous distribution, then with probability one the first three points are noncollinear [Efron (1965)].

After finding the extreme points of a set S, in $O(N)$ time we may find a point q that is internal to the hull. It only remains to sort the extreme points by polar angle, using q as origin. We may do this by transforming the points to polar coordinates in $O(N)$ time, then using $O(N \log N)$ time to sort, but the explicit conversion to polar coordinates is not necessary. Since sorting can be performed by pairwise comparisons, we need only determine which of two given angles is greater; we do not require their numerical values. Let us consider this question in more detail because it illustrates a simple but important geometric "trick" that is useful in many applications: Given two points p_1 and p_2 in the plane, which one has greater polar angle? This question is readily answered by considering the signed area of a triangle. Indeed given two points p_1 and p_2, p_2 forms a strictly smaller polar angle with the real axis than p_1 if and only if Triangle (O, p_2, p_1) has strictly positive signed area (see Figure 3.5).

The details of our first planar convex hull algorithm are now complete. We have shown that the problem can be solved in $O(N^4)$ time using only arithmetic operations and comparisons.

3.3.2 Graham's scan

An algorithm that runs in $O(N^4)$ time will not allow us to process very much data. If improvements are to be made they must come either by eliminating redundant computation or by taking a different theoretical approach. In this section we explore the possibility that our algorithm may be doing unnecessary work.

To determine whether a point lies in *some* triangle defined by a set of N

---

[8] The centroid of a finite set of points $p_1, \ldots, p_N$ is their arithmetic mean $(p_1 + \cdots + p_N)/N$.

[9] That is, if the interior is nonempty. [Benson (1966), exercise 25.15] "Interior" refers to the relative (subspace) topology. The convex hull of two distinct points in E^3 is a line segment, whose interior is empty in the metric topology of E^3, but nonempty in the relative topology.

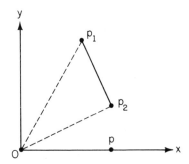

Figure 3.5 Comparing polar angles by means of the signed area of Triangle (O, p_2, p_1).

points, is it necessary to try *all* such triangles? If not, there is some hope that the extreme points can be found in less than $O(N^4)$ time. R. L. Graham, in one of the first papers specifically concerned with finding an efficient geometric algorithm [Graham (1972)] showed that performing the sorting step first enables the extreme points to be found in linear time. The method he uses turns out to be a very powerful tool in computational geometry.

Suppose that we have already found an internal point and trivially transformed the coordinates of the others so that this point is at the origin. We now sort the N points lexicographically by polar angle and distance from the origin. In performing this sort we do not, of course, compute the actual distance between two points, since only a magnitude comparison is required. We can work with distance *squared*, avoiding the square root, but this case is even simpler. The distance comparison only need be done if two points have the same polar angle, but then they and the origin are collinear and the comparison is trivial.

After arranging the sorted points into a doubly-linked circular list we have the situation depicted in Figure 3.6. Note that if a point is not a vertex of the convex hull, then it is internal to some triangle (Opq) where p and q are

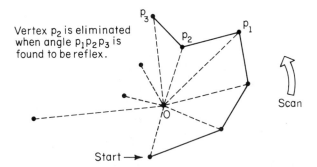

Figure 3.6 Beginning the Graham Scan.

consecutive hull vertices. The essence of Graham's algorithm is a single scan around the ordered points, during which the internal points are eliminated. What remains are the hull vertices in the required order.

The scan begins at the point labelled START, which may be taken as the rightmost smallest-ordinate point of the given set and hence is certainly a hull vertex. We repeatedly examine triples of consecutive points in counter-clockwise order to determine whether or not they define a reflex angle (one that is $\geq \pi$). If internal angle $p_1 p_2 p_3$ is reflex, then $p_1 p_2 p_3$ is said to be a "right turn," otherwise it is a "left turn." This can be determined easily by applying equation (3.4). It is an immediate consequence of convexity that in traversing a convex polygon we will make only left turns. If $p_1 p_2 p_3$ is a right turn, then p_2 cannot be an extreme point because it is internal to triangle $(Op_1 p_3)$. This is how the scan progresses, based on the outcome of each angle test:

1. $p_1 p_2 p_3$ *is a right turn.* Eliminate vertex p_2 and check $p_0 p_1 p_3$.
2. $p_1 p_2 p_3$ *is a left turn.* Advance the scan and check $p_2 p_3 p_4$.

The scan terminates when it advances all the way around to reach START again. Note that START is never eliminated because it is an extreme point. A simple argument shows that this scan only requires linear time. An angle test can be performed in a constant number of operations. After each test we either advance the scan (case 2), or delete a point (case 1). Since there are only N points in the set, the scan cannot advance more than N times, nor can more than N points be deleted. This method of traversing the boundary of a polygon is so useful that we shall refer to it as the *Graham scan*. A more precise description of the algorithm is given below, where S is a set of N points in the plane.

procedure GRAHAMHULL(S)

1. Find an internal point q.
2. Using q as the origin of coordinates, sort the points of S lexicographically by polar angle and distance from q and arrange them into a circular doubly-linked list, with START pointing to the initial vertex. (*The RLINK and LLINK associated with a node point respectively to its successor and predecessor in the list as in [Knuth (1968), p. 278].*)

3. (Scan)
begin $v :=$ START;
      while (RLINK$[v] \neq$ START) **do**
          if (the three points v, RLINK$[v]$, RLINK[RLINK$[v]$] form a left turn)
              then $v :=$ RLINK$[v]$
          else begin DELETE RLINK$[v]$;
              $v :=$ LLINK$[v]$
          end
end.

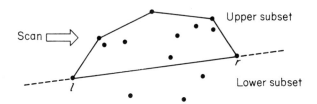

Figure 3.7   The left and right extremes partition the set into two subsets.

At the completion of execution, the list contains the hull vertices in sorted order.

Theorem 3.7. *The convex hull of N points points in the plane can be found in $O(N \log N)$ time and $O(N)$ space using only arithmetic operations and comparisons.*

PROOF. From the above discussion, only arithmetics and comparisons are used in Graham's algorithm. Steps 1 and 3 take linear time, while the sorting step 2, which dominates the computation, uses $O(N \log N)$ time. $O(N)$ storage suffices for the linked list of points.   □

By recalling the lower bound discussed in Section 3.2, we see that this simple and elegant algorithm has optimal running time. There is, however, one detail that may be a reason of concern to the readers: the use of polar coordinates. This, indeed, involves coordinate transformations that may be cumbersome in systems having a restricted set of primitives. As it happens, some researchers were equally concerned about this feature and proposed refinements that avoid it. We shall briefly describe the modification due to Andrew (1979); another noteworthy modification is due to Akl and Toussaint (1978).

Given a set of N points in the plane, we first determine its left and right extremes l and r (see Figure 3.7), and construct the line passing by l and r. We then partition the remaining points into two subsets (lower and upper) depending upon whether they lie below or above this line. The lower subset will give rise to a polygonal chain (*lower-hull* or L-*hull*) monotone with respect to the x-axis; similarly the upper subset gives rise to an analogous chain (*upper-hull* or U-*hull*) and the concatenation of these two chains is the convex hull.

Let us consider the construction of the upper-hull. The points are ordered by increasing abscissa and the Graham scan is applied to this sequence. In this manner, trigonometric operations are avoided. Notice, however, that this is nothing but a specialization of the original Graham method, where the reference point q—the pole—is chosen at $-\infty$ on the y-axis of the plane, whereby the ordering of abscissae is consistent with the ordering of polar angles.

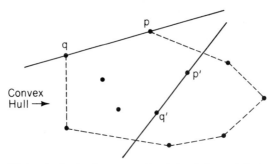

$\overline{pq}$ is a hull edge because all points of the set lie to one side of it.
$\overline{p'q'}$ is not a hull edge because there are points on both sides of it.

Figure 3.8   A hull edge cannot separate the set.

Even though we have shown that Graham's algorithm is optimal, there are still many reasons for continuing to study the convex hull problem.

1. The algorithm is optimal in the *worst-case* sense, but we have not yet analyzed its expected performance.
2. Because it is based on Theorem 3.6, which applies only in the plane, the algorithm does not generalize to higher dimensions.
3. It is not on-line since all points must be available before processing begins.
4. For a parallel environment we would prefer a recursive algorithm that allows the data to be split into smaller subproblems.

Before closing, we note that Graham's method makes explicit use of sorting; one could hardly find a more direct connection to the sorting problem. Just as the study of sorting reveals that no single algorithm is best for all applications, we will find the same to be true of hull-finding. Let us continue to explore combinatorial geometry, looking for ideas that may lead to other hull algorithms.

3.3.3 Jarvis's march

A polygon can equally well be specified by giving its edges in order as by giving its vertices. In the convex hull problem we have concentrated so far on isolating the extreme points. If we try instead to identify the hull edges, will a practical algorithm result? Given a set of points, it is fairly difficult to determine quickly whether a specific point is extreme or not. Given *two* points, though, it is straightforward to test whether the line segment joining them is a hull edge.

Theorem 3.8. *The line segment l defined by two points is an edge of the convex hull if and only if all other points of the set lie on l or to one side of it* [Stoer–Witzgall (1970), Theorem 2.4.7]. (See Figure 3.8.)

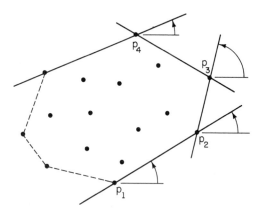

Figure 3.9 The Jarvis' march for constructing the convex hull. The algorithm of Jarvis finds successive hull vertices by repeatedly turning angles. Each new vertex is discovered in $O(N)$ time.

There are $\binom{N}{2} = O(N^2)$ lines determined by all pairs of N points. For each of these lines we may examine the remaining $N - 2$ points and apply relation (3.4) to determine in linear time whether the line meets the criteria of the theorem. Thus in $O(N^3)$ time we are able to find all pairs of points that define hull edges. It is then a simple matter to arrange these into a list of consecutive vertices.

Jarvis has observed that this algorithm can be improved if we note that once we established that segment $\overline{pq}$ is a hull edge, then another edge must exist with q as an endpoint [Jarvis (1973)]. His paper shows how to use this fact to reduce the time required to $O(N^2)$ and contains a number of other ideas that are worth treating in detail. We shall report here a version which includes a modification, due to Akl (1979), obviating a minor incorrectness.

Assume that we have found the lexicographically lowest point p_1 of the set as in Section 3.3.2. This point is certainly a hull vertex, and we wish to find the next consecutive vertex p_2 on the convex hull. This point p_2 is the one that has the least polar angle with respect to p_1 as origin. Likewise, the next point p_3 has least polar angle with respect to p_2 as origin, and each successive point can be found in linear time. Jarvis's algorithm marches around the convex hull, (hence, the appropriate name of *Jarvis's march*), finding extreme points in order, one at a time. (Refer to Figure 3.9.) In this manner one correctly constructs the portion of the convex hull (a polygonal chain) from the lexicographically lowest point (p_1 in Figure 3.9) to the lexicographically highest point (p_4 in the same figure). At this point, one completes the convex hull polygon by constructing the other chain from the lexicographically highest point to the lexicographically lowest. Due to the symmetry of the two steps, one must reverse the direction of the two axes, and refer now to least polar angles with respect to the *negative* x-axis.

As we have already seen in Section 3.3.1 the smallest angle can be found with arithmetics and comparisons alone, without actually computing any polar angles.

Since all N points of a set may lie on its convex hull, and Jarvis's algorithm expends linear time to find each hull points, its worst-case running time is $O(N^2)$, which is inferior to Graham's. If h is the actual number of vertices of the convex hull, Jarvis's algorithm runs in $O(hN)$ time, which is very efficient if h is known in advance to be small. For example, if the hull of the set is a polygon of any constant number of sides, we can find it in *linear* time. This observation is quite important in connection with the average-case analysis of convex hull algorithms to be presented in the next chapter.

Another appropriate remark is that the idea of finding successive hull vertices by repeatedly turning angles is intuitively suggestive of wrapping a two-dimensional package. Indeed, Jarvis's method can be viewed as a two-dimensional specialization of the "gift-wrapping" approach, proposed by Chand and Kapur (1970) even before the appearance of Jarvis's paper. The gift-wrapping approach applies also to more than two dimensions and will be outlined in Section 3.4.1.

3.3.4 QUICKHULL techniques

The sorting problem is a source of inspiration of ideas for the convex hull problem. For example, the central idea of QUICKSORT can be easily recognized in some techniques which—with minor variants—have been proposed independently and almost at the same time [Eddy (1977); Bykat (1978); Green and Silverman (1979); Floyd (private communication, 1976)]. Due to this close analogy with the QUICKSORT algorithm, we choose to refer to them as QUICKHULL techniques.

To better elucidate the analogy, let us briefly recall the mechanism of QUICKSORT [Hoare (1962)]. We have an array of N numbers and we aim at partitioning it into a left and a right subarray, such that each term of the first is no larger than each term of the second. This is done by establishing two pointers to array cells, which are initially placed respectively at the two extremes of the array and move, one cell at a time, toward each other. Any time the two terms being addressed by the pointers violate the desired order, an exchange occurs. The two pointers move alternately—one moving and one halting—the roles being exchanged at each effected exchange. Where the two pointers collide the array is split into two subarrays, and the same technique is separately applied to each of them. As is well known, this approach is very efficient (resulting in $O(N \log N)$ time) if each array partition is approximately balanced.

The corresponding QUICKHULL technique partitions the set S of N points into two subsets, each of which will contain one of two polygonal chains whose concatenation gives the convex hull polygon. The initial partition is determined by the line passing through the two points l and r with

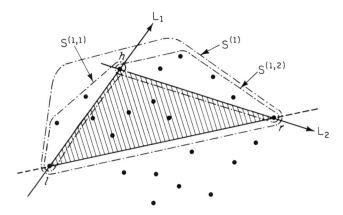

Figure 3.10   Points l, r, and h subdivide the set $S^{(1)}$ and eliminate all points in the shaded triangle from further consideration.

smallest and largest abscissae (refer to Figure 3.10), as in Andrew's variant of Graham's algorithm (see Section 3.3.2). Let $S^{(1)}$ be the subset of the points on or above the line through l and r; $S^{(2)}$ is symmetrically defined as the subset of points on or below the same line. (Technically, $\{S^{(1)}, S^{(2)}\}$ is not a partition of S, since $S^{(1)} \cap S^{(2)} \supseteq \{l, r\}$; this minor detail, which supplies a convenient symmetry, should in no way trouble the reader.)

Each successive step operates on sets like $S^{(1)}$ and $S^{(2)}$ in the following manner (we refer for concreteness to $S^{(1)}$ in Figure 3.10). We determine a point $h \in S^{(1)}$ such that the triangle (hlr) has maximum area among all triangles $\{(plr): p \in S^{(1)}\}$, and among all triangles having this maximum area—if there are more than one—the angle $\angle (hlr)$ is maximum. Note then, that point h is guaranteed to belong to the convex hull. Indeed, if we trace a line parallel to segment $\overline{lr}$ and passing through h, there are no points of S above this line; there may be other points of S on this line besides h, but by our choice, h is the leftmost of them. Thus h cannot be expressed as the convex combination of two other points of S.

Next we construct two lines, one L_1 directed from l to h, the other L_2 from h to r. Each point of $S^{(1)}$ is then tested with respect to each of these lines: there is clearly no point lying simultaneously to the left of both L_1 and L_2, while all those to the right of both of them are in the interior of the triangle (lrh) and can therefore be eliminated from further consideration. The points not to the left of L_2 but lying on or to the left of L_1 form a set $S^{(1,1)}$; similarly a set $S^{(1,2)}$ is formed. The newly formed sets $S^{(1,1)}$ and $S^{(1,2)}$ are then passed to the subsequent level of recursion.

The primitives used by the technique are evaluations of areas of triangles and discrimination of points with respect to lines. Both types of primitives require a few additions and multiplications.

Now that the approach has been explained at an intuitive level, we mention a way to unify with the general step the apparent anomaly of the initial partition: the corresponding choice $\{l_0, r_0\}$ of l and r has $l_0 = (x_0, y_0)$ as a point of S with smallest abscissa, while r_0 is chosen as $(x_0, y_0 - \varepsilon)$, with arbitrarily small positive ε. This effectively selects the initial partition line as the vertical line by l_0. At the completion of execution, r_0 is eliminated (setting $\varepsilon = 0$ identifies l_0 and r_0).

We can now give a less informal description of the technique. Here S is assumed to contain at least two points and FURTHEST $(S; l, r)$ is a function computing point $h \in S$ as explained above; also, QUICKHULL returns an ordered list of points, and "$*$" denotes "list concatenation."

    function   QUICKHULL$(S; l, r)$
1. **begin if** $(S = \{l, r\})$ **then return** (l, r) (*a single convex hull directed edge*)
2.   **else begin** $h := $ FURTHEST$(S; l, r)$;
3. $S^{(1)} := $ points of S on or to the left of $\overrightarrow{lh}$;
4. $S^{(2)} := $ points of S on or to the left of $\overrightarrow{hr}$;
5.   **return** QUICKHULL$(S^{(1)}; l, h)*($QUICKHULL$(S^{(2)}; h, r) - h)$
             end
   end.

Thus, once the function QUICKHULL is available, the following simple program carries out the desired task.

begin $l_0 = (x_0, y_0) := $ point of S with smallest abscissa;
       $r_0 := (x_0, y_0 - \varepsilon)$;
       QUICKHULL$(S; l_0, r_0)$;
       delete point r_0 (*this is equivalent to setting $\varepsilon = 0*$)
end.

The extraction from S of $S^{(1)}$ and $S^{(2)}$—with the implicit elimination of the points internal to the triangle (lrh)—as effected by lines 2, 3, and 4 above, is carried out with $O(N)$ operation. This is followed by recursive calls on $S^{(1)}$ and $S^{(2)}$. Now, if each of the latter has cardinality at most equal to a constant fraction of the cardinality of S and this holds at each level of recursion, then the algorithm runs in time $O(N \log N)$. However, in the worst case QUICKHULL—in spite of its simplicity—suffers from the same disability as QUICKSORT, resulting in an $O(N^2)$ running time.

3.3.5 Divide-and-conquer algorithms

The QUICKHULL technique described in the preceding section is not only of very simple implementation but attempts to achieve one of the goals set forth at the end of Section 3.3.2 (goal 4, parallelizability). Indeed, it subdivides the original problem into two subproblems, each of which can be solved independently, and thus simultaneously, in a parallel environment. In addition the

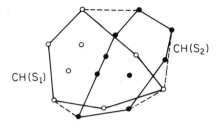

Figure 3.11   Forming the hull by divide-and- conquer. By dividing S into two subsets and finding their hulls recursively we can reduce the problem to finding the hull of the union of two convex polygons.

combination of the results of the two subproblems—the so-called "merge" step—is the very simple concatenation of the two results. Unfortunately, this simplicity has as a price the inability to accurately control the sizes of the two subproblems and therefore to guarantee an acceptable worst-case performance.

Therefore, although an example of the divide-and-conquer paradigm, which is so fundamental in algorithm design, QUICKHULL has obvious shortcomings. Indeed, the central feature of divide-and-conquer is to invoke the principle of balancing [Aho–Hopcroft–Ullman (1974), p. 65], which suggests that a computational problem should be divided into subproblems of nearly equal size. Suppose that in the convex hull problem we have split the input S into two parts, S_1 and S_2, each containing half of the points. If we now find $CH(S_1)$ and $CH(S_2)$ separately but recursively, how much additional work is needed to form $CH(S_1 \cup S_2)$, that is, the hull of the original set? To answer this we may use the relation

$$CH(S_1 \cup S_2) = CH(CH(S_1) \cup CH(S_2)). \qquad (3.7)$$

While at first glance equation (3.7) seems to involve more work than just finding the hull of the set directly, it is crucial to note that $CH(S_1)$ and $CH(S_2)$ are convex *polygons*, not just unordered sets of points. (See Figure 3.11.)

PROBLEM CH4 (HULL OF UNION OF CONVEX POLYGONS). Given two convex polygons P_1 and P_2 find the convex hull of their union.

Apart from its inherent interest, this problem is important because it is the merge step of a divide-and-conquer procedure and is thus a fundamental geometric tool. We cannot hope that the final hull algorithm will be efficient unless the subproblem solutions can be combined quickly.

procedure MERGEHULL(S)
1. If $|S| \leq k_0$ (k_0 is a small integer), construct the convex hull directly by some method and stop; else go to Step 2.
2. Partition the original set S arbitrarily into two subsets S_1 and S_2 of approximately equal cardinalities.

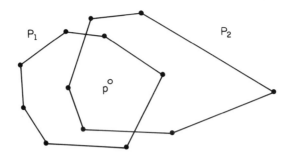

Figure 3.12   Point p lies inside P_2. Since p lies inside both polygons, the vertices of P_1 and P_2 occur in sorted order about p and can be merged in linear time.

3. Recursively find the convex hulls of S_1 and S_2.
4. Merge the two hulls together to form CH(S).

Let $U(N)$ denote the time needed to find the hull of the union of two convex polygons, each having $N/2$ vertices. If $T(N)$ is the time required to find the convex hull of a set of N points, then applying equation (3.7) gives

$$T(N) \le 2T(N/2) + U(N). \tag{3.8}$$

The following "merge" algorithm is due to M. I. Shamos [Shamos (1978)].

procedure HULL OF UNION OF CONVEX POLYGONS(P_1, P_2)

1. Find a point p that is internal to P_1. (For example, the centroid of any three vertices of P_1. This point p will be internal to CH($P_1 \cup P_2$).)
2. Determine whether or not p is internal to P_2. This can be done in $O(N)$ time by the method of Section 2.2.1. If p is not internal, go to step 4.
3. p is internal to P_2 (see Figure 3.12). By Theorem 3.6, the vertices of both P_1 and P_2 occur in sorted angular order about p. We may then merge the lists in $O(N)$ time to obtain a sorted list of the vertices of *both* P_1 and P_2. Go to step 5.
4. p is not internal to P_2 (see Figure 3.13). As seen from p, polygon P_2 lies in a wedge whose apex angle is $\le \pi$. This wedge is defined by two vertices u and v of P_2 which can be found in linear time by a single pass around P_2. These vertices partition P_2 into two chains of vertices that are monotonic in polar angle about p, one increasing in angle, the other decreasing. Of these two chains, the one convex toward p can be immediately discarded, since its vertices will be internal to CH($S_1 \cup S_2$). The other chain of P_2 and the boundary of P_1 constitute two sorted lists that contain a total of at most N vertices. They can be merged in $O(N)$ time to form a list of vertices of $P_1 \cup P_2$, sorted about p.
5. The Graham scan (step 3 of GRAHAMHULL) can now be performed on the obtained list, which uses only linear time. We now have the hull of $P_1 \cup P_2$.

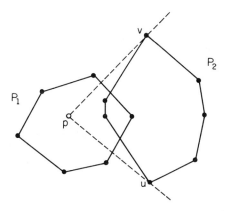

Figure 3.13   Point p is external to P_2. As seen from p, polygon P_2 lies in a wedge defined by vertices u and v, which partition P_2 into two chains of vertices. One may be discarded, the other merged with the vertices of P_1 in linear time.

If polygon P_1 has m vertices and P_2 has n vertices, this algorithm runs in $O(m + n)$ time, which is certainly optimal, so we know that $U(N) = O(N)$, giving $T(N) = O(N \log N)$ as the solution of recurrence (3.8). Thus we have

Theorem 3.9. *The convex hull of the union of two convex polygons can be found in time proportional to their total number of vertices.*

   A byproduct of the described merge technique is the calculation of the "supporting lines," when they exist, of two convex plygons. A *supporting line* of a convex polygon P is a straight line l passing through a vertex of P and such that the interior of P lies entirely on one side of l (in a sense, the notion of supporting line is analogous to the notion of tangent). Clearly, two convex polygons P_1 and P_2, with n and m vertices respectively, such that one is not entirely contained in the other, have common supporting lines (at least two and as many as $2 \min(n, m)$). Once the convex hull of the union of P_1 and P_2 has been obtained, the supporting lines are computed by scanning the vertex list of $CH(P_1 \cup P_2)$. Any pair of consecutive vertices of $CH(P_1 \cup P_2)$, originating one from P_1 and one from P_2, identifies a supporting line.
   An alternative technique to find the union of two disjoint convex polygons was independently developed by Preparata–Hong (1977) and is based on finding in linear time the two supporting lines of these polygons. This method, however, will not be illustrated here.

3.3.6 Dynamic convex hull algorithms

Each of the convex hull algorithms we have examined thus far requires all of the data points to be present before any processing begins. In many geometric applications, particularly those that run in real-time, this condition cannot be met and some computation must be done as the points are being received. In general, an algorithm that cannot look ahead at its input is referred to as *on-line*, while one that operates on all the data collectively is termed *off-line*.

A general feature of on-line algorithms is that no bound is placed on the update time, or equivalently, a new item (point) is input on request as soon as the update relative to the previously input item has been completed. We shall refer to the time interval between two consecutive inputs as the *interarrival delay*. A more demanding case of on-line applications occurs when the interarrival delay is outside the control of the algorithm. In other words, inputs are not acquired on request by the algorithm, but occur with their own independent timing; we shall assume, however, that they be *evenly* spaced in time. For this situation, the update must be completed in time no greater than the constant interarrival delay. Algorithms for this mode of operation are appropriately called in *real-time*. It should be pointed out that frequently known on-line algorithms are globally less efficient than the corresponding off-line algorithms (some price must be paid to acquire the on-line property).

The purpose of this section is to develop on-line convex hull algorithms tailored to specific requirements.

PROBLEM CH5 (ON-LINE CONVEX HULL). Given a sequence of N points in the plane, $p_1, \ldots, p_N$, find their convex hull in such a way that after p_i is processed we have $CH(\{p_1, \ldots, p_i\})$.

PROBLEM CH6 (REAL-TIME CONVEX HULL). Given a sequence of N points in the plane $p_1, \ldots, p_N$, find their convex hull on-line assuming constant interarrival delay.

The algorithm must maintain some representation of the hull and update it as points arrive; the question is whether this can be done without sacrificing $O(N \log N)$ worst-case running time for processing the entire set.

It is not a challenge to exhibit an on-line hull algorithm if execution time is not a consideration. For example, as each point is received we could run Graham's algorithm and obtain $O(N^2 \log N)$ behavior. We can be slightly more clever by recalling that Graham's algorithm consists of separate sort and scan steps.

1. Input points until three noncollinear ones are found. Their centroid will be internal to the final hull and is thus a suitable origin for sorting by polar angle as in algorithm GRAHAMHULL.
2. Maintain a linked list of the ordered extreme points. As point p_i is received, insert it temporarily into the list, according to its polar angle, in $O(i)$ time.
3. Perform a Graham scan of the linked list. Since the Graham scan is linear, this requires only $O(i)$ time. There are three possible outcomes of this step:

   a. p_i is an extreme point so far, and its inclusion causes some other points to be eliminated.

b. p_i is an extreme point, but no others are eliminated.

c. p_i is internal to the developing hull and is removed.

In any case, the size of the list increases by at most one. The total time used by this algorithm is $O(N^2)$ in the worst case, which occurs if each new point survives as an extreme point.

It may appear that the above procedure could be improved if we were able to perform *binary* insertion at Step 2 instead of linear insertion, and an equally efficient search (i.e., logarithmic) at Step 3 instead of a linear Graham scan.

Following this lead, Shamos (1978) designed an algorithm for an N-point set which globally runs in time $\theta(N \log N)$, and is therefore optimal, but exhibits an update time $O(\log^2 N)$. To analyze it with reference to the real-time property, we note that lower bounds for off-line algorithms apply equally well to on-line algorithms. This fact can be used to give lower bounds on the processing time that an on-line algorithm must use between successive inputs. Let $T(N)$ be the worst-case time used by an on-line algorithm in solving a problem with N inputs. Let $U(i)$ be the time used between inputs i and $i + 1$. If $L(N)$ is a lower bound on the time sufficient to solve the problem, then

$$T(N) = \sum_{i=1}^{N-1} U(i) \geq L(N). \tag{3.9}$$

which gives a lower bound on $U(i)$. In the present case, Theorem 3.2 and Equation (3.9) together lead to the following theorem, which, for given N, establishes a lower bound to the constant interarrival delay mentioned in the formulation of Problem CH6.

Theorem 3.10. *Any on-line convex hull algorithm must spend* $\Omega(\log N)$ *processing time between successive inputs, in the worst case.*

We therefore conclude that the algorithm mentioned above fails to achieve the real-time property. Subsequently, however, Preparata (1979) developed an on-line algorithm with the same global running time but whose update time is $O(\log N)$, thereby matching the bound on the interarrival delay established by Theorem 3.10. The two algorithms are closely related, and we shall limit our discussion to the latter.

The key to an efficient on-line algorithm is the observation that all we need is to be able to construct rapidly the two supporting lines (see Section 3.3.5) from a point to a convex polygon. Specifically, if points are processed in the sequence $p_1, p_2, \ldots$, let p_i be the current point and C_{i-1} be the convex hull of $\{p_1, p_2, \ldots, p_{i-1}\}$. We must obtain the supporting lines from p_i to C_{i-1} if they exist (i.e., if p_i is external to C_{i-1}); they do not exist if and only if p_i is internal to C_{i-1}. In the latter case, p_i is eliminated; in the former case, the appropriate chain of vertices contained between the two vertices of support is to be eliminated, while p_i is to replace them. The situation is illustrated in Figure 3.14. For ease of reference we shall call the two supporting lines as *left* and *right*, as they appear to an observer placed in p_i and facing C_{i-1}.

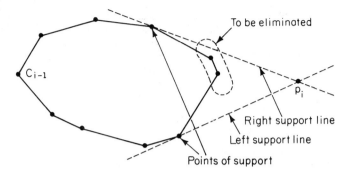

Figure 3.14   Supporting lines from a point p_i to a convex polygon C_{i-1}.

We shall therefore tackle the construction of the supporting lines from a point p to a convex polygon C. (The data structure in which the vertices of C are to be held has not yet been specified. We shall choose one after deciding which operations it must support.) The significant notion for our purpose is the following classification of each vertex v of C with respect to the segment $\overline{pv}$ (see Figure 3.15). We say that vertex v is *concave* (we should add with respect to segment $\overline{pv}$, but we shall normally imply this qualification) if the segment $\overline{pv}$

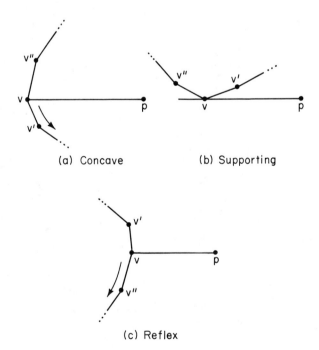

Figure 3.15   Classifications of a vertex v of a polygon C with respect to a segment $\overline{pv}$. (The arrows denote the directions of march when seeking a left supporting line.)

intersects the interior of C; otherwise, if the two vertices adjacent to v lie on the same side of the line containing $\overline{pv}$, v is *supporting*; in the remaining case, v is *reflex*. We leave it as an exercise to show that a vertex v can be classified in constant time.

If vertex v is supporting, our task has been completed. Otherwise, to make a concrete instance, suppose we are seeking a left supporting line. Then we must march (or, hopefully, jump!) on the polygon C counterclockwise or clockwise from v depending upon whether v is concave or reflex (see Figure 3.15). In this manner we can determine the two points of support (if they exist). Once this is completed, we must be able to delete from the cycle of the vertices of C a (possibly empty) string of vertices and insert p into the occurring gap.

It is now clear what must be expected of the on-line data structure. We must be able to efficiently perform the following:

  i. SEARCH an ordered string of item (the hull vertex cycle) to locate the supporting lines from p_i;
 ii. SPLIT a string into two substrings and CONCATENATE two strings (SPLICE);
iii. INSERT one item (the current p_i).

The data structure which nicely fits these specifications is well-known and is called a *concatenable queue*.[10] It is realized by means of a height-balanced search tree and each of the above operations can be performed in $O(\log i)$ worst-case time, where i is the number of nodes of the tree. Of course, in the tree data structure, called T, the cycle of vertices appears as a chain, where the first and last items are to be thought of as adjacent. In T, two vertices will be used as references: m, the leftmost member, and M, the root member. In addition, we shall make use of the angle α defined as $\angle (mp_iM)$, which is classified as *convex* ($\leq \pi$) or *reflex* ($> \pi$).

Depending upon the classifications of vertices m and M (concave, supporting, or reflex) and angle α, we can have a total of eighteen cases. However these cases can be conveniently reduced to eight (which cover all possibilities) as is summarized in Table I and illustrated in Figure 3.16. The diagrams of Figure 3.16 are to be read as follows: The "circle" on which points M and m lie stands for the polygon P; the ordered sequence of vertices starts at m and runs counterclockwise on the circle; $L(M)$ and $R(M)$ are the vertex sequences stored in the left and right subtree of the root of T.

Each of the cases illustrated in Figure 3.16 requires a distinct action to locate the left and right points of support, referred to as l and r respectively.

Let us consider first cases 2, 4, 6, and 8. Here, l and r are known to exist (because p cannot be internal) and are to be found in separate subtrees of the root of T (one of these subtrees is extended to include the root itself). Thus, l

---

[10] See Section 1.2.3, [Aho–Hopcroft–Ullman (1974)], and [Reingold–Nievergelt–Deo (1977)] for further details.

Table I

Case	α	m	M
1	convex	concave	concave
2	convex	concave	nonconcave
3	convex	nonconcave	reflex
4	convex	nonconcave	nonreflex
5	reflex	reflex	reflex
6	reflex	reflex	nonreflex
7	reflex	nonreflex	concave
8	reflex	nonreflex	nonconcave

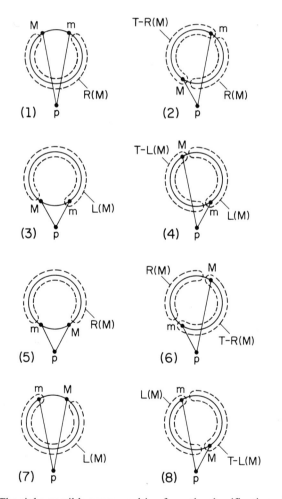

Figure 3.16   The eight possible cases resulting from the classification of m, M, and α.

and r are to be found by analogous functions. For example, l is found by the following:

function LEFTSEARCH(T)
Input: a tree T, describing a sequence of vertices
Output: a vertex l
1. **begin** $c := \text{ROOT}(T)$;
2.   **if** ($\overline{pc}$ is supporting) **then** $l := c$
3.   **else begin if** (c is reflex) **then** $T := \text{RTREE}(c)$;
4. $l := \text{LEFTSEARCH}(T)$
          end;
5.   **return** l
  end.

Clearly, LEFTSEARCH involves tracing a path of T, spending a bounded amount at each node to classify the point associated with it, and so does the analogous RIGHTSEARCH.

Considering now cases 1, 3, 5, and 7, here both l and r are to be found in the same subtree of the root, if they exist (note however, that in cases 1 and 7 p could be internal to P). Thus in each case, the search must call itself recursively in a subtree of the current tree[11] (the one corresponding to the boundary sequence encircled in Figure 3.16) until one of the cases 2, 4, 6, or 8 occurs; at this point separate LEFT and RIGHTSEARCHes are initiated. If we reach a leaf of T without verifying any of those cases, then p is internal to the polygon. In conclusion, in the most general case, the determination of the supporting lines from p can be visualized as tracing an initial path from the root of T down to a node c, at which two separate paths originate. Since the tree T is balanced and contains at most $i < N$ vertices, and a bounded amount of work is expended at each node, this activity uses $O(\log i)$ time.

To complete our description, we consider the restructuring of C_{i-1} which is in order when p_i is external to it. The vertices between l and r must be deleted and p_i inserted in their place. Slightly different actions (refer to Figure 3.17) are required depending upon whether l precedes r in T or not. In the first case we must split twice and splice once; in the second, only two splittings occur.

As mentioned earlier both SPLIT and SPLICE are performable in $O(\log i)$ time. In summary, since the convex hull update can be done in time $O(\log i)$ we have

Theorem 3.11. *The convex hull of a set of N points in the plane can be found on-line in $\theta(N \log N)$ time with $\theta(\log N)$ update time, that is, in real-time.*

---

[11] To ensure that each recursive call be done in $O(1)$ time, it is important that the elements m and M of the subtree be readily available. This is trivial for M, the root of the subtree. However, m of a right subtree is obtainable from the current root by "threading" the tree, i.e., by introducing a pointer NEXT, linking vertex v with its successor v_{i+1} on the boundary of the polygon.

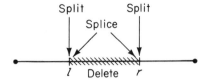

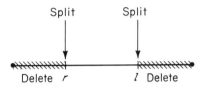

Figure 3.17   Different actions are required to delete the points becoming internal to the convex hull depending upon the relative positions of l and r.

3.3.7 A generalization: dynamic convex hull maintenance

The technique described in the preceding section could be viewed as the maintenance of a data structure, describing the convex hull of a set of points, when the only operations permitted are *insertions*. The most natural question arising at this point is: Can we design a data structure, organizing a set of points in the plane and describing their current convex hull, under the hypothesis that not only *insertions* but also *deletions* are permitted?

   This question, as may be expected, has no simple answer. Indeed, whereas in the on-line convex hull algorithm of Section 3.3.6 points found to be internal to the current convex hull are definitively eliminated from consideration, in this new situation we must carefully organize *all* points currently in the set because the deletion of a current hull point may cause several internal points to resurface on the hull boundary (refer to Figure 3.18).

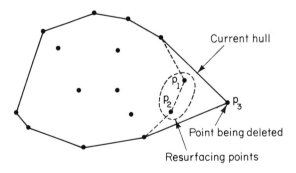

Figure 3.18   Points p_1 and p_2 resurface on the hull when p_3 is deleted.

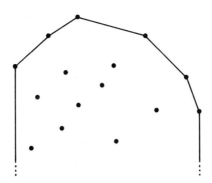

Figure 3.19   The U-hull of a point set.

This was the problem considered by Overmars and van Leeuwen, which can be formalized as follows.

PROBLEM CH7 (HULL MAINTENANCE). Given an initially empty set S and a sequence of N points $(p_1, p_2, \ldots, p_N)$ each of which corresponds either to an insertion or to a deletion from S (of course, a point can be deleted only if present in S), maintain the convex hull of S.

We shall now illustrate the very interesting solution of this problem proposed by Overmars and van Leeuwen (1981).

First, we exploit the fact that the convex hull boundary is the union of two (convex) monotone chains. Typically, one may consider the two chains comprised between two points of smallest and largest x-coordinate, respectively. These chains are upward and downward convex, and have been appropriately called the U-hull and L-hull of the point set, respectively (U for upper, L for lower). For example, the U-hull of a point set S is obtained (see Figure 3.19) as the ordinary convex hull of $S \cup \{\infty_L\}$, where ∞_L is the point $(0, -\infty)$ in the plane.[12] After observing that the convex hull of S is obtained by trivially intersecting (concatenating) its U-hull and L-hull, we may restrict our analysis to one of the latter, say the U-hull.

Not surprisingly, both the on-line convex hull technique of Section 3.3.6 and the present fully dynamic technique use trees as data structures. However, there is a basic difference in the representation of the hull in the two approaches. In the former, each tree node represents a point. In the latter, only leaves are used to represent points, while each internal node represents the U-hull of its leaves.

The data structure T is organized as follows. Its skeletal component is a height-balanced binary search tree T (with some care, a 2-3 tree [Aho–Hopcroft–Ullman (1974), p. 146] could be used instead) whose leaves are

---

[12] This notion was also used in Andrew's variant of Graham's algorithm (see Section 3.3.2).

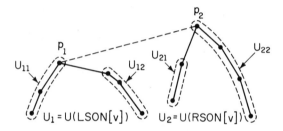

Figure 3.20   To obtain the U-hull of the union of U_1 and U_2 we must construct the common supporting segment (bridge) $\overline{p_1 p_2}$.

designed to store the points in the current set. The search is intended to operate on the abscissa x, so that the left-to-right leaf traversal gives the point set in sorted horizontal order. Note now that the sequence of points on the U-hull (the *vertices* of it) is also ordered by increasing abscissa, so that it is a *subsequence* of the global point sequence stored in the leaves.

Let v denote a node of T, with LSON[v] and RSON[v] denoting its left and right offsprings, respectively. We wish to be able to construct the U-hull of the (points stored in the) leaves of the subtree rooted at v. Since we shall have to perform splitting and splicing of chains of vertices, we assume that each such chain is represented as a *concatenable queue* (see Section 3.3.6). Notationally, let $U(v)$ denote the U-hull of the set of points stored in the leaves of the subtree rooted at v. Now, assume inductively that $U(\text{LSON}[v])$ and $U(\text{RSON}[v])$ are available, i.e., the U-hulls pertaining to the offsprings of node v. How can we construct $U(v)$? Referring to Figure 3.20, all we need is to identify the two support points p_1 and p_2 of the single common supporting segment of the two hulls. To this end we need a function BRIDGE(U_1, U_2) to produce the support segment of two U-hulls U_1 and U_2. BRIDGE effectively allows us to split U_1 into the ordered pair of chains (U_{11}, U_{12}) (refer to Figure 3.20) and similarly U_2 into (U_{21}, U_{22}). We stipulate that support point $p_1 \in U_1$ be assigned to U_{11}, and point $p_2 \in U_2$ be assigned to U_{22} (i.e., in both cases to the "external" subchain). At this stage, by splicing U_{11} to U_{22} we obtain the desired U-hull of $U_1 \cup U_2$. It is natural to have each node v of T point to a concatenable queue representing that portion of $U(v)$ which does not belong to $U(\text{FATHER}[v])$.

Suppose we wish to perform the converse operation, i.e., reassemble $U(\text{LSON}[v])$ and $U(\text{RSON}[v])$ assuming that $U(v)$ is available. All that is needed in this case is, with the above notation, knowledge of the bridging edge $\overline{p_1 p_2}$, i.e., a single integer $J[v]$ denoting the position of p_1 in the vertex chain $U(v)$. With this information $U(v)$ can be split into chains U_{11} and U_{22}, which in turn can be spliced with the chains stored at LSON[v] and RSON[v], respectively. In conclusion, the data structure T is completed by adding to each node v of T the following items.

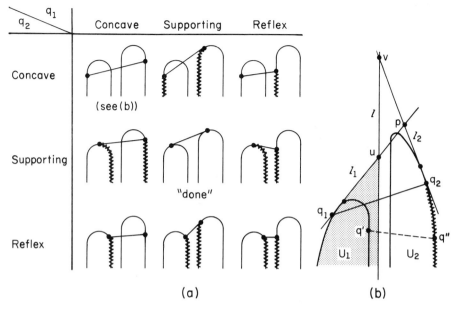

Figure 3.21   (a) All possible cases arising when choosing a vertex in U_1 and a vertex in U_2. (b) Illustration of the (concave, concave) case.

i. A pointer to a concatenable queue $Q[v]$, storing the portion of $U(v)$ not belonging to $U(\text{FATHER}[v])$ (if v is the root, then $Q[v] = U(v)$).
ii. An integer $J[v]$ denoting the position of the left support point on $U(v)$.

This interesting data structure uses only $O(N)$ space, where N is the size of the current point set. Indeed, the skeletal tree T has N leaves and $N - 1$ internal nodes, while the points stored in the concatenable queues represent a partition of the point set.

Since the operations of splitting and splicing concatenable queues are standard, we shall concentrate on the operation **BRIDGE** for which Overmars and van Leeuwen (1981) propose the following solution.

Lemma 3.1. *The bridging of two separated convex chains of N points (in total) can be done in $O(\log N)$ steps.*

PROOF. Given two U-hulls U_1 and U_2 and two vertices $q_1 \in U_1$ and $q_2 \in U_2$, each of these two vertices can be readily classified with respect to the segment $\overline{q_1 q_2}$ as either reflex, or supporting, or concave. (See Section 3.3.6 for an explanation of these terms.) Depending upon this classification there are nine possible cases, which are schematically illustrated in Figure 3.21(a). The wiggly subchains are those which can be eliminated from further contention for containing a support point. All cases are self-explanatory, except the case

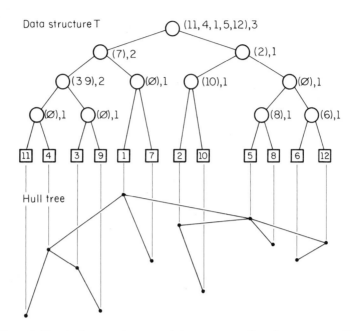

Figure 3.22   A planar point set and the corresponding data structure T.

$(q_1, q_2) = $ (concave, concave), which is further illustrated in Figure 3.21(b). Let line l_1 contain q_1 and its right neighbor on U_1; similarly let l_2 contain q_2 and its left neighbor on U_2, and let p be the intersection of l_1 and l_2. Recall that U_1 and U_2 are separated, by hypothesis, by a vertical line l. Assume, at first that p is to the right of l. We observe that support point p_1 can only belong to the shaded region, and that u has a lower ordinate than v. This implies that each vertex q'' on the subchain to the right of q_2 appears concave with respect to the segment $\overline{q'q''}$, where q' is any vertex of U_1. This shows that the chain to the right of q_2 can be eliminated from contention, but no similar statement can be made for the chain to the left of q_1. If intersection p is to the left of l, then we can show that the chain to the left of q_1 can be eliminated.

In all cases a portion of one or both hulls is eliminated. If this process starts from the roots of both trees representing U_1 and U_2, respectively, since these trees are balanced, BRIDGE(U_1, U_2) will run in time $O(\log N)$, where N is as usual the total number of vertices in the two hulls.   □

With the function BRIDGE at our disposal, we may analyze the dynamic maintenance of the planar convex hull. A typical situation is illustrated in Figure 3.22. Here points are indexed by their order of insertion into the set. In the data structure T, each leaf corresponds to a point, whereas each nonleaf node corresponds to a bridge and is shown with the pair $Q[v]$, $J[v]$. It is immediate to realize that data structure T describes a free tree on the set of

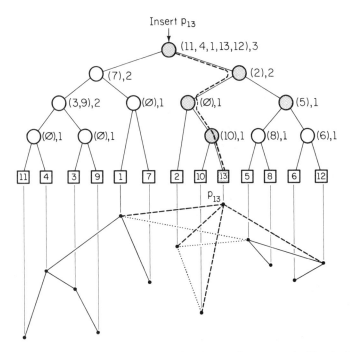

Figure 3.23   Insertion of point p_{13}. Shaded nodes are those perturbed by the insertion.

points currently being maintained and is appropriately referred to as the *hull tree*. It is suggested that the reader spend a little time to fully absorb the details of Figure 3.22.

Suppose now we wish to insert a new point p. The insertion must not only produce a height-balanced T by employing the usual rebalancing techniques [Reingold–Nievergelt–Deo (1977)], but must also perform whatever is necessary to ensure that the concatenable queue associated with each vertex still conforms with the definition of the data structure T. For simplicity, let us assume that no rebalancing is necessary (all rebalancing activities simply increase in no significant way the number of nodes to be processed, with no basic conceptual difference). Therefore let p_{13}, the point to be inserted in the point set of Figure 3.22, be as shown in Figure 3.23. In this figure, we have also shown the state of the data structure T after the insertion of p_{13}. Note that p_{13} uniquely identifies a path from the root of T to a leaf (where p_{13} is to be inserted). While we trace this path from the root, at each node v we visit we assemble the U-hull $U(v)$ pertaining to that node and subsequently we disassemble it—using the parameter $J[v]$—in order to pass the appropriate portions to its two offsprings. In this manner we have at the sibling u of each node on the path the complete representation—as a concatenable queue $U[u]$—of $U(u)$. Less informally this descent in the tree T for the insertion of a point p is carried out by a call DESCEND(root(T), p), where DESCEND(v, p) is the following recursive procedure using the functions SPLIT and SPLICE discussed earlier.

procedure DESCEND(v, p)
begin if $(v \neq \text{leaf})$ **then**
      begin $(Q_L, Q_R) := \text{SPLIT}(U[v]; J[v])$
          $U[\text{LSON}[v]] := \text{SPLICE}(Q_L, Q[\text{LSON}[v]]);$
          $U[\text{RSON}[v] := \text{SPLICE}(Q[\text{RSON}[v]], Q_R);$
          if $(x(p) \leq x[v])$ **then** $v := \text{LSON}[v]$ **else** $v := \text{RSON}[v];$
          DESCEND(v, p)
      end
end.

At this point, we insert the new leaf and begin tracing the same path of T toward the root. At each node v on this path we arrive with the complete U-hull pertaining to that node. We have just shown that at v also the U-hull of the sibling of v is available. So we now bridge these two hulls, effectively splitting the U-hull $U[v]$ into two portions Q_1 and Q_2, and, correspondingly, $U[\text{SIBLING}[v]]$ into Q_3 and Q_4. The portions Q_2 and Q_3 are to be kept at v and SIBLING$[v]$, respectively, while the portiom Q_1 is to be passed to the father of v, where it will be spliced with the analogous portion Q_4 forwarded by the sibling of v. In this manner we have obtained the U-hull of the father of v and the ascent toward the root can proceed. Less informally again, we use the following procedure.

procedure ASCEND(v)
begin if $(v \neq \text{root})$ **then**
      begin $(Q_1, Q_2, Q_3, Q_4; J) := \text{BRIDGE}(U[v], U[\text{SIBLING}[v]]);$
          $Q[\text{LSON}[\text{FATHER}[v]]] := Q_2;$
          $Q[\text{RSON}[\text{FATHER}[v]]] := Q_3;$
          $U[\text{FATHER}[v]] := \text{SPLICE}(Q_1, Q_4);$
          $J[\text{FATHER}[v]] := J;$
          ASCEND(FATHER$[v]$)
      end;
    else $Q[v] := U[v]$
end.

From the performance standpoint, we recall that SPLIT and SPLICE each run in time $O(\log k)$, where k is the size of the queue, before splitting or after splicing. Since $k \leq N$, we see that each node visit in DESCEND costs $O(\log N)$ time. Since the depth of T is also $O(\log N)$, DESCEND has a worst-case running time $O(\log^2 N)$. As to ASCEND, we have shown earlier that BRIDGE also runs in time $O(\log N)$, whereby the same bound applies to this case.

A similar analysis applies to the deletion of a point from the current set, and we leave it as a challenging exercise for the reader. We may therefore summarize this section with the following theorem.

Theorem 3.12. *The* U-*hull and the* L-*hull of a set of N points in the plane can be dynamically maintained at the worst-case cost of* $O(\log^2 N)$ *per insertion or deletion.*

Notice that if we use this technique to construct the convex hull of an N-point set on-line—that is, requiring only insertions—we achieve an $O(N \log^2 N)$ running time against the $O(N \log N)$ of the less powerful technique. Clearly, we are paying a price for using a technique which is more powerful than demanded by the application.

3.4 Convex Hulls in More Than Two Dimensions

We are now ready to consider the problem of constructing the convex hull of a finite set of points in more than two dimensions. We have already seen that such a hull is a convex polytope; we have also seen that polytopes in many dimensions are not as simple geometric objects as their two-dimensional counterparts, convex polygons. We recall, however, that in three dimensions the numbers v, e, and f of vertices, edges and faces of the hull boundary (a polyhedral surface) are related by Euler's formula $v - e + f = 2$.

In the higher-dimensional cases, the following terminology is particularly useful; we say that a point p is *beneath* a facet F of a polytope P if the P lies in the open half-space determined by hyperplane aff(F) and containing P. (In other words aff(F) is a supporting hyperplane of P, and v and P belong to the same half-space bounded by it.) Point p is *beyond* F if p lies in the open half-space determined by aff(F) and *not* containing P. See Figure 3.24 for an illustration of the notions in two dimensions.

3.4.1 The gift-wrapping method

As we mentioned earlier (Section 3.1.2), no reasonably efficient convex hull algorithms of a finite set of points is known that does not produce a complete description of the boundary (facial graph) of the convex hull polytope. In the d-dimensional case, one must therefore seek a careful organization of the computation of the facets of the convex hull in order to cut the likely overhead. A significant attempt in this direction is the "gift-wrapping" method proposed by Chand and Kapur (1970). The analysis of the technique was produced by Bhattacharya (1982) about a decade later.

The basic idea is to proceed from a facet to an adjacent facet, in the guise in which one wraps a sheet around a plane-bounded object. The simplest instance of this principle is Jarvis's march, which we discussed earlier (Section 3.3.3): in that case the material used to wrap is better likened to a rope than to a sheet of paper. The latter idea is more attuned to three dimensions, and we

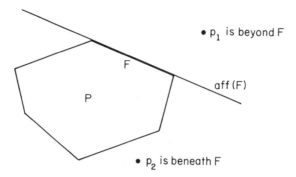

Figure 3.24   Two-dimensional illustration of the "beneath/beyond" notions.

shall frequently refer to this case to appeal to the reader's intuition. It must be stressed, however, that the three-dimensional instantiation is motivated by purely presentational needs and does not impair the generality of the approach. All the ensuing discussion, however, is based on the assumption that the resulting polytope be *simplicial* (see Section 3.1); we shall later comment on the general case.

For a simplicial *d*-polytope, we recall that each facet, which is a $(d - 1)$-simplex, is determined by exactly *d* vertices. In addition we have the following straightforward theorem.

Theorem 3.13. *In a simplicial polytope, a subfacet is shared by exactly two facets and two facets F_1 and F_2 share a subfacet e if and only if e is determined by a common subset, with $(d - 1)$ vertices, of the sets determining F_1 and F_2 (F_1 and F_2 are said adjacent on e).*

This theorem is the basis of the method, which uses a subfacet *e* of an already constructed facet F_1 to construct the adjacent facet F_2, which shares *e* with F_1. In this sense, subfacets are the fundamental geometric objects of the technique.[13]

Let $S = \{p_1, p_2, \ldots, p_N\}$ be a finite set of points in E^d, and assume that a facet *F* of CH(S) is known, with all its subfacets. The mechanism to advance from *F* to an adjacent facet *F'*—sharing with *F* the subfacet *e*—consists in determining, among all points in *S* which are not vertices of *F*, the point *p'* such that all other points are beneath the hyperplane aff($e \cup p'$). In other words, we seek among all hyperplanes determined by *e* and a point of *S* not in *F*, the one which forms the "largest angle" (in some appropriate sense) with aff(*F*). The situation is illustrated for $d = 3$ in Figure 3.25(a). Here we consider the

---

[13] Indeed, the gift-wrapping method is referred to as "edge-based," to mean "subfacet-based" in our terminology.

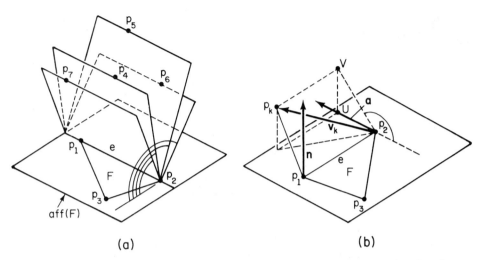

(a)   (b)

Figure 3.25   (a) Illustration of the half-plane determined by e and p_6 forming the largest convex angle with the half-plane containing F. (b) Illustration of the calculation of the cotangent.

collection of half-planes sharing the line through e and seek the half-plane which forms the largest angle $< \pi$ (*convex angle*) with the half-plane containing F: in the illustration the half-plane containing point p_6 is the result of our search. The angle comparison is carried out by comparing cotangents. Let $\mathbf{n}$ be the unit normal to F (in the "beneath" half-space of aff(F)), and let $\mathbf{a}$ be a unit vector normal to both edge e and vector $\mathbf{n}$ (so that $\mathbf{a}$ is oriented like $\mathbf{n} \times \overrightarrow{p_2 p_1}$). Also let $\mathbf{v}_k$ denote the vector $\overrightarrow{p_2 p_k}$. The cotangent of the angle formed by the half-plane containing F with the half-plane containing e and p_k is given by the ratio $-|\overline{Up_2}|/|\overline{UV}|$, where $|\overline{Up_2}| = \mathbf{v}_k \cdot \mathbf{a}^T$ and $|\overline{UV}| = \mathbf{v}_k \cdot \mathbf{n}^T$. Thus for each p_k not in F, we compute the quantity

$$\rho_k \triangleq -\mathbf{v}_k \cdot \mathbf{a}^T / \mathbf{v}_k \cdot \mathbf{n}^T \qquad (3.10)$$

and select ρ_i so that

$$\rho_i = \max_k \rho_k. \qquad (3.11)$$

The latter equation has a unique solution in the simplicial case.

Once we have visualized the situation in our familiar three-dimensional space, it is relatively easy to show the validity of (3.10) and (3.11) for arbitrary dimension d as a solution to our problem. The computational cost of this primitive operation is readily assessed. Assuming that vector $\mathbf{n}$ is known (the unit normal to aff(F)), let subfacet e be determined by vertices $\{p'_1, p'_2, \ldots, p'_{d-1}\}$. Therefore vector $\mathbf{a}$ is orthogonal to both $\mathbf{n}$ and each of the $(d-2)$ vectors $\overrightarrow{p'_i p'_{d-1}}$, and it can be determined by solving a system of $(d-2)$

equations in $(d-1)$ unknowns, and by normalizing the result to achieve unit length. This computation involves $O(d^3)$ arithmetic operations. The computation of each ρ_k uses $O(d)$ arithmetic operations (see (3.10)), where the selection of ρ_i uses $O(Nd)$ operations. At this point, we can construct the unit normal to the new facet F', which is proportional to $-\rho_i\mathbf{a} + \mathbf{n}$. Thus, the overall cost of advancing from F to an adjacent F' (*gift-wrapping step*) is $O(d^3) + O(Nd)$.

Once we have understood the mechanism of the "gift-wrapping" primitive, we can describe the overall organization of the algorithm. The algorithm starts from a given initial facet (we shall see later how to obtain it), and, for each subfacet of it, it constructs the adjacent facet. Next, it moves to a new facet and proceeds until all facets have been generated. For each newly obtained facet we construct all of its subfacets, and keep a pool of all subfacets which are candidates for being used in the gift-wrapping step. (Note that a subfacet e, shared by facets F and F', can be such a candidate only if either F or F', *but not both*, have been generated in the gift-wrapping exploration.) The orderly visit of all facets is best organized by means of an ordinary *queue* Q of facets and of a file $\mathscr{T}$, the "pool" of subfacets.

procedure GIFTWRAPPING$(p_1, \ldots, p_N)$
1. **begin** $Q := \varnothing;\ \mathscr{T} := \varnothing;$
2. $F :=$ find an initial convex hull facet;
3. $\mathscr{T} \Leftarrow$ subfacets of F;
4. $Q \Leftarrow F$;
5.   **while** $(Q \neq \varnothing)$ **do**
6.   **begin** $F \Leftarrow Q$ (*extract front element from queue*);
7. $T :=$ subfacets of F;
8.   **for each** $e \in T \cap \mathscr{T}$ **do** (*e is a gift-wrapping candidate*)
9.   **begin** $F' :=$ facet sharing e with F; (*giftwrapping*)
10.   insert into $\mathscr{T}$ all subfacets of F' not yet present and delete all those already present;
11. $Q \Leftarrow F'$
            end;
12.   output F
        end
    end.

This algorithm encompasses several major activities whose implementation must now be analyzed in detail. They are

Step 2: Find an initial convex hull facet;
Step 7: Generate the subfacets of facet F;
Step 8: Check if subfacet e is a candidate;

Step 9 is the gift-wrapping step, which has been discussed earlier. We now consider each of the above three operations, and evaluate their performance.

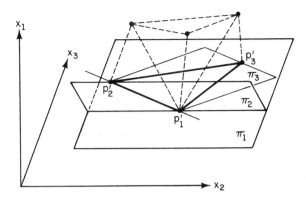

Figure 3.26   The sequence (π_1, π_2, π_3) of hyperplanes, so that π_3 contains the initial facet of the process.

Step 2. "Find an initial convex hull facet." The idea is to obtain a hyperplane containing a facet of the convex hull by successive approximations, that is, by constructing a sequence of d supporting hyperplanes, $\pi_1, \pi_2, \ldots, \pi_d$, each of which shares with the convex hull one more vertex than the preceding one. In essence, the technique is an adaptation of the gift-wrapping mechanism, where at the j-th of the d iterations the hyperplane π_j contains a $(j-1)$-face of the convex hull. Thus we begin by determining a point of least x_1-coordinate (call it p_1'); this point is certainly a vertex (a 0-face) of the convex hull. Therefore, hyperplane π_1 is chosen orthogonal to vector $(1, 0, \ldots, 0)$ and passing by p_1'. After this initialization, at the generic j-th iteration ($j = 2, \ldots, d$) hyperplane π_{j-1} has normal $\mathbf{n}_{j-1}$ and contains vertices $p_1', p_2', \ldots, p_{j-1}'$. The vector $\mathbf{a}_j$ is chosen to be normal to $\mathbf{n}_{j-1}$, to each of the $(j-2)$ vectors $\overrightarrow{p_1'p_2'}, \overrightarrow{p_1'p_3'}, \ldots, \overrightarrow{p_1'p_{j-1}'}$, and to each of the coordinate axes $x_{j+1}, x_{j+2}, \ldots, x_d$. These $(d-1)$ conditions determine $\mathbf{a}_j$ and therefore the gift-wrapping primitive can be applied with vectors $\mathbf{a}_j$ and $\mathbf{n}_{j-1}$. The sequence of hyperplanes is illustrated for $d = 3$ in Figure 3.26. The computational work of each iteration is essentially due to the construction of $\mathbf{a}_j$ and to the gift-wrapping selection of the next vertex p_j'. The latter is of complexity $O(Nd)$, while only $O(d^2)$ work is needed to determine $\mathbf{a}_j$, since $\mathbf{a}_j$ is obtained by adding just *one* linear constraint to those determining $\mathbf{a}_{j-1}$. Thus, since we have d iterations, the construction of the initial face is completed in time $O(Nd^2) + O(d^3) = O(Nd^2)$, since $N \geq d$.

Step 7. "Generate the subfacets of facet F." Due to the simplicial polytope assumption, each facet is determined by exactly d vertices and each subset of $(d-1)$ vertices of F determines a subfacet. Thus the subfacets of F can be generated in a straightforward manner in time $O(d^2)$. Each facet will be described by a d-component vector of the indices of its vertices, while a subfacet will be described by an analogous $(d-1)$-component vector.

Step 8. "Check if a subfacet e is a candidate." A subfacet is a candidate if it is contained in *just one* facet generated by the algorithm, therefore if we keep a file $\mathcal{T}$ of such subfacets, a search of this file is the required test. As noted above, a subfacet is described by a $(d-1)$-component vector of integers (the indices of the vertices determining it). We can now arrange the file $\mathcal{T}$ lexicographically and store it as a height-balanced binary search tree. Given a subfacet e, we may access $\mathcal{T}$ in two different circumstances: either to test e for being a candidate (line 8 of algorithm) or to update the file (line 10). In the first case, we search the file $\mathcal{T}$ for membership of e; in the second case, we also search $\mathcal{T}$ for membership of e, but, in addition, if e is already present we delete it, otherwise we insert it. Both searches are carried out in time $O(d \log M)$, where M, the size of the file $\mathcal{T}$, is bounded above by $SF(d, N)$, the maximum number of subfacets.[14]

We can now estimate the overall performance of the algorithm. Initialization (lines 1–4) has a cost $O(Nd^2)$. Steps 6, 11, and 12 (addition and removal from the queue, or output of a facet) can be treated together; denoting by φ_{d-1} and φ_{d-2} the actual numbers of facets and subfacets of the polytope, respectively, their overall complexity is $O(d) \times \varphi_{d-1}$. The overall complexity of Step 7—generation of all subfacets—is $O(d) \times \varphi_{d-2}$, for each subfacet is constructed in time $O(d)$ and is generated twice. The test embodied by line 8 as well as the file update in line 10, are carried out in time $O(d \cdot \log \varphi_{d-2})$ per subfacet, whence their overall complexity is $(d \log \varphi_{d-2}) \times \varphi_{d-2}$. Finally, the overall complexity of Step 9—gift-wrapping—is $\varphi_{d-1} \times (O(d^3) + O(Nd))$. Recalling that the maximum values of both φ_{d-1} and φ_{d-2} are $O(N^{\lfloor d/2 \rfloor})$, we have the following conclusion.

Theorem 3.14. *The convex hull of a set of N points in d-dimensional space can be constructed by means of the gift-wrapping technique in time $T(d, N) = O(N \cdot \varphi_{d-1}) + O(\varphi_{d-2} \cdot \log \varphi_{d-2})$. Thus in the worst-case $T(d, N) = O(N^{\lfloor d/2 \rfloor + 1}) + O(N^{\lfloor d/2 \rfloor} \log N)$.*

Finally, we cannot ignore the *caveat* that the above technique is applicable under the simplicial polytope assumption. What happens in the general, albeit improbable case, of more than d vertices per facet? If a facet F is not a simplex, equation (3.9) has more than one solution, i.e., there are several points of the set S realizing the condition. All of these points belong to the hyperplane $\mathrm{aff}(F)$; however, to determine the face structure of F (i.e., the $(d-2)$-faces, ..., 0-faces) we must recursively solve a $(d-1)$-dimensional convex hull problem in $\mathrm{aff}(F)$. Thus the generation of the subfacets—which was straightforward in the simplicial case—becomes considerably more involved. The recursive definition of the algorithm for the general case, of course, makes its analysis substantially more difficult.

---

[14] $SF(d, N) = O(N^{\lfloor d/2 \rfloor})$ [Grünbaum (1967), 9.6.1].

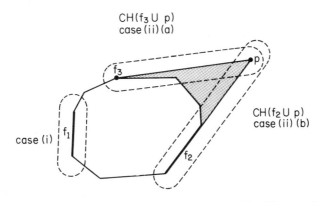

Figure 3.27 Illustration of the cases contemplated by Theorem 3.15.

3.4.2 The beneath-beyond method

For a long time the gift-wrapping method was the only known general technique to compute convex hull of a finite point set in d-dimensional space. A new technique, called the "beneath-beyond" method has been proposed [Kallay (1981)]. This technique exhibits, in addition, the on-line property.

The basic idea is very different from that of gift-wrapping and is akin to the dynamic two-dimensional convex hull algorithm described in Section 3.3.6. Informally, the technique considers one point p at a time and, if p is external to the current hull P, it constructs the supporting "cone" of P from p and removes the portion of P which falls in the "shadow" of this cone. (The intuitive ideas of cone and shadow are supplied by the three-dimensional case.) The method is based on the following theorem [McMullen–Shepard (1971), p. 113].

Theorem 3.15. *If P is a polytope and p a point in E^d, let $P' = CH(P \cup p)$. Then each face of P' is of one of the following types:*
(i) *a face f of P is also a face of P' if and only if there exists a facet F of P such that $f \subset F$ and p is beneath F.*
(ii) *if f is a face of P, then $f' = CH(f \cup p)$ is a face of P' if and only if either*
    (a) *among the facets of P containing f there is at least one such that p is beneath it and at least one such that p is beyond it, or*
    (b) *$p \in \mathrm{aff}(f)$.*

Intuitively, case (i) concerns the faces of P which are not in the cone shadow and become faces of P', the updated hull. (See Figure 3.27, where all cases are illustrated in the simple, but nonetheless general, two-dimensional case.) Case (ii) concerns the faces of the cone; specifically, subcase (ii)(a) identifies the faces of support of the cone, while (ii)(b) identifies the degenerate instance of case (a). Note that in subcase (ii)(a), the dimension of the face $f' = CH(f \cup p)$

of P' is one higher than the dimension of f of P. Although the objective may be, in general, the construction of just the facets of the convex hull, the dimensional step-up encompassed by case (ii)(a) forces us to maintain *an adequate description of all faces* of the current polytope.

Since the algorithm is on-line—i.e., it constructs the convex hull by introducing one point at a time—it will take at least $(d + 1)$ iterations before the polytope reaches full dimensionality d; that is, we obtain first a 0-face, then a 1-face, and so on. The update mechanism, however, is uniform, and we shall describe it in the general case. For simplicity, we assume that the points are in general positions, so that no degeneracy occurs.

An adequate description of the current polytope P is its facial graph $H(P)$ (Hasse diagram of the set of faces of P under the relation of inclusion, Section 3.1), from which the node corresponding to P itself is removed as redundant when P has reached full dimensionality. As a data structure, we could use $(d + 1)$ lists $L_{-1}, L_0, L_1, \ldots, L_{d-1}$, where L_j is the list of j-faces. Each record in L_j pertains to a j-face f and contains:

(1) The affine base $BASE(f)$ of f;
(2) Pointers $SUPER(f)$ to each $(j + 1)$-face containing f and pointers $SUB(f)$ to each $(j - 1)$-face contained in f:
(3) Pointers $FACETS(f)$ to each facet containing f.

Pointers (3) are crucial to carry out the beneath-beyond test (case (ii) (a) in Theorem 3.15): the other data are needed to obtain the necessary information should a face step-up its dimension to ultimately evolve into a facet.

The update of P due to the introduction of a new point p can be effected in more than one way. The following approach is particularly simple to describe, although it is susceptible of several algorithmic improvements.

Let P be described by its facial graph $H(P)$. We now create another graph $H_p(P)$, isomorphic to $H(P)$, whose vertices are defined as

$$\{f' : f' = CH(f \cup p), \quad \text{for each } f \text{ of } P\}$$

and, for each f of P we also establish an edge to represent the inclusion $f \subset f'$ (see Figure 3.28(b)). All that is involved in the construction of f' is the addition to $BASE(f)$ of the point p, since the points are assumed in general position (which guarantees $p \notin aff(f)$). We call $H(P, p)$ the graph thus obtained.

Clearly $H_p(P)$ contains the supporting "cone" of P from p; indeed, when the dimension of P is less than d, $H_p(P)$ is exactly the supporting cone, and the update is completed at this point with $H(P, p) = H(CH(P \cup p))$.

We now consider the update mechanism when the dimension of P is d. Assuming that p is external to P the desired $H(CH(P \cup p))$ is a *refinement* of $H(P, p)$. Specifically we have:

(i) $H(P)$ properly contains a portion falling in the cone shadow. Indeed, each facet of P such that, for each facet F containing f, p is beyond F, belongs to the shadow and is to be deleted (in our example of Figure 3.28 this holds for p_3, e_{34}, and e_{23}). We make use of the following straightforward property.

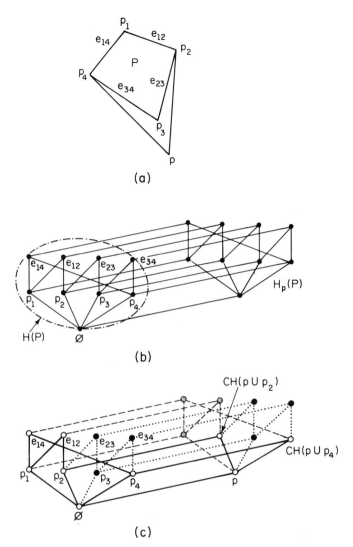

Figure 3.28   Illustration of the update of P. (a) Geometry of the update. (b) Creation of $H_p(P)$ as a "translation" of $H(P)$. (c) Refinement of the diagram of (b) to obtain $H(CH(P \cup p))$.

Property 1. *If $f \notin CH(P \cup p)$, and $f \subset f'$, then $f' \notin CH(P \cup p)$.*

Application of Property 1 enables us to delete all ascendants of f in the graph $H(P, p)$. (The faces to be so deleted are shown as solid circles in Figure 3.28(c).) For each deleted face f, we remove also its incident edges (i.e., both SUPER(f) and SUB(f)).

(ii) $H_p(P)$ contains the supporting cone. Indeed, each face f of P such that, for each facet F containing f, p is beneath F, is also a face of $CH(P \cup p)$ (Case i of Theorem 3.15), so that $CH(f \cup p) \in H_p(P)$ is to be deleted (in our example, this holds for $f = p_1, e_{14}, e_{12}$). We make use of the following straightforward property.

Property 2. *If* $CH(f \cup p) \notin CH(P \cup p)$, *and* $f \subset f'$, *then* $CH(f' \cup p) \notin CH(P \cup p)$.

Application of Property 2 enables us to delete all ascendants of $CH(f \cup p)$ in $H_p(P)$. (The faces to be so deleted are shown as shaded circles in Figure 3.28(c).) Next for each term of $H_p(P)$ so deleted, we remove also its incident edges. The resulting graph, obtained as a refinement of $H(P, p)$ is the facial graph of $CH(P \cup p)$. We also note that when p is not external to P, Property 2 holds for each f of P, so that the entire subgraph $H_p(P)$ is deleted and $CH(P \cup p) = CH(P)$, correctly.

From an implementation viewpoint, the fundamental part of the technique is the classification of each of the N vertices of P *with respect to p*, as *concave*, *reflex*, or *supporting* (according to a generalization of the terminology of Section 3.3.6). Specifically, we say

vertex v is $\begin{cases} \textit{concave} \text{ if for each facet } F \text{ containing } v, p \text{ is beneath } F. \\ \textit{reflex} \text{ if for each facet } F \text{ containing } v, p \text{ is beyond } F. \\ \textit{supporting, otherwise.} \end{cases}$

We leave as an interesting exercise for the reader (Exercise 3.9 at the end of this chapter) to devise a method to perform this classification in time $O(\varphi_{d-1}) + O(Nd)$, where φ_{d-1} is, as usual, the actual number of facets of P. Once this classification is obtained, Property 1 is applied to all ascendants of each reflex vertex, and Property 2 to all ascendants of concave vertices in $H_p(P)$: this operation can be done in time proportional to the total numbers of vertices and edges of $H(P)$; it is not hard to show that the latter is $O(\varphi_{d-1})$. Thus the total update time is $O(\varphi_{d-1})$ and, since there are N updates, we have the following theorem.

Theorem 3.16. *The convex hull of a set of N points in d-dimensional space can be constructed on-line by the beneath-beyond technique in time*

$$T(d, N) = O(N^{\lfloor d/2 \rfloor + 1}).$$

Two remarks are now in order. First, special care must be taken to handle the degenerate cases, when p belongs to the affine hull of some face of P. Second, when P is not simplicial, it can be shown that P can be perturbed without significantly increasing the face complexity or the complexity of the inclusion relation. As a consequence, the time bound developed for the simplicial case applies to the general case.

In summary, the beneath-beyond technique is attractive because it has a performance comparable to that of the gift-wrapping method and, in addition, it has the desirable property of being on-line.

3.4.3 Convex hulls in three dimensions

Among all the higher dimensional convex hull problems, the three-dimensional instance has enormous importance for its unquestionable relevance to a host of applications, ranging from computer graphics to design automation, to pattern recognition, to operations research. It is a fortunate circumstance—and a most refreshing finding—that this significant applicational need is matched by what appears to be a "lull" in the computational difficulty. We shall now try to elaborate on this rather vague notion.

We have already noted that the objective of a convex hull algorithm is the description of the convex hull polytope. Thus, we know from Section 3.1 that $\Omega(N^{\lfloor d/2 \rfloor})$ is a trivial worst-case lower bound—the size of the output—to the running time of any convex hull algorithm on N points in E^d. We have also seen (see the preceding section) that the beneath-beyond technique exceeds this lower-bound by a factor $O(N)$, and that for $d = 3$ (d is odd) this is the best we can expect from an on-line technique. If we renounce the on-line property, we may expect to reduce this factor, possibly by exchanging $O(N)$ with something like $O(\log N)$: in three dimensions this would lead to a cost $O(N^{\lfloor 3/2 \rfloor} \log N) = O(N \log N)$. On the other hand, the general lower bound (Section 3.2) is exactly $\Omega(N \log N)$, so the best we can hope for is an off-line algorithm with running time $\theta(N \log N)$. This objective can be achieved [Preparata–Hong (1977)].

As usual, we have a set $S = \{p_1, p_2, \ldots, p_N\}$ of N points in E^3. We assume for simplicity, that for any two points p_i and p_j in S we have $x_k(p_i) \neq x_k(p_j)$, for $k = 1, 2, 3$. This simplification helps bring out the basic algorithmic ideas, while the modifications required for the unrestricted case are straightforward. With the target of an $O(N \log N)$ performance, we now try to develop a technique of the divide-and-conquer type.

As a preliminary step we sort the elements of S according to the coordinate x_1 and relabel them if necessary so that we may assume $x_1(p_i) < x_1(p_j) \Leftrightarrow i < j$. We then have the following simple recursive algorithm (given as a function).

> **function** CH(S)
> 1. **begin if** $(S \leq k_0)$ **then**
>   **begin** construct CH(S) by brute force;
>   **return** CH(S)
>   **end**
> 2.   **else begin** $S_1 := \{p_1, \ldots, p_{\lfloor N/2 \rfloor}\}$; $S_2 := \{p_{\lfloor N/2 \rfloor + 1}, \ldots, p_N\}$;
> 3. $P_1 := $ CH(S_1); $P_2 := $ CH(S_2);
> 4. $P := $ MERGE(P_1, P_2);
> 5.   **return** P
>   **end**
> **end**.

The initial sorting of the x_1 coordinates of the elements of S uses $O(N \log N)$ operations. Notice that, because of this sorting and of Step 1, the sets $CH(S_1)$ and $CH(S_2)$ are two nonintersecting 3-dimensional polytopes. Now, if the "merging" of two convex hulls with at most N vertices in total, i.e., the construction of the convex hull of their union, can be done in at most $M(N)$ operations, an upper bound to the number $T(N)$ of operations used by algorithm CH is given by the equation

$$T(N) = 2T(N/2) + M(N).$$

(Note that we have assumed that N be even for simplicity, but practically without loss of generality.) Thus, if we can show that $M(N)$ is $O(N)$, we shall obtain that $T(N)$ is $O(N \log N)$, and, taking into account the initial sorting pass, an overall complexity $O(N \log N)$ results for the convex hull determination.

Clearly the MERGE function is the crucial component of the method. We begin by considering the data structure suited to represent a convex 3-polytope P. All we need to describe is the boundary of P, which in three dimensions has the topology of a planar graph. As we noted in Section 1.2.3.2, the natural structure to represent a planar graph embedded in the plane is the doubly-connected-edge-list (DCEL). The interesting property of the DCEL is that it contains explicitly only the edges of the graph, but it allows for the extraction of the vertex and face sets in optimal time.[15]

Assume now that the polytopes $P_1 = CH(S_1)$ and $P_2 = CH(S_2)$ have been recursively obtained. It is important to stress that—due to the initial sorting and to the chosen partition of the resulting point set (line 1 of Algorithm $CH(S)$)—P_1 and P_2 are *nonintersecting*. The "merge" of P_1 and P_2 may be obtained by the following operations (see Figure 3.29 for an intuitive illustration):

1. Construction of a "cylindrical" triangulation $\mathcal{T}$,[16] which is supporting P_1 and P_2 along two circuits E_1 and E_2, respectively.
2. Removal both from P_1 and P_2 of the respective portions which have been "obscured" by $\mathcal{T}$.

Here, the terms "cylindrical" and "obscured" have not been formally defined; rather, they have been used in their intuitive connotations, as suggested by Figure 3.29. Still on an intuitive level, the construction of $\mathcal{T}$ may be viewed as a gift-wrapping operation: indeed we will "wrap" both P_1 and P_2 in the same "package" by means of $\mathcal{T}$. Although $\mathcal{T}$ may have $O(N)$ facets and each gift-wrapping step takes in general $O(N)$ work, the nature of the three-dimensional polytope allows for a crucial simplification, as we shall see.

---

[15] Here the term "face" is used in the conventional terminology of planar graphs. However, for the rest of this section, we shall use the term "facet" to emphasize their nature as $(d-1)$-dimensional sets.

[16] We are implicity assuming that at each stage in the execution of the algorithm we are dealing with simplicial (triangulated) polytopes. As we shall see later this is done with no essential sacrifice of generality.

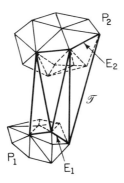

Figure 3.29   Illustration of the principle of the method. A triangulation $\mathcal{T}$ is "wrapped-around" the convex hulls of P_1 and P_2.

The initial step in the construction of $\mathcal{T}$ is the determination of a facet or edge of it. Although there is more than one way to carry out this task, an expedient way is to maintain projections of each convex polytope as a polygon in a coordinate plane (say, the plane (x_1, x_2) as shown in Figure 3.30). Thus, let P_i' be the projection of P_i on the (x_1, x_2)-plane $(i = 1, 2)$. We can now use our favorite algorithm (see, e.g., Section 3.3.5) to construct a common support line of P_1' and P_2' and obtain a planar edge e' which is the projection of an edge e of $\mathcal{T}$. Once e is available, a supporting plane through e and parallel to the x_3-axis can be used to start the construction of $\mathcal{T}$.

The advancing mechanism uses as reference the last constructed facet of $\mathcal{T}$. We shall denote by the letter "a" vertices of P_1, while "b" is used for P_2. As shown in Figure 3.30, let (a_2, b_2, a_1) be the reference facet (shaded area) for the current step. We must now select a vertex $\hat{a}$, connected to a_2, such that the facet $(a_2, b_2, \hat{a})$ forms the largest convex angle with (a_2, b_2, a_1) among the facets (a_2, b_2, v), for $v \neq a_1$ connected to a_2; similarly we select $\hat{b}$ among the

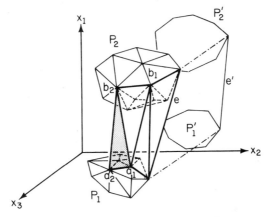

Figure 3.30   The initial step and the generic step in the construction of $\mathcal{T}$.

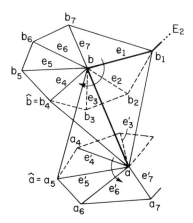

Figure 3.31   The advancing step in the construction of $\mathcal{T}$.

vertices connected to b_2. For reasons to become apparent later, we call these comparisons of Type 1.

Next, once the "winners" $(a_2, b_2, \hat{a})$ and $(a_2, b_2, \hat{b})$ have been selected, we have a run-off comparison, called of Type 2. If $(a_2, b_2, \hat{a})$ forms with (a_2, b_2, a_1) a larger convex angle than $(a_2, b_2, \hat{b})$, then $\hat{a}$ is added to $\mathcal{T}$ ($\hat{b}$ is added in the opposite case) and the step is complete. The described schedule of comparisons is an efficient way, as we shall see, to implement the gift-wrapping step around edge (a_2, b_2).

The efficient implementation of the just described advancing mechanism rests crucially on the following considerations. In the DCEL, the edges incident on a vertex can be efficiently scanned in either clockwise or counterclockwise order. Referring to Figure 3.31 for concreteness of illustration, suppose now that b and (b, a) are the most recently added vertex and edge of $\mathcal{T}$, respectively, and let (b_1, b) be the edge of E_2 reaching b. Without loss of generality, we may assume that the numbering of the edges incident on b and of their termini $b_1, b_2, \ldots, b_k$ be as shown in Figure 3.31 (where $k = 7$) and analogously for the edges incident on vertex a. Let (b_s, b, a) be the facet which forms the largest convex angle with (b_1, b, a) among the facets (b_i, b, a), for $i = 2, \ldots, k$ (in our example $s = 4$). A first important observation is that *any* (b, b_i) *for* $1 < i < s$ *becomes internal to the final hull* $CH(P_1 \cup P_2)$ *and need not be further considered*. As regards vertex a, we note that the set of its incident edges had been partially scanned at some earlier stage in the process (since vertex a was reached earlier than vertex b by the algorithm). As a consequence of the preceding observation the scan around a should resume from the last visited edge since all others have been eliminated (this edge is shown as (a, a_4) in Figure 3.31).

The second observation concerns the numbers of angle comparisons, each to be performed in constant time, by specializing the general technique outlined in Section 3.4.1 for the d-dimensional case. Each Type 1 comparison definitively eliminates one edge of either P_1 or P_2 from further consideration. Let v_i and m_i denote, respectively, the numbers of vertices and edges of $P_i (i = 1, 2)$. Thus, the number of Type 1 comparisons is upper-bounded by $(m_1 + m_2)$. Next, each Type 2 comparison adds a new vertex to either E_1 or E_2. Since the numbers of vertices of E_1 and E_2 are at most v_1 and v_2, respectively, the number of Type 2 comparisons is bounded by $(v_1 + v_2 - 1)$. *Due to the planar graph structure of the polytopal surface*, $m_i \leq 3v_i - 6$, whence the number of angle comparisons grows no faster than linearly in the total number of vertices of P_1 and P_2.

The update of the DCEL can be performed dynamically during the construction of $\mathscr{T}$. Specifically, refer to Figure 3.31 for an illustration. Assume that we have advanced from edge (a, b_1) to edge (a, b), "pivoting" on vertex a. Edge scans will take place around both b and a, although the edges incident on the latter (the pivot) have been already partially scanned. First a new DCEL node for edge (a, b) is created, and (a, b) becomes the successor of (a, b_1) around a, while (b_1, b) becomes its successor around b_1. At this point the clockwise scan of the edges around b begins starting from e_1 and stops at e_4: the nodes of edges e_2 and e_3 have been deleted[17] (as e_2 and e_3 have become internal), while (a, b) is made the successor of e_1. A similar scanning takes place around a from some edge e'_4 to some other edge $e' = (a, a_5)$. Then, if (a, b, b_4) is the winner of the run-off comparison we set (a, b_4) and (b, b_4) as the successors of (a, b) around a and b, respectively; otherwise (a, a_5) and (b, a_5) play these roles. This effects the update of the DCEL, as a linked data structure. We therefore conclude that the overall running time of the MERGE algorithm is $O(N)$, yielding the following theorem.

Theorem 3.17. *The convex hull of a set of N points in three-dimensional space can be computed in optimal time $\theta(N \log N)$.*

Finally, we must discuss how to handle degeneracies, that is, the case of nonsimplicial polytopes. The most expedient way is to let the method introduce an "artificial" triangulation of nontriangular facets, so that at each stage the polytope appears simplicial. Specifically, consider the scan of the set of edges incident on vertex b, as illustrated in Figure 3.32. Under the simplicial hypothesis, the scan stops at an edge e which is shared by two facets F and F' so that vertex a is beyond F and beneath F' (see the beginning of Section 3.4 for the beneath-beyond terminology): this is the only possible situation, since no four vertices are coplanar by hypothesis. Removing this assumption, we modify the criterion as follows: the scan stops at an edge e shared by F and F'

---

[17] Note that these deletions may not entirely remove from the DCEL all edges which have become internal to the convex hull. However, the not yet deleted edges form a planar graph on a set V' of vertices, which are themselves internal to the hull. Thus, the number of nondeleted edges is proportional to $|V'|$, and the total memory occupancy of the DCEL is still proportional to $|S|$. Note that the graph on vertex set V' is disconnected from the convex hull graph.

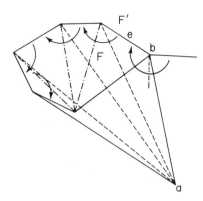

Figure 3.32   Illustration of a triangulation of a nontriangular facet.

so that vertex a is nonbeneath F and beneath F'. In this manner (see Figure 3.32) if vertex a and facet F are coplanar, the resulting facet $CH(F \cup a)$ appears triangulated by a fan of edges issuing from a. We conclude that $\mathscr{T}$ will appear triangulated, although adjacent facets may be coplanar. The removal of the unnecessary edges can be easily accomplished at the end by a single pass over the DCEL, by marking any edge that is coplanar with its two neighbors incident on the same vertex, and then deleting all edges so marked.

3.5 Notes and Comments

In this chapter we have repeatedly alluded to the intimate connection between sorting and hull finding in the plane. In particular, any ordered hull algorithm—i.e., a solution of problem CH1 in the plane—no matter what technique it employs, should be a sorting algorithm in disguise. We have already illustrated this analogy with incidental remarks whenever appropriate. This section affords us the opportunity to more systematically review the algorithms of this chapter by discovering their corresponding sorting methods.

As a general comment, the analogy is quite strong if all the given points are thought of as being hull points. (Of course, the sorting problem has no analog for points internal to the convex hull.)

Graham's algorithm uses sorting explicitly at the start. The first algorithm of Section 3.3.1, which checks each pair of points to determine whether they define a hull edge (and is so inefficient that we did not even dignify it with a name), is analogous to the following curious sorting algorithm.

1. To sort $x_1, \ldots, x_N$, find the smallest value x_m and exchange it with x_1. (∗Time: $O(N)$∗).
2. Now find the successor of x_1 in the final sorted list by examining $x_2, \ldots, x_N$ as possible candidates. x_j is the successor of x_1 if and only if no other element lies between them. (∗Time: $O(N^2)$∗).
3. Find the successors of $x_2, \ldots, x_{N-1}$ similarly. (∗Time to find each successor: $O(N^2)$. Total time = $O(N^3)$∗).

This procedure can obviously be improved by speeding up the successor search. To determine which element follows x_i, we need only find the smallest element among

$x_{i+1}, \ldots, x_N$. With this change, the above algorithm becomes SELECTION SORT [Knuth (1973)], and corresponds to Jarvis's algorithm (Section 3.3.3).

The divide-and-conquer hull algorithms (Section 3.3.5), which split the set in two, find the hulls of the subproblems, and combine them in linear time, are the geometric analog of MERGESORT. The real time algorithm is an insertion sort. The analogs of QUICKSORT have been discussed in Section 3.3.4.

Although $\Omega(N \log N)$ is a lower bound to the worst-case running time of any convex hull algorithm on an N-point set S, it is plausible that smaller running times may be achievable when the number h of extreme points is substantially smaller than N. This lead was successfully explored by Kirkpatrick and Seidel who developed an $O(N \log N)$ algorithm they characterized as "the ultimate," and also proved it to be asymptotically optimal [Kirkpatrick–Seidel (1983)]. Their algorithm constructs separately the upper and lower hulls. Referring to the upper hull, they partition S into two equal-size vertically separated subsets; but instead of recursively constructing the upper hull of the two halves and computing their common supporting line (referred to as their *upper bridge*), they first construct this bridge and then separately construct the upper hulls of the subsets respectively to the left and to the right of the leftmost and rightmost endpoints of the bridge. The key of the technique is an efficient way to construct the bridge. Recognizing that the bridge can be defined as the solution of a linear programming problem, they resort to the linear time technique recently proposed for linear programming by Megiddo and Dyer to be presented in detail in Section 7.2.5. Denoting by h_1 and h_2 ($h_1 + h_2 = h$) the numbers of hull vertices to the left and right of the bridge respectively, and by $T(N, h)$ the running time of the algorithm, it is easy to verify that the recurrence relation (c is a constant)

$$T(N, h) = cN + \max_{h_1 + h_2 = h} (T(N/2, h_1) + T(N/2, h_2))$$

is solved by $T(N, h) = O(N \log h)$.

Only recently considerable attention has been given to the problem of the convex hull in E^d. Bhattacharya's analysis of the Chand–Kapur algorithm and the beneath-beyond method of Kallay are two examples of this renewed interest. An algorithm analogous to Kallay's was discovered, independently and almost simultaneously, by Seidel (1981). Seidel employs an analogous approach, only in the dual space (see Section 1.3.3). The main idea is illustrated in Figure 3.33 for $d = 2$, which shows that

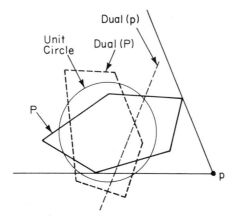

Figure 3.33   The construction of the supporting lines is equivalent to intersection in the dual space.

the construction of the supporting lines from p to P is equivalent, in the dual space, to the intersection of dual(P) with dual(p). An efficient algorithm to compute this intersection in E^d achieves $O(N^{\lfloor(d+1)/2\rfloor})$ running time which improves over Kallay's $O(N^{\lfloor d/2+1\rfloor})$, and is indeed optimal ($O(N^{\lfloor d/2\rfloor})$ in the size of the output) for even d.

3.6 Exercises

1. *Simple Polygonal Path.* Given N points in the plane, construct a simple polygon having them as its vertices.
   (a) Show that $\Omega(N\log N)$ is a lower bound to running time.
   (b) Design an algorithm to solve the problem.
   (*Hint*: Modify Graham's scan.)

2. Two disjoint convex polygons P_1 and P_2 in the plane are given.
   (a) How many common supporting lines do P_1 and P_2 have?
   (b) Construct the supporting lines by an algorithm based exclusively on the angles between polygon edges and straight lines through vertices.

3. Given a point p and a vertex v of a convex polygon P in the plane, give an algorithm to classify in constant time vertex v with respect to $\overline{pv}$ (as *concave* or *supporting* or *reflex*).

4. Referring to the real-time planar convex hull algorithm of Section 3.3.6:
   (a) Prove that in Cases 2, 4, 6, and 8 of Figure 3.16 point p is necessarily external to polygon P.
   (b) Assuming that p is internal to P, discuss how the search procedure will detect this situation.

5. Referring to the dynamic convex hull maintenance technique (Section 3.3.7), solve the following problems:
   (a) Give a Pigdin Algol program for the function BRIDGE(U_1, U_2) that produces the support segment of two U-hulls U_1 and U_2.
   (b) Develop a twin pair of procedures ASCEND–DESCEND to process the deletion of a point from the planar set.

6. *Keil–Kirkpatrick.* Let S be a set of N points in the plane with integer coordinates between 1 and N^d, where d is a constant. Show that the convex hull of S can be found in linear time.

7. *Seidel.* Let $S = \{p_1, \ldots, p_N\} \subset E^3$ with $x(p_1) < x(p_2) < \cdots < x(p_n)$. Show that in spite of knowing the order of S in the x-direction, it still takes $\Omega(N\log N)$ to find the convex hull of S.
   (*Hint*: Use transformation from the 2-dimensional convex hull problem.)

8. *Seidel.* In Chand–Kapur's gift-wrapping algorithm one needs to maintain a data structure for the set Q of $(d-2)$-faces that can be "gift-wrapped over" next. In this algorithm facets can be printed out immediately after they are found and need not be stored. Thus the size of the data structure for Q determines the space complexity

of the gift-wrapping algorithm. Let S be a set of n points in E^d in general position. The gift-wrapping algorithm is to be applied to construct the convex hull of S.

(a) Show that there is an order for the gift-wrapping steps such that Q is never larger than the maximum number of facets of a polytope with N vertices in E^d.

(*Hint*: Use the idea of "shelling" a polytope developed in H. Bruggesser and P. Mani "Shellable decompositions of cells and spheres," *Math. Scand.* **29** (1971) pp. 197–205.)

(b) Develop a variant of the gift-wrapping algorithm with space complexity

$$O(\varphi_{d-1}) = O(N^{\lfloor (d-1)/2 \rfloor}),$$

where φ_{d-1} is the number of facets of the polytope.

9. In the beneath-beyond algorithms for the d-dimensional convex hull (Section 3.4.2), show that the classification of each of the N vertices of the polytope P as concave, reflex, or supporting can be done in time $O(\varphi_{d-1}) + O(Nd)$, where φ_{d-1} is the number of facets of P.

CHAPTER 4
Convex Hulls: Extensions and Applications

This chapter has two objectives. The first is the discussion of variants and special cases of the convex hull problem, as well as the average-case performance analysis of convex hull algorithms. The second objective is the discussion of applications that use the convex hull. New problems will be formulated and treated as they arise in these applications. Their variety should convince the reader that the hull problem is important both in practice and as a fundamental tool in computational geometry.

4.1 Extensions and Variants

4.1.1 Average-case analysis

Referring to the two-dimensional convex hull algorithms discussed in Section 3.3 of the preceding chapter, we note that Graham's convex hull algorithm always uses $O(N \log N)$ time, regardless of the data, because its first step is to sort the input. Jarvis's algorithm, on the other hand, uses time that varies between linear and quadratic, so it makes sense to ask how much time it can be *expected* to take. The answer to this question will take us into the difficult but fascinating field of stochastic geometry, where we will see some of the difficulties associated with analyzing the average-case performance of geometric algorithms.

Since Jarvis's algorithm runs in $O(hN)$ time, where h is the number of hull vertices, to analyze its average-case performance we need only compute $E(h)$, the expected value of h. In order to do this, we must make some assumption about the probability distribution of the input points. This problem brings us into the province of stochastic geometry, which deals with the properties of random geometric objects and is an essential tool for dealing with expected-time analysis.[1] We would like to be able to say, "Given N points chosen uniformly in the plane ...," but technical difficulties make this impossible— elements can be chosen uniformly only from a set of bounded Lebesgue measure [Kendall–Moran (1963)] so we are forced to specify a particular figure from which the points are to be selected. Fortunately, the problem of calculating $E(h)$ has received a good deal of attention in the statistical literature, and we quote below a number of theorems that will be relevant to the analysis of several geometric algorithms.

Theorem 4.1 [Rényi–Sulanke (1963)]. *If N points are chosen uniformly and independently at random from a plane convex r-gon, then as $N \to \infty$,*

$$E(h) = \left(\frac{2r}{3}\right)(\gamma + \log_e N) + O(1), \qquad (4.1)$$

where γ denotes Euler's constant.

Theorem 4.2 [Raynaud (1970)]. *If N points are chosen uniformly and independently at random from the interior of a k-dimensional hypersphere, then as $N \to \infty$, $E(f)$, the expected number of facets of the convex hull, is given asymptotically by*

$$E(f) = O(N^{(k-1)/(k+1)}). \qquad (4.2)$$

This implies that

$$E(h) = O(N^{1/3}) \text{ for } N \text{ points chosen uniformly in a circle, and}$$

$$E(h) = O(N^{1/2}) \text{ for } N \text{ points chosen uniformly in a sphere.}$$

Theorem 4.3 [Raynaud (1970)]. *If N points are chosen independently from a k-dimensional normal distribution, then as $N \to \infty$ the asymptotic behavior of $E(h)$ is given by*

$$E(h) = O((\log N)^{(k-1)/2}). \qquad (4.3)$$

Theorem 4.4 [Bentley–Kung–Schkolnick–Thompson (1978)]. *If N points in k dimensions have their components chosen independently from any set of continuous distributions (possibly different for each component), then*

$$E(h) = O((\log N)^{k-1}). \qquad (4.4)$$

Many distributions satisfy the conditions of this theorem, including the uniform distribution over a hypercube.

---

[1] Consult [Santaló (1976)] for a monumental compilation of results in this field.

The surprising qualitative behavior of the hulls of random sets can be understood intuitively as follows: for uniform sampling within any bounded domain F, the hull of a random set tends to assume the shape of the boundary of F. For a polygon, points accumulating in the "corners" cause the resulting hull to have very few vertices. Because the circle has no corners, the expected number of hull vertices is comparatively high, although we know of no elementary explanation of the $N^{1/3}$ phenomenon in the planar case.[2]

It follows directly from these results that the expected time used by Jarvis's algorithm can be described by the following table.

Table I Average-Case Behavior of Jarvis's Algorithm

Distribution	Average-Case Time
Uniform in a convex polygon	$O(N \log N)$
Uniform in a circle	$O(N^{4/3})$
Normal in the plane	$O(N(\log N)^{1/2})$

Note that for the normal distribution Jarvis's algorithm can be expected to take slightly less time than Graham's.[3]

All of the distributions considered in this section have the property that the expected number of extreme points in a sample of size N is $O(N^p)$ for some constant $p < 1$. We shall refer to these as N^p-*distributions*.

We now turn out attention to the planar divide-and-conquer algorithm described in Section 3.3.5, which we already showed to have optimal worst-case performance. This algorithm, with a few crucial implementation *caveats*, achieves linear expected time, as shown by Bentley and Shamos (1978), and perhaps holds the potential of being generalized to higher dimensional cases. We briefly recall it for the reader's benefit. If N, the number of given points, is less than some constant N_0, then the procedure calculates the hull by some direct method and returns. If N is large, though, the procedure first divides, *in an arbitrary way*, the N points into two subsets of approximately $N/2$ points each. It then finds the convex hulls of the two subproblems recursively. The result of each of the recursive calls is a convex polygon. The hull of the given set is now just the hull of the union of the hulls found in the subproblems. The latter can be found in time proportional to the total number of vertices of both.

To adapt this method for linear expected time, the crux is the "division" step of the divide-and-conquer scheme. Indeed, in our previous formulation in

---

[2] The expected number of hull vertices of a set of N points can be expressed as the integral of the Nth power of the integral of a probability density [Efron (1965)]. What is of interest is not the value of this quantity but its asymptotic dependence on N.

[3] The bivariate normal is often used to model the distribution of projectiles aimed at a target.

Section 3.3.5, this division could have taken linear time, such as required by the mere copying of the subproblems. This approach, however, would by itself lead to $O(N \log N)$ overall time. Thus a different approach is in order, to achieve a two-fold objective:

(i) the division step should use time $O(N^p)$, for $p < 1$;
(ii) the points in each of the two subproblems must obey the same probability distribution as the original points.

Objective (ii) is rather easily accomplished by initially subjecting the original points to an arbitrary permutation and arranging them in an array (this is achieved by storing the points in this array in the order of their input). This ensures that any subsequence of the resulting sequence is random, i.e., it obeys the given distribution. A subproblem concerns the set of points in a subarray of contiguous cells, and is completely specified by two delimiting pointers. Thus division is accomplished by splitting a subarray at its mid-cell (in constant time), and by passing only two pointers in each subroutine call. This strategy achieves Objective (i). This objective is implicitly achieved if we provide an iterative (rather than recursive) formulation of the algorithm. Such implementation would work bottom-up on sets of four points, then eight, etc.

Thus, the running time of the merge step is proportional to the sum of the sizes of the convex hulls of each of the two subsets. Let $C(N)$ denote the latter quantity; also, $T(N)$ is as usual the running time of the algorithm on a set of size N. We readily have the recurrence relation

$$T(N) \leq 2T(N/2) + C(N).$$

If we now let $T_{\text{ave}}(N) \triangleq E[T(N)]$ (E is the standard expectation operator), since E is linear, we obtain the new recurrence

$$T_{\text{ave}}(N) \leq 2T_{\text{ave}}(N/2) + E[C(N)]. \tag{4.5}$$

The solution depends upon the form of $E[C(N)]$. Specifically if

$$E[C(N)] \begin{cases} = O(N), & \text{then } T_{\text{ave}}(N) = O(N \log N). \\ = O(N/\log N), & \text{then } T_{\text{ave}}(N) = O(N \log \log N). \\ = O(N^p), \ p < 1, & \text{then } T_{\text{ave}}(N) = O(N). \end{cases} \tag{4.6}$$

The quantity $E[C(N)]$, due to the linearity of $E[\ ]$, is at most twice the expected value of the size $h(N/2)$ of the convex hull of a set of $N/2$ points. Since for all the distributions discussed above have the property that $E[h(N/2)] = O((N/2)^p)$ for some $p < 1$, then $E[C(N)] = O(N^p)$. This is summarized in the following theorem.

Theorem 4.5. *For an N^p-distribution, the convex hull of a sample of N points in two dimensions can be found in $O(N)$ expected time.*

It is tempting to say that the above analysis applies to the three-dimensional algorithm of Preparata–Hong, described in Section 3.4.3, which is also of the divide-and-conquer type. However, the merge step (at least in the presented formulation), makes crucial use of the fact that the two convex hulls to be merged are disjoint. Unfortunately, this hypothesis invalidates the condition that throughout the execution of the algorithm each subset be random and drawn from the same distribution.

4.1.2 Approximation algorithms for convex hull

An alternative to choosing a convex hull algorithm on the basis of its average-case performance (discussed in the preceding section), is the design of algorithms that compute *approximations* to the actual convex hull, in exchange for simplicity and performance. Such an algorithm would be particularly useful for applications that must have rapid solutions, even at the expense of accuracy. Thus, it would be appropriate, for example, in statistical applications where observations are not exact but are known within a well-defined precision.

We shall now discuss one such algorithm for the plane [Bentley–Faust–Preparata (1982)]. Its basic, simple idea is to sample some subset of the points and then use the convex hull of the sample as an approximation to the convex hull of the points. The particular sampling scheme we shall choose is based, however, on a computation model different from the one considered so far. Indeed, in this section we shall assume that the *floor function* "$\lfloor \ \rfloor$" has been added to our set of primitive operations.[4] Referring to Figure 4.1(a), the first step is to find the minimum and the maximum values in the x dimension, and then divide the vertical strip between them into k equally spaced strips. These k strips from a sequence of "bins" in which we shall distribute (here is where the floor function is needed!) the N points of the given set S. Next, in each strip, we find the two points in that strip with minimum and maximum y-value [see Figure 4.1(b)]. We also select the points with extreme x-values; if there are several points with extreme x-values, we select both their maximum and minimum y-value. The resulting set, which we call S^*, therefore has a maximum of $2k + 4$ points. Finally, we construct the convex hull of S^* and use that as the approximate hull of the set; see Figure 4.1(c). Note that the resulting hull is indeed only an approximation: in Figure 4.1(c) one of the points of the set lies outside it.

The outlined method is extremely easy to implement. The k strips are set up as a $(k + 2)$-element array (the zero-th and $(k + 1)$-st lines contain the two points with extreme x-values $x_{\min}$ and $x_{\max}$, respectively). To assign a point p to a bin, we subtract the minimum x-value from $x(p)$, divide that difference by $(1/k)$-th of the difference of the extreme x-values and take the floor of the ratio

---

[4] Notice that the decision-tree computation model, used in all convex hull algorithms presented so far, rules out the use of the floor function.

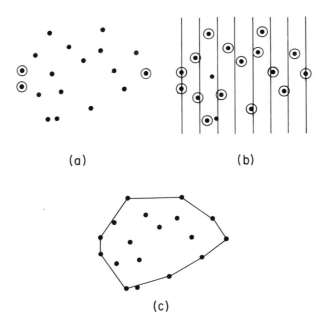

(a)　　　　　　　　　　(b)

(c)

Figure 4.1　Illustration of the approximate hull algorithm: (a) A given planar point set and its rightmost and leftmost members; (b) Partition into k strips and determination of vertical extremes in each strip; (c) Approximate convex hull.

as the strip number. Concurrently we can find the minimum and maximum in each strip, since for each point we check to see if it exceeds either the current maximum or minimum in the strip, and, if so, update a value in the array. Finally we can use a suitable convex-hull algorithm: clearly the most appropriate one is the modified version of the Graham scan proposed by Andrew (see Section 3.3.2). Note that the points in S^* are nearly sorted by x-value; indeed, sorting is completed by comparing, in each strip, the x-values of the two points of S^* in that strip.

The performance of this technique is also easy to analyze. The space that it takes is proportional to k (if we assume the input points are already present). Finding the minimum and maximum x-values requires $\theta(N)$ time. Finding the extremes in each strip requires $\theta(N)$ time, and the convex hull of S can be found in $O(k)$ time (because the (at most) $2k + 4$ points in S are already sorted by x-value). The running time of the entire program is therefore $\theta(N + k)$.

The simplicity and the efficiency of the described algorithm would be of little value if the result were grossly inaccurate. We now wish to estimate the accuracy of the approach. First of all, because it uses only points in the input set as hull points, it is a conservative approximation in the sense that every point within the approximate hull is also within the true hull. The next question is: how far outside of the approximate hull can a point of S be? The answer is provided by the following straightforward proposition.

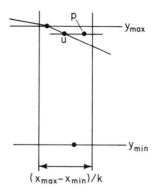

$(x_{max} - x_{min})/k$

Figure 4.2   Analysis of the accuracy of the approximation algorithm.

Theorem 4.6. *Any point $p \in S$ that is not inside the approximate convex hull is within distance $(x_{max} - x_{min})/k$ of the hull.*

PROOF. Consider the strip where p falls (see Figure 4.2). Because p is outside the approximate hull, it cannot have either the largest or smallest y-value of points in the strip; so $y_{min} \leq y(p) \leq y_{max}$. If u is the intercept on the approximate hull of a horizontal line through p, the length of $\overline{pu}$ bounds above the distance of p from the hull, and is in turn bounded above by the width $(x_{max} - x_{min})/k$ of the strip.   □

The approximation just described yields an approximate hull that is a subset of the true hull, but it is not hard to modify it to yield a superset of the true hull. To do this, we just replace every maximal point p in a strip with two outermost points in the strip having the same ordinate. Obviously, the resulting hull contains the true convex hull. An error analysis similar to the one above shows that any point in the approximate hull but not in the true hull must be within distance $(x_{max} - x_{min})/k$ of the true hull.

Can we apply this approach to higher dimensions? We limit ourselves to the discussion of the three-dimensional algorithm, which is an extension of the planar algorithm just described. An initial sampling pass finds the minimum and maximum values in both the x and the y dimensions; the resulting rectangle $[x_{min}, x_{max}] \times [y_{min}, y_{max}]$ is then partitioned into a grid of (at most $(k + 2) \times (k + 2)$ squares by means lines parallel to the x-and y-axes. (We use k of the squares in each dimension for "actual" points in squares, and the two squares on the end for the extreme values.) Note that one of the dimensions can have fewer than $(k + 2)$ intervals if the point set does not happen to occupy a square region. In each square of the grid we find the points with minimum and maximum z-values; the resulting set of points, called S^*, consists of at most $2(k + 2)^2$ points. Finally, we construct the convex hull of S and use it as the approximate hull of the original set. The hull can be constructed using Preparata–Hong's general hull algorithm for point sets in 3-space (Section 3.4.3). Notice the striking difference with respect to the two-dimensional method, where we were able to take advantage of the known ordering of the points in the x dimension: we do not exploit the grid structure of the points to yield a faster algorithm, but rather use a general algorithm. An intriguing question is whether the regular placement of the points of S in a two-dimensional grid can be used to obtain an algorithm more efficient than the general one.

The analysis of the three-dimensional algorithm is somewhat different from that of the two-dimensional algorithm. The set S^* can be found in time $\theta(N + k^2)$, but computing the convex hull of S^* using the general algorithm requires time $O(k^2 \log k)$ rather than $O(k^2)$ since we are unable to exploit any natural ordering of the grid points as we were in the two-dimensional case. Thus, the total running time of the algorithm is $O(N + k^2 \log k)$.

To estimate the accuracy of the technique, we note that the length of the diagonal of a horizontal section of a cell is an upper bound to the distance from the approximate hull of any point of S external to it. The latter is readily found to be $\sqrt{(x_{max} - x_{min})^2 + (y_{max} - y_{min})^2}/k$.

An interesting feature of this approach is that the accuracy of the result can be made to grow quickly with the expended computational work (by increasing the integer k). Specifically, the number of points in the sample controls the accuracy of the approximation, while the way the sample is selected controls the approximation type (liberal or conservative).

4.1.3 The problem of the maxima of a point set

This section is devoted to the discussion of a problem that bears an intriguing resemblance to that of the convex hull and yet on a deeper level, to be elucidated later, is fundamentally different from it. This problem occurs in a large number of applications, in statistics, economics, operations research, etc. Indeed, it was originally described as the "floating-currency problem"[5]: In Erehwon, every citizen has a portfolio of foreign currencies; these foreign currencies wildly fluctuate in their values, so that every evening a person whose portfolio has the largest cumulative value is declared the King of Erehwon. Which is the smallest subset of the population that is certain to contain all potential kings?

The problem can now be precisely formulated. As usual, all points belong to the d-dimensional space E^d, with coordinates $x_1, x_2, \ldots, x_d$.[6]

---

[5] This amusing formulation is due to H. Freeman (1973, unpublished).

[6] In a somewhat more general formulation, given d totally ordered sets $U_1, \ldots, U_d$, each element is a vector $v \in V$, where V is the cartesian product $U_1 \times \cdots \times U_d$.

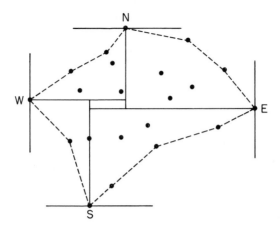

Figure 4.3   In the quadrants defined by the four extreme points N, S, E, and W, the maxima provide a crude representation of the boundary of the point set.

Definition 4.1. Point p_1 *dominates* point p_2 (denoted by $p_2 \prec p_1$) if $x_i(p_2) \le x_i(p_1)$ for $i = 1, 2, \ldots, d$. (The relation "$\prec$" is naturally called *dominance*.)

Given a set S of N points in E^d the relation dominance on set S is clearly a partial ordering on S for $d > 1$. A point p in S is a *maximal element* (or, briefly a *maximum*) of S if there does not exist q in S such that $p \ne q$ and $q \succ p$. The "maxima problem" consists of finding *all* the maxima of S under dominance.

We now illustrate the relationship between the maxima and the convex hull problems (See also [Bentley–Kung–Schkolnick–Thompson (1978)].) The essence of this similarity is that both the maxima and the convex hull are representations of the "boundary" of S, although the former provides a cruder representation. Specifically, given a point set S we can formulate as many maxima problems as there are orthants in E^d (i.e., 2^d such problems; see Figure 4.3 for an instantation to $d = 2$ where there are four orthants—called quadrants). Each of these problems is obtained by our assignment of signs $+$ and $-$ to each of the coordinates of the points of S (the original formulation corresponds to the assignment $(+ + \cdots +)$). The desired relationship is expressed by the following theorem.

Theorem 4.7. *A point p in the convex hull of S is a maximum in at least one of the assignments of signs to the coordinates.*

PROOF. Assume for a contradiction there is a hull point p which is not a maximum in any assignment. Imagine the origin of E^d placed in p and consider the 2^d orthants of E^d. Since p is not a maximum in any assignment, in each such orthant there is at least another point of S. Let S^* be the set of these points: clearly conv (S^*) contains p in its *interior*, and since conv $(S^*) \subseteq$ conv (S), we contradict the hypothesis that p is a hull point.   □

We shall return to this interesting connection later in this section. Before describing algorithms for the maxima problem, it is appropriate to estimate its complexity by establishing a lower bound to the running time of any maxima algorithm in the computation-tree model. In particular, the argument to be presented below refers to a slightly weaker instance of the linear decision-tree model, known as the comparison-tree model, although the extension of the result to the stronger version is possible. (In the comparison-tree model the linear functions are restricted to two variables.) As we already did for the convex hull (Section 3.1), we shall seek a lower bound for the two-dimensional case. Obviously, since any maxima problem in E^{d-1} can be viewed as a problem in E^d, such lower bound will hold for all dimensions (unfortunately, we do not know how tightly for $d > 3$, as we shall see). Not surprisingly, as it frequently happens for nontrivial low-complexity lower bounds, we shall resort to a familiar reduction:

$$\text{SORTING} \propto_N \text{MAXIMA.}$$

Let $\mathscr{A}$ be an algorithm, described by a comparison-tree T(see Section 1.4), which is claimed to solve any two-dimensional N-point maxima problem. The execution of $\mathscr{A}$ on any problem instance corresponds to tracing a path from the root to a leaf of T. The reduction from sorting works as follows. Let x_1, $x_2, \ldots, x_N$ be N real numbers. We now form, in linear time, a set of points in the plane $S = \{p_i: i = 1, \ldots, N, p_i = (x_i, y_i), x_i + y_i = \text{const}\}$. The condition $x_i + y_i = \text{const.}$ for $i = 1, \ldots, N$ is equivalent to

$$x_i < x_j \Leftrightarrow y_i > y_j. \tag{4.7}$$

Note that, by this construction, each element (x_i, y_i) of S is maximal. To determine that (x_i, y_i) is maximal, algorithm $\mathscr{A}$ must be able to conclude that "there is no $j \neq i$ such that $x_i < x_j$ and $y_i < y_j$"; but this is equivalent to saying that "for all $j \neq i$ either $x_i > x_j$ or $y_i > y_j$, that is, either $x_i > x_j$ or $x_i < x_j$" (by (4.7)). In other words, for each pair $\{x_i, x_j\}$ we know the relative ordering of its elements, i.e., each leaf is associated with a unique ordering of $\{x_1, x_2, \ldots, x_N\}$ which yields a solution to the sorting problem. Therefore, we have proved

Theorem 4.8 [Kung–Luccio–Preparata (1975)]. *In the comparison-tree model, any algorithm that solves the maxima problem in two dimensions requires time* $\Omega(N \log N)$.

With the gauge provided by the lower bound, we now consider the design of maxima algorithms. This problem affords us the excellent opportunity of comparing the relative merits and efficacies of two basic paradigms of computational geometry: divide-and-conquer and dimension-sweep (Section 1.2.2).

We shall begin with the latter. The first step of a sweep technique is to sort all the points of S according to a selected coordinate, say x_d; this coordinate becomes the dimension along which to sweep. With the terminology introduced in Section 1.2.2, the "event-point schedule" is a queue Q of the points arranged by *decreasing* x_d; the nature of the "sweep-section status" structure will be examined later. By the simplified notation $p \prec U$ we mean that there is a point $q \in U$ such that $p \prec q$. We can now outline the following general algorithm:

function MAXIMA1(S, d)

1. **begin** $M := \varnothing$;
2. $M^* := \varnothing$; (*M and M^* are sets of maxima, of dimensions d and d-1 respectively*)
3.   **while** $(Q \neq \varnothing)$ **do**
          begin $p \Leftarrow Q$ (*extract element form queue*);
5. $p^* :=$ project p on $(x_1, \dots, x_{d-1})$;
6.   **if** $(p^* \not\prec M^*)$ **then**
7.   **begin** $M^* =$ maxima $(M^* \cup \{p^*\})$;
8. $M := M \cup \{p\}$
            end
          end;
9.   **return** M
   end.

Descriptively, MAXIMA1 sweeps the point set S by decreasing x_d and for each point p it determines (lines 5–6) whether or not it is dominated on coordinates $x_1, x_2, \dots, x_{d-1}$ by any of the previously scanned points (which already dominate p on x_d by the organization of the sweep). This establishes the correctness of the approach whose crucial component is clearly lines 6 and 7: "*if $p^* \not\prec M$ then $M^* :=$ maxima $(M^* \cup \{p^*\})$.*" Basically this activity is nothing but the incremental (i.e., one point at a time, on-line) construction of the maxima of a $(d-1)$-dimensional set. This is by no means a trivial problem; efficient solutions for it are known only for $d = 2, 3$.

Indeed, for $d = 2$ the set M^* is one-dimensional (i.e., it contains only one element) and the test "$p^* \prec M^*$?" reduces to "$x_1(p) \leq \bar{x}_1 = \max_{q \in M} x_1(q)$?," while the update $M^* :=$ maxima $(M^* \cup \{p^*\})$ becomes $\bar{x}_1 := x_1(p)$; both these operations can be carried out in constant time. For $d = 3$, the situation is naturally more complex but still manageable. Here M^*—a two-dimensional set of maxima (see Figure 4.4)—has a *total ordering* and can be organized as a height-balanced search tree on coordinate x_2. The test "$p^* \prec M^*$?" is a search of $x_2(p^*)$ in this tree: if $x_2(q)$ is the successor of $x_2(p^*)$, then we have the immediate equivalence $(p^* \prec M^*) \Leftrightarrow x_1(p^*) \leq x_1(q)$. The update of M^*, when $p^* \not\prec M^*$, involves a traversal of the tree starting from $x_2(p^*)$ in the direction of decreasing x_2, deleting all elements until the first q' is found for which $x_1(q') > x_1(p^*)$. The computational work globally expended for this task in the sweep is readily found to be $O(N \log N)$, since each test "$p^* \prec M^*$?" (and possible insertion) uses time $O(\log N)$, while each deletion required in the update traversal can be done in constant time by trivially modifying the search tree (threading). Since there are N tests/insertions and at most N deletions, the claim is proved.

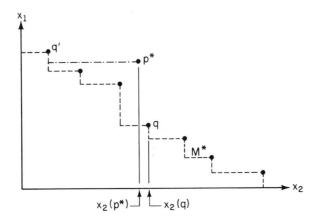

Figure 4.4   The set of maxima M^* has a total ordering and the test "$p^* \prec M^*$?" consists of comparing the x_1-coordinate of p^* with that of the point of M^* immediately to its right.

Recalling that the sweep technique requires a preliminary sorting operation, we conclude that for $d = 2, 3$ the algorithm MAXIMA1 runs in optimal time $\theta(N \log N)$.

Unfortunately this technique yields—to the best of our knowledge—time $O(N^2)$ already for $d = 4$. Thus, alternative approaches must be explored. One such approach is divide-and-conquer. As we shall see, the algorithm contemplates partitioning point sets according to the values of the coordinates x_d, $x_{d-1}, \ldots, x_2$, in turn. So it is convenient to introduce a preliminary step to facilitate these frequent operations. This step is a sorting of the elements of the set S on each of the coordinates. The resulting data structure is a multi-pointer list $((d-1)$-tuply-threaded list), each node of which is threaded in each of the single-pointer lists corresponding to the chosen coordinates. This organization is obtained at an overall cost $O(dN \log N)$; however, it affords straightforward and efficient set splitting. To avoid clumsy language, we say that a set S is "split at x_j into (S_1, S_2)" if for each $p \in S_1$ we have $x_j(p) \le x_j$ and for each $q \in S_2$ we have $x_j(q) > x_j$; moreover, we say that S is "equipartitioned on x_j into (S_1, S_2)" if S is split at $\bar{x}_j$ into (S_1, S_2) and $\bar{x}_j$ is the median of the x_j-coordinates of the points of S (so that $|S_1| \simeq |S_2|$).

The first step of the maxima algorithm—referred to later as MAXIMA2—is the equipartition of the given S on x_d into (S_1, S_2). We then recursively apply the algorithm to the sets S_1 and S_2 and obtain the two sets, max(S_1) and max(S_2), of their respective maxima. To combine the result of the recursive calls, we first note that all elements of max(S_2) are also maxima of S, while the elements of max(S_1) are not necessarily so. Indeed an element of max(S_1) is a maximal element of S if and only if it is not dominated by any element of max(S_2). The situation is illustrated in Figure 4.5 for $d = 3$. Therefore if U and V are two point sets such that for all $p \in U$ and $q \in V$ we have $x_d(p) \le x_d(q)$, it is

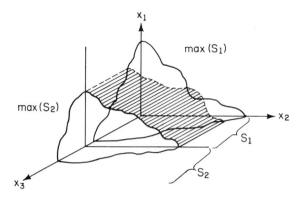

Figure 4.5   The "merge" step of the divide-and-conquer approach for $d = 3$.

convenient to define a new set, filter$(U|V) \subseteq U$, as

$$\text{filter}(U|V) \triangleq \{p: p \in U \text{ and } p \not< V\}.$$

Clearly, the "merge step" of the divide-and-conquer is the computation of filter(max(S_1)|max(S_2)). Thus we have the following simple algorithm:

> **function** MAXIMA2(S, d)
> **begin if** $(|S| = 1)$ **then return** S
>   **else begin** $(S_1, S_2) :=$ equipartition S on x_d;
> $M_1 :=$ MAXIMA2 (S_1); $M_2 :=$ MAXIMA2(S_2);
>   **return** $M_2 \cup$ filter$(M_1|M_2)$
>   **end**
> **end.**

If we denote by $T_d(|S|)$ the running time of MAXIMA2(S) and by $F_{d-1}(|S_1|, |S_2|)$, the time needed to compute filter$(S_1|S_2)$,[7] we have the straightforward recurrence (assume that N is even)

$$T_d(N) \le 2T_d(N/2) + F_{d-1}(N/2, N/2) + O(N), \tag{4.8}$$

where $O(N)$ is the time used to equipartition S into S_1 and S_2.

Clearly, the crux of the technique is the computation of "filter." Again, let U and V be two point sets in E^d as defined earlier, with $|U| = u$ and $|V| = v$. We resort once again to the divide-and-conquer approach. Thus we equipartition V on x_{d-1} into (V_1, V_2), and let $\bar{x}_{d-1}$ be the largest value of x_{d-1} for all points in V_1. Next we split U at $\bar{x}_{d-1}$ into (U_1, U_2). The situation is illustrated in Figure 4.6, where we have projected all points of U and V on the (x_d, x_{d-1})-plane. Notice that $|V_1| \simeq v/2$ and $|V_2| \simeq v/2$, while $|U_1| = m$, and $|U_2| =$

---

[7] To justify the subscripts of $T(\ )$ and $F(\ ,\ )$, we note that S is a set of points in E^d. However, since each point of S_2 dominates each point of S_1 in the coordinate x_d, the computation of filter$(S_1|S_2)$ is a $(d-1)$-dimensional problem.

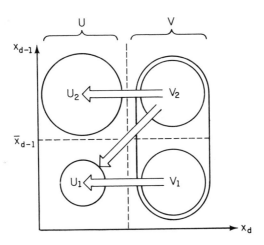

Figure 4.6 Divide-and-conquer computation of filter$(U|V)$ (transformation into three simpler subproblems).

$u - m$, for some integer m. With this subdivision, the original problem of computing filter$(U|V)$ has been replaced by the four subproblems of computing filter$(U_1|V_1)$, filter$(U_1|V_2)$, filter$(U_2|V_1)$, and filter$(U_2|V_2)$. Note, however, that filter$(U_2|V_1)$ is trivial, since for any $p \in U_2$ and $q \in V_1$ we have $x_d(p) \le x_d(q)$ and $x_{d-1}(p) > x_{d-1}(q)$. Moreover, each element of V_2 dominates each element of U_1 in both coordinates x_d and x_{d-1}, so that the computation of filter$(U_1|V_2)$ is really a $(d - 2)$-dimensional problem. The two remaining subproblems are still $(d - 1)$-dimensional, but involve smaller sets. Descriptively, computation proceeds on the tracks of a double recursion: one on set sizes, the other on the dimension. The first one stops when the cardinality of one of the two sets of the pair becomes small (one), the second one when we reach a dimension where a direct efficient approach is applicable (which happens for dimension 3, by resorting to a modification of the dimension-sweep technique described earlier). More formally we have (notice that in algorithm MAXIMA2 the set filter$(M_1|M_2)$ is computed by the call FILTER$(M_1, M_2; d - 1)$)

$\quad$ **function** FILTER$(U, V; d)$
begin if $(d = 3)$ **then** $A := $ FILTER2(U, V);
$\quad\quad$ **if** $(V = \{q\})$ **then** $A := \{p: p \in U \text{ and } p \not\prec q\}$;
$\quad\quad$ **if** $(U = \{p\})$ **then**
$\quad\quad\quad\quad$ **begin if for each** $q \in V: p \not\prec q$ **then** $A := \{p\}$
$\quad\quad\quad\quad\quad$ **else** $A := \emptyset$
$\quad\quad\quad$ **end**
$\quad\quad$ **else begin** $(V_1, V_2) := $ equipartition V on x_d;
$\quad\quad\quad\quad\quad$ $\bar{x}_d := \max\{x_d(p): p \in V_1\}$;
$\quad\quad\quad\quad\quad$ $(U_1, U_2) := $ split U at $\bar{x}_d$;

$$A := \text{FILTER}(U_2, V_2; d) \cup (\text{FILTER}(U_1, V_1; d) \cap \text{FILTER}$$
$$(U_1, V_2; d - 1))$$

         end;
      return A
end.

We must still describe the procedure FILTER2. This procedure is of the sweep-type and is similar to MAXIMA1 for $d = 2$. The points in $U \cup V$ are scanned by descreasing x_2 and placed in a queue Q; if the current p belongs to V, then M is updated, otherwise p is either accepted (as a maximal element) or discarded. Thus we have

      function FILTER2(U, V)
begin $Q := \text{sort } U \cup V$ by decreasing x_2;
    $M := \emptyset$;
    $x_1^* := 0$;
    while $(Q \neq \emptyset)$ **do**
      begin $p \Leftarrow Q$;
        if $(x_1(p) > x_1^*)$ **then**
          begin if $(p \in U)$ **then** $M := M \cup \{p\}$ (*p is found to be
             maximal*)
           else $x_1^* := x_1(p)$
        end
      end;
      return M
end.

To analyze the performance, let $F_d(u, v)$ be the running time of FILTER$(U, V; d)$. An inspection of the procedure FILTER2 yields the straightforward result that $F_2(u, v) = O(u + v)$, (if both U and V have been presorted on x_2). In general, we obtain from the procedure FILTER the recurrence

$$F_d(u, v) \leq F_d(m, v/2) + F_d(u - m, v/2) + F_{d-1}(m, v/2), \qquad (4.9)$$

for some $0 \leq m \leq u$. A tedious calculation,[8] and the obtained expression of F_2, show that the right-side is maximized as $O((u + v) \log u (\log v)^{d-3})$. Therefore, inequality (4.8) becomes

$$T_d(N) \leq 2T_d(N/2) + O(N(\log N)^{d-3}) + O(N)$$

whence

$$T_d(N) = O(N(\log N)^{d-2}) \qquad d \geq 2. \qquad (4.10)$$

When we take into consideration the time used by the presorting operation, we obtain the following theorem.

---

[8] The reader is referred to [Kung–Luccio–Preparata (1975)] for a proof.

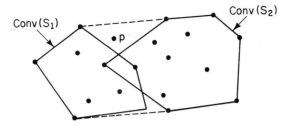

Figure 4.7   A point p is outside both conv(S_1) and conv (S_2) but it is inside the convex hull of their union.

Theorem 4.9. *The maxima of a set of N points in E^d, $d \geq 2$, can be obtained in time $O(N(\log N)^{d-2}) + O(N \log N)$.*

This remarkable success of the divide-and-conquer approach, which failed for $d > 3$ in the convex hull problem (see Section 3.4) is the clue to the elucidation of the deep difference between the two problems. The key notion is closely related to, but not quite the same as, that of "decomposability," introduced by Bentley (1979). Specifically, let us consider the searching problems (see Chapter 2) associated with the maxima and convex hull problems, i.e., "test for maximum" and "convex hull inclusion," respectively. Suppose that a set S of points is arbitrarily partitioned into $\{S_1, S_2\}$. The query "Is p a maximum in $S \cup \{p\}$?" is answered affirmatively if and only if affirmative answers are obtained for both "Is p a maximum in $S_1 \cup \{p\}$?" and "Is p a maximum in $S_2 \cup \{p\}$?". However, it is not hard to partition S so that "Is p outside the convex hull of S?" is answered negatively, while both "Is p outside the convex hull of S_1?" and "Is p outside the convex hull of S_2?" receive affirmative answers (refer to Figure 4.7).

Before closing this section, we consider the average-case performance of the above algorithm for a rather general probability distribution. The result we need is expressed in the following theorem.

Theorem 4.10 [Bentley–Kung–Schkolnick–Thompson (1978)]. *The average number $\mu(N, d)$ of maxima of a set S of N points in E^d, under the hypothesis that the coordinates of each point of S are independent and drawn from an identical, continuous distribution is*

$$\mu(N, d) = O((\log N)^{d-1}). \tag{4.11}$$

We now observe that the set splitting contemplated by algorithm MAXIMA2 preserves randomness, so that the same probability distribution applies to each of the two subproblems. Also, set splitting can be done in constant time by the "pointer-passing" strategy outlined in Section 4.1.1. With these observations, relation (4.8) is replaced in the average-case analysis by

$$E[T_d(N)] \leq 2E[T_d(N/2)] + E[F_{d-1}(m_1, m_2)] + O(1), \qquad (4.12)$$

where m_1 and m_2—the sizes of the sets of maxima of S_1 and S_2—are now random variables, both with average equal to $\mu(N/2, d)$. By the preceding discussion, we have $F_{d-1}(m_1, m_2) \leq K_1(m_1 + m_2) \log m_1 (\log m_2)^{d-4}$, for some constant K_1.

Since both m_1 and m_2 are upperbounded by $N/2$, we have $F_{d-1}(m_1, m_2) \leq K_1(m_1 + m_2)(\log N)^{d-3}$, whence

$$E[F_{d-1}(m_1, m_2)] \leq K_1(\log N)^{d-3}(E[m_1] + E[m_2])$$

$$= K_1(\log N)^{d-3} 2\mu\left(\frac{N}{2}, d\right)$$

$$= O((\log N)^{d-3} \cdot (\log N)^{d-1}) = O((\log N)^{2d-4}).$$

Substituting this value in (4.12), we obtain the following result.

Theorem 4.11. *Under the hypothesis that the coordinates of each point are independent and drawn from an identical continuous distribution, the maxima of a set of N points in E^d can be found in average time*

$$E[T_d(N)] = O(N). \qquad (4.13)$$

4.1.4 Convex hull of a simple polygon

Whenever there exists a constraint on the formulation of a given problem, the set of problem instances one must consider is generally reduced. A most intriguing question in such cases is whether this restriction affords the development of an *ad hoc* technique which is inherently simpler than the general one. Of course not always can the situation be decisively settled, either in the form of the discovery of a simpler algorithm or in the establishment of a lower bound analogous to that for the general case. In all cases, the investigation is worthwhile and challenging.

One such constrained situation is the problem of the convex hull of a simple polygon. Notice that given a set S of N points, we can always think of it as a polygon, by simply arranging the points of S as a sequence and assuming the presence of an edge between adjacent (in the cyclic sense) points of this sequence. Such a polygon, however, is in general not simple, that is, it may have intersecting edges. What happens, then, if we know that the polygon is simple? (See Section 1.3. for a definition of simple polygon.)

The lower-bound arguments—even the one for the ordered hull—are inapplicable to the situation, so one must settle for the trivial $\Omega(N)$ lower bound. It is thus natural to seek algorithms with a better than $O(N \log N)$ worst-case time performance, and indeed several have been proposed all with an $O(N)$ performance. Unfortunately some of these may fail in very special cases [Sklansky (1972); Shamos (1975a)], while others correctly solve the

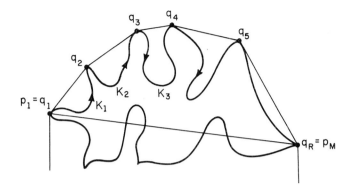

Figure 4.8   The upper-hull of a simple polygon.

problem. Of the latter we mention the rather complicated one by McCallum–Avis (1979), which makes use of two stacks. Two new algorithms have been since proposed, which use only one stack, one by D. T. Lee (1983a) the other by Graham–Yao (1983). The algorithm described below is an original variant of Lee's algorithm.

Let p_1 be the leftmost vertex of the given simple polygon P and let $(p_1, p_2, \ldots, p_N)$ be its directed vertex cycle (obviously, with p_1 following p_N). We assume that the interior of P lies to the right of the boundary, i.e., the cycle is directed clockwise (see Figure 4.8). On this cycle, let p_M be the rightmost vertex. Clearly both p_1 and p_M lie on the convex hull of P. Moreover, they partition the vertex cycle into two chains, one from p_1 to p_M, the other from p_M to p_1. It is sufficient to consider the hull of the chain $(p_1, p_2, \ldots, p_M)$ which, according to our previous terminology (Section 3.3.2) we shall call *upper-hull*. A *subsequence* $(q_1, q_2, \ldots, q_R)$, with $q_1 = p_1$ and $q_R = p_M$, of $(p_1, p_2, \ldots, p_M)$ is the desired convex hull polygon. Each of the edges $\overline{q_i q_{i+1}}$ $(i = 1, \ldots, r - 1)$ could be intuitively viewed as the "lid" of the "pocket" K_i, where pocket K_i is a *substring* of $(p_1, p_2, \ldots, p_M)$, whose first and last vertices are q_i and q_{i+1}, respectively.

The algorithm marches along $(p_1, \ldots, p_M)$ and, in this process, constructs in succession the lids of all pockets. The distinguished event in the execution of the algorithm is the identification of a vertex q that forms a lid with the last found q vertex (for ease of reference, we call q an *event vertex*). Note, however, that each event vertex is not necessarily a vertex of the final hull, but simply a candidate for that status. Basically, what the algorithm declares before its completion is that the so-far discovered sequence $(q_1, q_2, \ldots)$ of event vertices is consistent with the assumption that polygon P is simple (we shall give a more technical characterization below). Thus, we shall now concentrate on the activity leading from one event vertex to the next one. We call this the *advancing step*.

Suppose that the boundary has been scanned from p_1 to p_s $(s \leq M)$ and that

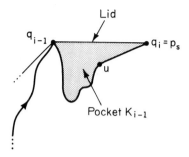

Figure 4.9   A pocket and its lid.

$p_s = q_i$ is an event vertex. The situation appears as in Figure 4.9, where pocket K_{i-1} has been closed by lid $\overline{q_{i-1}q_i}$. The march now proceeds beyond p_s. Calling u the vertex preceding q_i on the boundary of P, we distinguish two cases, depending upon the position of u relative to the oriented segment $\overrightarrow{q_M q_i}$.

1. *u is on or to the right of* $\overrightarrow{q_M q_i}$. In this case we identify in the vertical strip determined by q_1 and q_M three regions, labelled ①, ②, and ③ in Figure 4.10(a). These regions are defined by the line passing by q_{i-1} and q_i, the ray obtained by extending $\overline{q_i u}$ and the portion of the boundary of P corresponding to pocket K_{i-1}.
2. *u is to the left of* $\overrightarrow{q_M q_i}$. In this case we identify one additional region, region ④, as shown in Figure 4.10(b).

Letting v denote the vertex following q_i on the boundary of P, we observe that v lies in one of the previously identified regions. If v lies in either ② or ③, then it is an event vertex, whereas if it lies in either ① or ④, then it is not an event vertex. Specifically we now discuss the four cases:

1. $v \in$ region ①. In this case, the boundary enters the pocket (and extends it). We follow this path until we reach the first boundary edge that has the

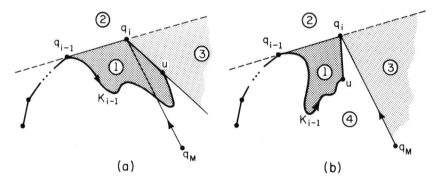

Figure 4.10   Regions where the vertex v following q_i may be (depending upon the relative positions of u and $\overrightarrow{q_M q_i}$).

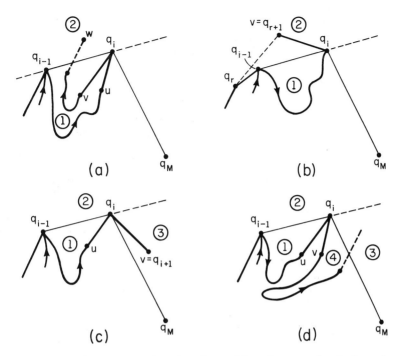

Figure 4.11   Illustration of the four situations arising when scanning the boundary of
P beyond event point q_i.

other vertex w outside of the pocket, i.e., in region ② [Figure 4.11(a)]. Since
polygon P is simple and the pocket with its lid also forms a simple polygon,
by the Jordan curve theorem the boudary of P necessarily crosses the lid.
Vertex w is then treated as v in the case $v \in$ region ② (discussed next).

2. $v \in$ region ②. Vertex v is an event vertex. The supporting line from v to the
   chain $(q_1, \ldots, q_{i-1})$ is found. If this line contains q_r ($r < i$), then vertices
   $q_{r+1}, \ldots, q_i$ are deleted and v becomes q_{r+1}. [Figure 4.11(b)]. Clearly v is a
   vertex of the convex hull (an event point), since it is external to the current
   hull $(q_1, q_2, \ldots, q_M)$. Tracing the supporting line is equivalent to the construction of $\text{conv}(q_1, \ldots, q_r, q_{r+1}, q_M)$.

3. $v \in$ region ③. Vertex v is an event vertex and v becomes q_{i+1}. Indeed [refer to
   Figure 4.11(c)], v is external to the current hull $(q_1, q_2, \ldots, q_i, q_M)$ and,
   being to the right of the line passing by q_{i-1} and q_i (and directed from q_{i-1}
   to q_i), the angle $(q_{i-1} q_i v)$ is convex. (Note that Figure 4.11(c) illustrates
   the case corresponding to Figure 4.10(b).)

4. $v \in$ region ④. The boundary enters the interior of the convex hull. As in
   Case 2 above, we follow this path until we reach the first boundary edge
   with the following property: one of its extremes is either external to ④ or it
   coincides with q_M. In the second case, the procedure terminates; in the first
   case, the external extreme lies either in ③ (handled by Case 3 above) or in

② (handled by Case 2 above) [Figure 4.11(d)]. Indeed, the portion C of the boundary being traced from q_i is a simple curve terminating at q_M; moreover, before it crosses $\overline{q_i q_M}$, it does not extend to the right of the vertical line by q_M, nor to the left of the vertical line by q_1. Since polygon P is simple, C cannot cross any of the boundaries of pockets $K_1, \ldots, K_{i-1}$, and so it cannot cross any of their lids. It follows that C can only cross $\overline{q_i q_M}$; if it does not cross it, then it just terminates at q_M.

The preceding discussion contains the proof of correctness of any implementation that handles Cases 1–4 in the manner described above. We shall now formalize one such algorithm. The most appropriate data structures are: (i) a queue P containing the sequence $(p_2, p_3, \ldots, p_M)$; (ii) a stack Q for the sequence $(q_0, q_1, q_2, \ldots)$—due to the last-in-first-out mechanism of insertion and deletion of hull points—where q_0 is a "dummy" sentinel vertex having the same abscissa as $p_1 = q_1$ and smaller ordinate. As mentioned earlier, u is the vertex preceding q_i on the boundary of P, and v is the currently visited vertex. Given an ordered triplet (stw) of vertices, we call it a *right turn* if w is to the right of the straight line passing by s and t and directed from s to t, and we call it a *left turn* otherwise. Also, while $U \Leftarrow v$ is as usual the operation "PUSH v into U", by POP U we mean: "$v \Leftarrow U$; ignore v"; q_i denotes the vertex at the top of the stack Q and FRONT(P) is the first element of queue P.

procedure POLYGON HULL$(p_1, \ldots, p_M)$

1. **begin** $P \Leftarrow (p_1, \ldots, p_M)$;
2. $Q \Leftarrow q_0$;
3. $Q \Leftarrow p_1$;
4.   **while** $(P \neq \varnothing)$ **do**
5.   **begin** $v \Leftarrow P$;
6.   **if** $((q_{i-1} q_i v)$ is right turn) (∗regions ①, ③, ④∗) **then**
7.   **if** $((u q_i v)$ is right turn) (∗regions ③, ④∗) **then**
8.   **if** $((q_M q_i v)$ is right turn) (∗region ③∗) **then** $Q \Leftarrow v$
9.   **else** (∗region ④∗)
10.   **while** (FRONT(P) is on or to left of $\overrightarrow{q_M q_i}$) **do** POP P
11.   **else** (∗region ①∗)
12.   **while** (FRONT(P) is on or to left of $\overrightarrow{q_i q_{i-1}}$) **do** POP P
13.   **else** (∗region ②∗)
14.   **begin while** $((q_{i-1} q_i v)$ is left turn) **do** POP Q;
15. $Q \Leftarrow v$
              end
        end
end.

The performance analysis of the algorithm POLYGON HULL is quite straightforward. After the initialization (lines 1–3), each vertex of the boundary of P is visited exactly once before it is either accepted (lines 8 or 15) or discarded (lines 10 or 12). Processing of each boundary vertex is done in constant time. The *while* loop of line 14, which constructs the support line, uses a constant amount of time per deleted vertex. Since both sequences $(p_1, \ldots, p_M)$ and $(q_1, \ldots, q_R)$ have $O(M)$ terms, and analogous arguments can be developed for the lower-hull of P, we conclude with the following theorem.

Theorem 4.12. *The convex hull of an N-vertex simple polygon can be constructed in optimal $\theta(N)$ time and $\theta(N)$ space.*

4.2 Applications to Statistics

The connection between geometry and statistics is a close one because a multivariate statistical sample can be viewed as a point in Euclidean space. In this setting, many problems in statistics become purely geometric ones. For example, linear regression asks for a hyperplane of best fit in a specified norm. Certain problems in voting theory are interpreted as finding which k-dimensional hemisphere contains the most points of a set. A survey of geometric techniques in statistics is given in [Shamos (1976)]. Determining the convex hull is a basic step in several statistical problems, which we treat separately in the next few paragraphs.

4.2.1 Robust estimation

A central problem in statistics is to estimate a population parameter, such as the mean, by observing only a small sample drawn randomly from the population. We say that a function t is an unbiased estimator of a parameter p if $E[t] = p$, that is, the expectation value of t is precisely p [Hoel (1971)]. While the sample mean is an unbiased estimator of the population mean, it is extremely sensitive to *outliers*, observations that lie abnormally far from most of the others. It is desirable to reduce the effects of outliers because they often represent spurious data that would otherwise introduce errors in the analysis. A related property that a good estimator should enjoy is that of *robustness*, or insensitivity to deviations from the assumed population distribution. Many such estimators have been proposed [Andrews (1972)]. An important class, known as the Gastwirth estimators [Gastwirth (1966)], are based on the fact that we tend to trust observations more the closer they are to the "center" of the sample.

Consider N points on the line (see Figure 4.12). A simple method of removing suspected outliers is to remove the upper and lower α-fraction of the points and taking the average of the remainder. This is known as the α-trimmed mean, and is a special case of the Gastwirth estimator,

$$T = \sum_{i=1}^{N} w_i x_{(i)}, \qquad \sum_{i=1}^{N} w_i = 1,$$

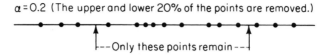

Figure 4.12   The α-trimmed mean.

where $x_{(i)}$ denotes the i-th smallest of the x_j. The α-trimmed mean is just the case

$$w_i = 1/(1 - 2a)N, \qquad \alpha N \le i \le (1 - \alpha)N.$$

Any trimmed mean can be computed in $O(N)$ time (using a linear selection algorithm) and any Gastwirth estimator in $O(N \log N)$ time (by sorting), but what are their analogs in higher dimensions? Tukey has suggested a procedure known as "shelling," or "peeling," which involves stripping away the convex hull of the set, then removing the convex hull of the remainder, and continuing until only $(1 - 2\alpha)N$ points remain [Huber (1972)]. This procedure motivates our next definition and problem.

Definition 4.2. The depth of a point p in a set S is the number of convex hulls (*convex layers*) that have to be stripped from S before p is removed. The depth of S is the depth of its deepest point. (See Figure 4.13.)

This motivates a geometric problem which is interesting in its own right.

PROBLEM CH8 (DEPTH OF A SET). Given a set of N points in the plane, find the depth of each point.

Theorem 4.13. *Any algorithm that determines the depth of each point in a set must make $\Omega(N \log N)$ comparisons, in the worst case.*

PROOF. By transformation from SORTING. Consider a one-dimensional set. Knowing the depth of each point, we can sort the set in only $O(N)$ additional comparisons.   □

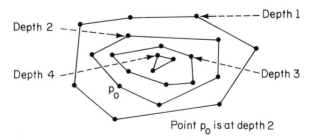

Figure 4.13   The depth of point in a set.

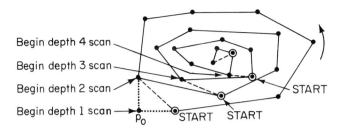

Begin depth 4 scan

Begin depth 3 scan

Begin depth 2 scan

Begin depth 1 scan

START

START   START

p_0

Figure 4.14   Finding depths by repeating the Jarvis march. Vertices which are successively labelled START are circled.

Only recently an algorithm attaining this bound has been developed. Before describing it, however, it is interesting to briefly review the history of the problem.

A brute force approach consists of finding first the convex hull of the given set S, subsequently the convex hull of the nonhull points, and so on, at a worst-case cost of $O(N^2 \log N)$ operations. (This is achieved when the hulls contain a set of $\lfloor N/3 \rfloor$ nested triangles.) A somewhat more subtle approach [Shamos (1978)] uses a repeated application of the Jarvis' march (Section 3.3.3) and achieves running time $O(N^2)$. Figure 4.14 illustrates the working of the algorithm. For reasons of uniformity we introduce a dummy point $p_0 = (x_0, y_0)$, whose coordinate are equal to the smallest abscissa and ordinate in S. The scan begins from p_0 and the first point reached is labelled START. The march then proceeds counterclockwise. Each visited point (except START) is removed from S. When the march cycles back to START, the depth 1 hull has been constructed. Point START is removed from S, the predecessor of START assumes the role previously held by p_0, and the process continues until all points are removed from S.

A more complex although more efficient algorithm is afforded by the planar configuration maintenance technique of Overmars–van Leeuwen (see Section 3.3.7). We just recall that the primary data structure is a binary search tree, the nodes of which point to secondary data structures, which are concatenable queues. Moreover, the queue pointed to by the root gives the upperhull of the point set, and each point deletion is carried out in $O(\log N)^2$ time. Thus, by simultaneously using the structures corresponding to the upper- and the lower-hull, one obtains the depth 1 hull. Each of the points on this hull is then deleted from both structures, which are now ready to deliver the depth 2 hull, and so on. The hulls are trivially obtained in linear time (traversal of a concatenable queue), while the global cost of the deletions is $O(N \log^2 N)$.

This idea has been further refined by Chazelle (1983b). Indeed, the general technique of Overmars–van Leeuwen maintains planar configurations under an arbitrary schedule of insertions of deletions. However, when constructing the convex layers, the schedule of deletions is under the control of the algorithm designer. Chazelle shows how to carefully batch these deletions to

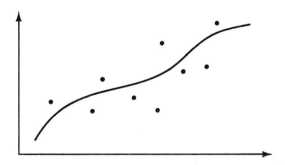

Figure 4.15   An isotone function (not best) approximating a finite point set.

efficiently obtain the layers. The reader is referred to the original work for a
proof of this interesting theorem.

Theorem 4.14. *The convex layers, and, therefore the depth of a set of N points in*
the plane can be computed in optimal time $\theta(N \log N)$.

4.2.2 Isotonic regression

Regression may be regarded as a problem of *best approximation in a subspace.*
We are given a finite set of points in E^d, considered as the values of a function f
of $(d-1)$ variables, and must specify both the allowed class of approximating
functions and the norm in which the error is to be measured.[9] A *regression*
function is some function f^* of the $(d-1)$ variables which minimizes the norm
$\| f - f^* \|$.

   The problem of isotonic regression is to find a best *isotone* (that is, monotone nonincreasing or nondecreasing) approximation to a finite point set. The
error norm usually chosen is L_2, or least-squares, because of its connection
with maximum likelihood estimation.[10] In other words, we are given a set of
points $\{(x_i, y_i): i = 1, \ldots, N\}$ and we are seeking an isotone function f that
minimizes

$$\sum_{i=1}^{N} (y_i - f(x_i))^2.$$

An isotone approximating function is shown in Figure 4.15.

   The above diagram is misleading because it reinforces the intuition that the
required function f should be smooth. A best least-squares isotone fit is a step
function, as illustrated in Figure 4.16. It "only" remains to determine the
locations and heights of the steps.

---

[9] In two variables the points (x_i, y_i) represent the function $y_i = f(x_i)$, and x_i is said to be the
independent variable. In k dimensions we have $x_k = f(x_1, \ldots, x_{k-1})$.

[10] Isotonic regression is discussed at length in [Barlow–Bartholomew–Bremner–Brunk (1972)].

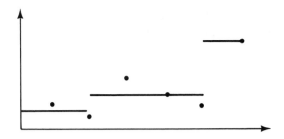

Figure 4.16 A best isotone fit is a step function. The number of steps and the points at which they break must both be determined.

Suppose the data have been ordered by x-coordinate. (In many experimental situations sorting is not necessary because the independent variable is time or the points are taken in order of increasing x.) Define the *cumulative sum diagram* (CSD) to be the set of points $p_j = (j, s_j)$, $p_0 = (0, 0)$, where s_j is the cumulative sum of the y's

$$s_j = \sum_{i=1}^{j} y_i$$

The slope of the line segment joining p_{j-1} to p_j is just y_j. Suppose now that we construct the lower-hull H (see Section 3.3.2) of the points set $\{p_0, p_1, \ldots, p_N\}$. There is a remarkable relationship between the lower-hull H just obtained and isotonic regression: indeed the isotonic regression of a point set is given by the slope of the lower-hull of its cumulative sum diagram [Barlow (1972)] (see Figure 4.17).

Thus we have the result

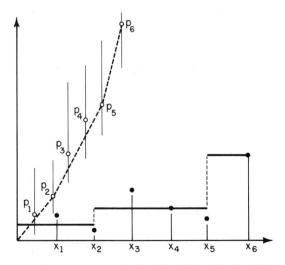

Figure 4.17 The lower convex hull of the CSD defines the isotonic fit.

Theorem 4.15. *Least-squares isotonic regression can be performed on a set of N points in the plane in $O(N \log N)$ time. If the points are ordered by abscissa, then linear time suffices.*

If the data are already ordered, the Graham scan can be run in linear time to find the lower hull. If the points are unordered, $O(N \log N)$ time will be used in the worst case.

4.2.3 Clustering (diameter of a point set)

To quote from [Hartigan (1975)], clustering is the *grouping of similar objects*. A clustering of a set is a partition of its elements that is chosen to minimize some measure of dissimilarity. Hartigan's book contains a large number of different such measures and procedures for clustering using them. We will focus on point data in two dimensions, where we assume that the x and y variables are scaled so that Euclidean distances are meaningful. A measure of the "spread" of a cluster is the maximum distance between any two of its points, called the *diameter* of the cluster. We feel intuitively that a cluster with small diameter has elements that are closely related, while the opposite is true of a large cluster. One formulation, then, of the clustering problem is

PROBLEM CH9 (MINIMUM DIAMETER K-CLUSTERING). Given N points in the plane, partition them into K clusters $C_1, \ldots, C_K$ so that the maximum cluster diameter is as small as possible.

It is difficult to imagine how to solve this problem unless we at least have an algorithm for determining cluster diameter. This motivates

PROBLEM CH10 (SET DIAMETER). Given N points in the plane, find two that are farthest apart.

This problem is seemingly so elementary that it is difficult to perceive that there is any real issue involved. After all, we can compute the distance between each of the $N(N - 1)/2$ pairs of points in a completely straightforward manner and choose the smallest of these to define the diameter. What is left to investigate? We certainly must wonder if this $O(N^2)$ procedure is the best possible algorithm. The complexity of problems of this type are basic questions in computational geometry, which we would have a duty to ask even in the absence of practical considerations.

An argument obtaining a nontrivial lower bound for the DIAMETER problem has been only recently developed.[11] It is based on transforming SET

---

[11] The original idea of the transformation can be found in K. Q. Brown's thesis (1979a). Apparently, D. Dobkin and I. Munro independently discovered the same mapping.

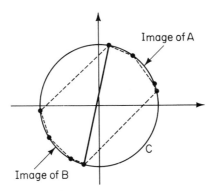

Figure 4.18   Mapping A and B to points on the unit circle C.

DISJOINTNESS (a problem for which a nontrivial lower bound is known) to SET DIAMETER. Let $A = \{a_1, \ldots, a_N\}$ and $B = \{b_1, \ldots, b_N\}$ be two sets of real nonnegative numbers. To test whether A and B do not share any elements (i.e., the disjointness of sets A and B) requires $\Omega(N \log N)$ comparisons [Reingold (1972)]. We now transform SET DISJOINTNESS to SET DIAMETER by mapping A and B to the first and third quadrant of the unit circle C in the plane as follows: a_j is mapped to the intersection of C with the line $y = a_j x$ in the first quadrant, while b_i is mapped to the analogous intersection in the third quadrant (Figure 4.18). Let S be the set of these $2N$ intersections. It is trivial to realize that the diameter of S equals 2 if and only if there are two diametrically opposed points on C, that is, $A \cap B \neq \emptyset$. So we have (with obvious validity in higher dimensions)

Theorem 4.16. *The computation of the diameter of a finite set of N points in E^d* *($d \geq 2$) requires $\Omega(N \log N)$ operations in the algebraic computation-tree model.*

Turning now our attention to the design of an algorithm for the DIAMETER problem, the key idea for possibly avoiding the examination of all pairs of points is provided by the following theorem.

Theorem 4.17 [Hocking–Young (1961)]. *The diameter of a set is equal to the diameter of its convex hull.* (See Figure 4.19.)

In the worst case, of course, all of the original points of the set may be vertices of the hull, so we will have spent $O(N \log N)$ time without eliminating anything.

Until further notice, we shall consider the two-dimensional problem. In this case the convex hull is a convex polygon, not just a set of points, so we have a different problem.

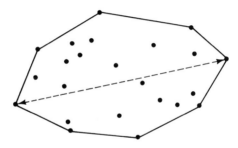

Figure 4.19   Diam(S) = Diam(CH(S)).

PROBLEM CH11 (CONVEX POLYGON DIAMETER). Given a convex poly-gon, find its diameter.

We have immediately that

$$\text{SET DIAMETER} \propto_{N \log N} \text{CONVEX POLYGON DIAMETER.}$$

If we can find the diameter of a convex polygon in less than quadratic time we will also be able to find the diameter of a set quickly.

While there is no *a priori* reason to suspect that convexity helps here, it is worthwhile to investigate further before giving up. We should at least examine alternative characterizations of the diameter. One of these is

Theorem 4.18. *The diameter of a convex figure is the greatest distance between parallel lines of support* [Yaglom–Boltyanskii (1961), p. 9].

Consult Figure 4.20 and notice that parallel lines of support cannot be made to pass through *every* pair of points. For example, no lines of support through vertices p_4 and p_6 can be parallel. This means that $\overline{p_4 p_6}$ is not a diameter. A pair of points that does admit parallel supporting lines will be

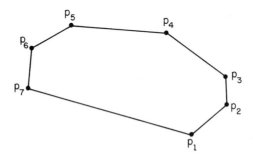

Figure 4.20   Not all vertex pairs are antipodal. Parallel lines of support cannot pass through p_4 and p_6 simultaneously. Thus $\overline{p_4 p_6}$ cannot be a diameter.

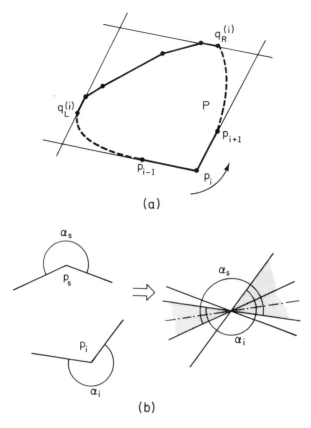

Figure 4.21   Characterization of antipodal pairs. The intersection of α_i and α_s is a pair of planar wedges sharing their vertex.

called *antipodal*. Because of Theorem 4.18, we need only consider antipodal pairs. The problem is to find them without examining *all* pairs of points.

Referring now to Figure 4.21(a), consider a vertex p_i of a convex polygon P (vertices are indexed in counterclockwise order). Suppose we traverse the counterclockwise chain of the boundary of P starting at p_i until we reach a vertex $q_R^{(i)}$ which is farthest from $\overline{p_{i-1}p_i}$: in case of ties—i.e., when P has parallel edges—$q_R^{(i)}$ is the first vertex encountered in this traversal with the desired property. Analogously, we define $q_L^{(i)}$ as farthest vertex from $\overline{p_ip_{i+1}}$ in the traversal of the clockwise boundary chain starting at p_i. We claim that the chain of vertices between $q_R^{(i)}$ and $q_L^{(i)}$ (extremes included) defines the set $C(p_i)$ of vertices, each of which forms an antipodal pair with p_i. Indeed, let $\alpha_i > \pi$ be the external angle formed by $\overline{p_{i-1}p_i}$ and $\overline{p_ip_{i+1}}$. Clearly, (p_i, p_s) is an antipodal pair if and only if there is a straight line in the intersection of α_s and α_i [refer to Figure 4.21(b)]. Since $C(p_i)$ is a convex chain, each vertex $p_s \in C(p_i)$ enjoys this property: moreover, the opposite holds for any other vertex of P not belonging to $C(p_i)$.

This property immediately affords an algorithm for the generation of all antipodal pairs. The primitive operation that is needed is a test for the farthest point from a segment $\overline{p_i p_{i+1}}$. This is readily accomplished by computing the "signed area," Area (p_i, p_{i+1}, p) (see Section 2.2.1), of the triangle (p_i, p_{i+1}, p). We can now describe a simple algorithm, where we assume that the vertices of P form a counterclockwise sequence $(p_1, p_2, \ldots, p_N)$ (in which, obviously, p_1 follows p_N) organized as a list governed by a pointer NEXT[ ].

                    procedure ANTIPODAL PAIRS
1. **begin** $p := p_N$;
2. $q := \text{NEXT}[p]$;
3.   **while** $(\text{Area}(p, \text{NEXT}[p], \text{NEXT}[q]) > \text{Area}(p, \text{NEXT}[p], q))$ **do**
                $q := \text{NEXT}[q]$;
          (∗march on P until you reach the first vertex farthest from
          $\overline{p\text{NEXT}[p]}$∗)
4. $q_0 := q$;
5.   **while** $(q \neq p_0)$ **do**
6.   **begin** $p := \text{NEXT}[p]$;
7.   print (p, q);
8.   **while** $(\text{Area}(p, \text{NEXT}[p], \text{NEXT}[q]) > \text{Area}(p,$
                     $\text{NEXT}[p], q))$ **do**
9.   **begin** $q := \text{NEXT}[q]$;
10.   **if** $((p, q) \neq (q_0, p_0))$ **then** print (p, q)
                  end;
11.   **if** $(\text{Area}(p, \text{NEXT}[p], \text{NEXT}[q]) = \text{Area}(p, \text{NEXT}[p], q))$
                  then
12.   **if** $((p, q) \neq (q_0, p_N))$ **then** print $(p, \text{NEXT}[q])$
                     (∗handling of parallel edges∗)
             end
       end.

The above algorithm is illustrated by an example in Figure 4.22.

Note that lines 1–3 locate the vertex $q_R^{(1)}$, in the preceding terminology (here referred to as q_0). The subsequent **while** loop, which constructs the set $C(p_i)$, uses two pointers p and q. These two pointers move counterclockwise around P, p advancing from p_N to $q_R^{(1)}$ and q advancing from $q_R^{(1)}$ to p_N. An antipodal pair is generated either each time any of these two pointers is advanced (lines 5–7 and lines 9–10) or each time we encounter a pair of parallel eges of P (lines 11–12). The first pair to be generated is clearly $(p, q) = (p_0, q_0)$ (the first time line 7 is executed). We claim that the last pair is $(p, q) = (q_0, p_N)$ since p_N is the last value assumed by q (otherwise the main **while** loop at line 5 aborts). Thus each antipodal pair is generated once, and since p advances from p_0 to q_0 and q advances from q_0 to p_N, in the **while** loop, we have a total of N moves, that is, N

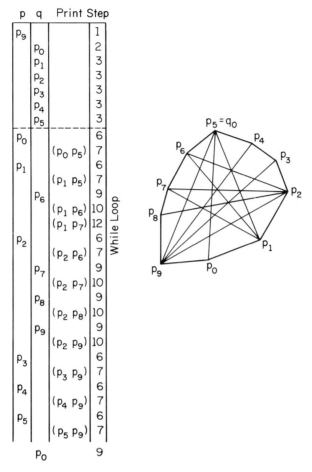

p	q	Print Step
p_9		1
	p_0	2
	p_1	3
	p_2	3
	p_3	3
	p_4	3
	p_5	3
p_0		6
	$(p_0\ p_5)$	7
p_1		6
	$(p_1\ p_5)$	7
	p_6	9
	$(p_1\ p_6)$	10
	$(p_1\ p_7)$	12
p_2		6
	$(p_2\ p_6)$	7
	p_7	9
	$(p_2\ p_7)$	10
	p_8	9
	$(p_2\ p_8)$	10
	p_9	9
	$(p_2\ p_9)$	10
p_3		6
	$(p_3\ p_9)$	7
p_4		6
	$(p_4\ p_9)$	7
p_5		6
	$(p_5\ p_9)$	7
	p_0	9

Figure 4.22 Illustration of ANTIPODAL PAIRS. Note that $\overline{p_1 p_2}$ and $\overline{p_6 p_7}$ are parallel edges.

antipodal pairs, in the absence of parallel edges. If there are pairs of parallel edges, their numbers is at most $\lfloor N/2 \rfloor$, so the total number of antipodal pairs generated by the algorithm is $\leq 3N/2$.

The algorithm could be modified to report only those antipodal pairs (exactly N) which are candidates for being the diameter: indeed, any time two polygon edges are parallel, only the diagonals of the trapezoid they determine need be considered.

Because enumeration of all antipodal pairs suffices to find the diameter of a polygon (by Theorem 4.18), we have the following results.

Theorem 4.19. *The diameter of a convex polygon can be found in time linear in the number of its vertices.*

Corollary 4.1. *The diameter of a set of N points in the plane can be found in optimal* $\theta(N \log N)$ *time.*

Theorem 4.20. *The diameter of a set of N points chosen from an N^p-distribution (Section 4.1.1) in the plane can be found in $O(N)$ expected time.*

PROOF. The convex hull can be found in linear expected time by Theorem 4.5 and then Theorem 4.19 applies. $\square$

Theorem 4.19 can be combined with Theorem 4.12 (Section 4.1.4) to give the following interesting result.

Corollary 4.2. *The diameter of a simple polygon can be found in linear time.*

It may superficially appear that some redundant work is performed in generating *all* the antipodal pairs to obtain the diameter. We shall now convince ourselves that—in the absence of parallel edges—this is not the case. The key notion is that of *diametral pair of P*, that is, a pair of vertices of P whose distance is exactly the diameter of the set itself. Since a diametral pair is also an antipodal pair, there are no more than N diametral pairs. Indeed, there is a classical result by Erdös (1946) that says

Theorem 4.21. *The maximum distance between N points in the plane can occur at most N times.*

Obviously a regular N-gon ($N \geq 3$) achieves this bound.

4.3 Notes and Comments

The problem of computing the diameter of a planar set of points has been optimally solved by the technique discussed in Section 4.2.3. It is therefore tempting to extend the approach to higher dimensional spaces. For E^3 a further inducement into this temptation is provided by the following result [Grünbaum (1956)] (originally known as Vaszonyi conjecture).

Theorem 4.22. *In three dimensions the maximum distance between N points can occur at most $2N - 2$ times.*

The strong similarity between Theorems 4.21 and 4.22 unfortunately conceals the fact that while in the plane the numbers of diametral pairs and antipodal pairs are both $O(N)$, the analogous notions in the space are $O(N)$ and $O(N^2)$, respectively. Indeed, it is not hard to constructively show that the number of antipodal pairs can be $O(N^2)$. This is illustrated in Figure 4.23, where a tetrahedron has been modified by replacing two nonadjacent edges by two "chains" of $\simeq N/2$ vertices. Any two vertices in distinct chains are obviously antipodal. Thus the generation of the antipodal pairs is an $O(N^2)$-

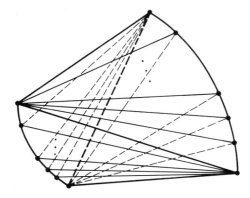

Figure 4.23   Construction of a set of N points in the 3-space with $O(N^2)$ antipodal pairs.

time task, perhaps even more consuming than the brute-force computation of the distances of all pairs of points. The diametral pairs are a subset of the antipodal pairs, but no efficient way to select them has been found to-date. In spite of its apparent simplicity, the computation of the diameter of a three-dimensional set has been a source of frustration to a great many workers.

In d dimensions the diameter of a set can always be found by rote in $O(dN^2)$ time by computing all interpoint distances. The possibility of doing better is darkened by the following theorem.

Theorem 4.23 [Erdös (1960)]. *The maximum distance between N points in d-space can occur $N^2/(2 - 2/\lfloor d/2 \rfloor) + O(N^{2-\varepsilon})$ times, for $d > 3$ and some $\varepsilon > 0$.*

A different approach to the diameter problem has been proposed by Yao (1982). This approach results in a $o(N^2)$-time algorithm. Specifically the time bound is $T(N, d) = O(N^{2-a(d)}(\log N)^{1-a(d)})$, where $a(d) = 2^{-(d+1)}$. For $d = 3$ the time bound can be improved to $O((N \log N)^{1.8})$. Whether or not the gap between $T(N, d)$ and $\Omega(N \log N)$ can be reduced is an open problem.

4.4 Exercises

1. Let S be a set of N points in the plane, so that both coordinates of each point are integers $\leq m$ (*lattice points*). Devise an algorithm to construct CH(S) whose running time is $O(N + m)$.

2. Let S be a set of $N = m^2$ points in E^3 defined $S = \{p_{ij}: x(p_{ij}) = i, y(p_{ij}) = j, z(p_{ij}) > 0, 0 < i, j \leq m\}$. Devise a specialized algorithm to construct the upper hull of S.

3. *Test for maximum in E^2*. Given a set of S of N points in the plane, devise a method for testing if a query point q is a maximum in $S \cup \{q\}$ in time $O(\log N)$. (Repetitive-

mode operation is assumed—Section 2.1) The search data structure should be constructed in time $O(N \log N)$ and use storage $O(N)$.

4. *Test for maximum in E^3.* Given a set S of N points in E^3, devise a method for testing if a query point q is a maximum in $S \cup \{q\}$ in time $O(\log N)$. (Repetitive-mode operation is assumed.) The search data structure should be constructed in time $O(N \log N)$ and use storage $O(N)$.
   (*Hint:* Use planar point location in a suitable planar subdivision.)

5. *Dominance depth.* Given a set S of N points in the plane, the *dominance hull* of S is the set $\max(S)$ of the maxima of S. The dominance depth of a point p in S is the number of dominance hulls that have to be stripped before p is removed; the dominance depth of S is the dominance depth of its deepest point.
   (a) Devise an $O(N \log N)$ algorithm to compute the dominance depth of each point in S.
   (b) Is this algorithm optimal?

CHAPTER 5
Proximity: Fundamental Algorithms

In Chapter 4 we illustrated an $O(N \log N)$ algorithm for finding the two farthest points of a plane set. One may think that finding the two closest points would be a simple extension, but it is not. The two farthest points are necessarily hull vertices and we may exploit convexity to give a fast algorithm; the two closest points do not necessarily bear any relation to the convex hull, so a new technique must be developed, which is the subject of this chapter. We will be concerned with a large class of problems that involve the proximity of points in the plane and our goal will be to deal with all of these seemingly unrelated tasks via a single algorithm, one that discovers, processes, and stores compactly all of the relevant proximity information. To do this, we revive a classical mathematical object, the Voronoi diagram, and turn it into an efficient computational structure that permits vast improvement over the best previously known algorithms. In this chapter, several of the geometric tools we have discussed earlier—such as hull-finding and searching—will be used to attack this large and difficult class—the *closest-point*, or *proximity problems*.

As suggested above, the status of this topic of computational geometry is no exception to the by now familiar current standard: in the plane, powerful and elegant techniques are available, while for the space—and even more in higher dimensions—very little is known and formidable difficulties lurk to suggest a negative prognosis.

Most of the problems studied in the previous chapters (the farthest-pair problem being an exception) involve exclusively the incidence properties of the geometric objects considered; thus, the corresponding results hold under the full linear group of transformations (Section 1.3.2). The problems treated in the next two chapters involve metric properties, and thus the validity of results is restricted to the Euclidean group of transformations (group of rigid motions).

5.1 A Collection of Problems

We begin by presenting a catalog of apparently diverse proximity problems which we shall then see to be intimately related in computational terms.

PROBLEM P.1 (CLOSEST PAIR). Given N points in the plane, find two whose mutual distance is smallest.[1]

This problem is so easily stated and important that we must regard it as one of the fundamental questions of computational geometry, both from the point of view of applications and from pure theoretical interest. For example, a real-time application is provided by the air-traffic control problem: with some simplification, the two aircraft that are in the greatest danger of collision are the two closest ones.

The central algorithmic issue is whether it is necessary to examine every pair of points to find the minimum distance thus determined. This can be done in $O(dN^2)$ time in d dimensions, for any d. In one dimension a faster algorithm is possible, based on the fact that any pair of closest points must be consecutive in sorted order. We may thus sort the given N real numbers in $O(N \log N)$ steps and perform a linear-time scan of the sorted sequence $(x_1, x_2, \ldots, x_N)$ computing $x_{i+1} - x_i$, $i = 1, \ldots, N - 1$. This algorithm, obvious as it is, will be shown later to be optimal.

PROBLEM P.2 (ALL NEAREST NEIGHBORS). Given N points in the plane, find a nearest neighbor of each.

The "nearest neighbor" is a relation on a set S of points as follows: point b is a nearest neighbor of point a, denoted $a \rightarrow b$, if

$$\text{dist}(a, b) = \min_{c \in S-a} \text{dist}(a, c).$$

The graph of this relation is pictured in Figure 5.1. Note that it is not necessarily symmetric, that is, $a \rightarrow b$ does not necessarily imply $b \rightarrow a$. Note also that a point is not the nearest neighbor of a unique point (i.e., "$\rightarrow$" is not necessarily a function).[2] The solution to Problem P.2 is a collection of ordered pairs (a, b), where $a \rightarrow b$.

---

[1] More than one pair may be closest. We will consider finding one such pair as a solution to the problem.

[2] Although a point can be the nearest neighbor of every other point, a point can have at most six nearest neighbors in two dimensions, and at most 12 in three dimensions. This maximum number of nearest neighbors in d dimensions is the same as the maximum number of unit spheres that can be placed so as to touch a given one. ([Saaty (1970)] states that this number is not known for d greater than 12.)

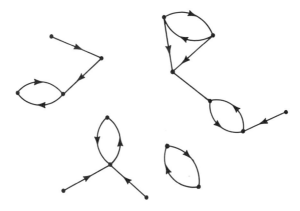

Figure 5.1   The nearest-neighbor relation "→" on a set of points.

A pair which does satisfy symmetry ($a \to b$ and $b \to a$) is called a *reciprocal* pair. If N points are chosen in the plane according to a Poisson point process, the expected fraction of reciprocal pairs is given by [Pielou (1977)]

$$6\pi/(8\pi + 3(3)^{1/2}) \sim 0.6215.$$

In mathematical ecology, this quantity is used to detect whether members of a species tend to occur in isolated couples. The actual number of reciprocal pairs is computed, and the ratio compared. Other studies involving nearest-neighbor computations arise in studying the territoriality of species, in which the distribution of nearest-neighbor distances is of interest, as well as in geography [Kolars–Nystuen (1974)] and solid-state physics. Obviously, in one dimension the same sorting algorithm which solves CLOSEST PAIR yields all the nearest neighbors. We shall later address the question of the respective difficulties of these two problems in higher dimensions.

PROBLEM P.3 (EUCLIDEAN MINIMUM SPANNING TREE). Given N points in the plane, construct a tree of minimum total length whose vertices are the given points.

By a solution to this problem we will mean a list of the $N - 1$ pairs of points comprising the edges of the tree. Such a tree is shown in Figure 5.2.

The Euclidean minimum spanning tree (EMST) problem is a common component in applications involving networks. If one desires to set up a communications system among N nodes requiring interconnecting cables, using the EMST will result in a network of minimum cost. A curious facet of federal law lends added importance to the problem: When the Long Lines Department of the Telephone Company establishes a communications hookup for a customer, federal tariffs require that the billing rate be proportional to the length of a minimum spanning tree connecting the customers' termini, the distance to be measured on a standard flat projection of the

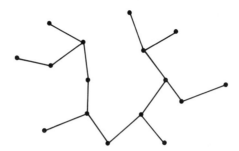

Figure 5.2   A minimum spanning tree on a planar point set.

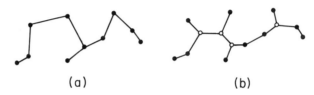

(a)   (b)

Figure 5.3   A Steiner Tree (b) may have smaller total length than the MST (a).

Earth's surface.[3] This is true regardless of the fact that the Earth is not flat and the Telephone Company may not choose to set up the actual network as an MST; nonetheless, the billing is based on a real Euclidean problem, one which must be solved hundreds of times daily, as television network configurations are constantly changing.

This law is a Solomon-like compromise between what is desirable and what is practical to compute, for the minimum spanning tree is not the shortest possible interconnecting network if new vertices may be added to the original set. With this restriction lifted, the shortest tree is called a Steiner Tree (Figure 5.3).[4]

The computation of Steiner trees has been shown by Garey, Graham and Johnson (1976) to be *NP*-hard, and we are unable with present technology to solve problems with more than about 20–25 points. It is therefore unreasonable for the FCC to require billing to be by Steiner tree.

The minimum spanning tree has been used as a tool in clustering [Gower–Ross (1969); Johnson (1967); Zahn (1971)], in determining the intrinsic dimension of point sets [Schwartzmann–Vidal (1975)], and in pattern recognition [Osteen–Lin (1974)]. It has also been used [Lobermann–Weinberger (1957)] to minimize wire length in computer circuitry, and provides the basis for several approximation algorithms for the traveling salesman problem, which we discuss in Section 6.1.

---

[3] Thanks go to Stefan Burr for providing this information.

[4] The Steiner tree in Figure 5.3 is taken from [Melzak (1973)].

The Minimum Spanning Tree problem is usually formulated as a problem in graph theory: Given a graph with N nodes and E weighted edges, find the shortest subtree of G that includes every vertex. This problem was solved independently by [Dijkstra (1959)], [Kruskal (1956)], and [Prim (1957)] and the existence of a polynomial time algorithm (which they all demonstrated) is a considerable surprise, because a graph on N vertices may contain as many as N^{N-2} spanning subtrees [Moon (1967)].[5] A great deal of work has been done in an attempt to find a fast algorithm for this general problem [Nijenhuis–Wilf (1975); Yao (1975)], and the best result to date [Cheriton–Tarjan (1976)] is that $O(E)$ time suffices if $E > N^{1+\varepsilon}$. (See also Section 6.1.)

In the Euclidean problem, the N vertices are defined by $2N$ coordinates of points in the plane and the associated graph has an edge joining every pair of vertices. The weight of an edge is the distance between its endpoints. Using the best known MST algorithm for this problem will thus require $\theta(E) = \theta(N^2)$ time, and it is easy to prove that this is a lower bound in an arbitrary graph because the MST *always contains a shortest edge of G.*[6] Indeed, since the edge weights in a general graph are unrestricted, an MST algorithm that ran in less than $O(N^2)$ time could be used to find the minimum of N^2 quantities in $o(N^2)$ time, which is impossible. It follows that any algorithm that treats a Euclidean MST problem as being embedded in the complete graph on N vertices is doomed to take quadratic time. What would then lead us to suspect that less time is sufficient? For one thing, the Euclidean problem only has $2N$ inputs (the coordinates of the points), while the graph problem has $N(N-1)/2$ inputs (the edge lengths). The Euclidean problem is therefore highly constrained, and we may be able to use its metric properties to give a fast algorithm.

PROBLEM P.4 (TRIANGULATION). Given N points in the plane, join them by nonintersecting straight line segments so that every region internal to the convex hull is a triangle. (See Section 1.3.1.)

Being a planar graph, a triangulation on N vertices has at most $3N - 6$ edges. A solution to the problem must give at least a list of these edges. A triangulation is shown in Figure 5.4.

This problem arises in the finite-element method [Strang–Fix (1973); Cavendish (1974)] and in numerical interpolation of bivariate data when function values are available at N irregularly-spaced data points (x_i, y_i) and an approximation to the function at a new point (x, y) is desired. One method of doing this is by piecewise linear interpolation, in which the function surface is represented by a network of planar triangular facets. The projection of each point (x, y) lies in a unique facet and the function value $f(x, y)$ is obtained by

---

[5] This was first proved by Cayley in 1889.

[6] This was shown by Kruskal and Prim.

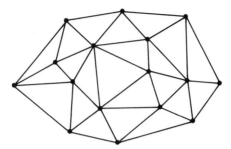

Figure 5.4   Triangulation of a point set.

interpolating a plane through the three facet vertices. Triangulation is the
process of selecting triples that will define the facets. Many criteria have been
proposed as to what constitutes a "good" triangulation for numerical purposes [George (1971)], some of which involve maximizing the smallest angle or
minimizing the total edge length. These conditions are chosen because they
lead to convenient proofs of error bounds on the interpolant, not because they
necessarily result in the best triangulation.

The preceding four problems—P.1 through P.4—are "one-shot" applications which concern the construction of some geometric object (closest pair,
all nearest neighbors, Euclidean minimum spanning tree, and triangulation).
Next we shall review two problems of the "search type" (see Chapter 2), that
is, problems to be executed in a repetitive mode, allowing preprocessing of the
initial data.

PROBLEM P.5 (NEAREST-NEIGHBOR SEARCH). Given N points in the
plane, with preprocessing allowed, how quickly can a nearest neighbor of a
new given query point q be found [Knuth (1973)]?

We may solve this problem in $O(dN)$ time in d dimensions, but we are
interested in using preprocessing to speed up the search. There are a multitude
of applications for such fast searching, possibly the most important of which is
the classification problem. One classification method is the nearest-neighbor
rule [Duda–Hart (1973)], which states that when an object must be classified
as being in one of a number of known populations, it should be placed in the
population corresponding to its nearest neighbor. For example, in Figure 5.5
the unknown point U would be classified "B."

A similar application occurs in information retrieval, where the record
that best matches the query record is retrieved [Burkhard–Keller (1973)];
[Friedman–Bentley–Finkel (1977)]. If many objects are to be processed,
either in classification tasks [Duda–Hart (1973)] (speech recognition, elementary particle identification, etc.) or in retrieval tasks (best match retrieval), we
must be able to perform nearest-neighbor searching quickly.

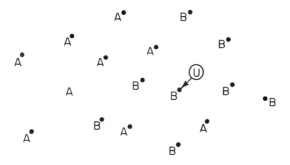

Figure 5.5   The nearest-neighbor rule.

PROBLEM P.6 (*k*-NEAREST NEIGHBORS). Given *N* points in the plane, with preprocessing allowed, how quickly can the *k* points nearest to a new given point *q* be found?

The *k*-nearest neighbors have been used for interpolation and contouring [Davis (1975)] and for classification (the *k*-nearest-neighbor rule is more robust than just looking at a single neighbor). Though the problem seems to be more difficult than P.5, we shall see later that they can both be solved by means of analogous geometric structures (Section 6.3.1).

5.2 A Computational Prototype: Element Uniqueness

In investigating the computational complexity of a problem, that is, lower bounds to some significant performance parameters such as running time and memory use, we frequently seek to transform (or to "reduce," as the jargon goes) a well-known problem, for which nontrivial lower bounds are known, to the problem under consideration (Section 1.4). Examples of this approach are now classical, and we have already encountered them in our study of computational geometry: recall, for example, the transformation of sorting—in the computation-tree model—to the problem of computing the ordered convex hull of a set of points. Another famous instance is SATISFIABILITY as regards *NP*-complete problems [Garey–Johnson (1979)]. It seems appropriate to refer to such archetypal problems, which act as fundamental representatives for classes of problems, as *computational prototypes*.

For the one-shot problems presented in the preceding section it is easy to develop straightforward polynomial time algorithms. It is therefore natural to try to resort to "sorting" as a possible prototype. Although we cannot rule out that sorting could be found to be an adequate prototype, so far the necessary transformation has not been found for all the problems P.1–P.4. Fortunately, however, we can resort to another prototype, ELEMENT UNIQUENESS [Dobkin–Lipton (1979)], which is stated as follows.

PROBLEM (ELEMENT UNIQUENESS). Given N real numbers, decide if any two are equal.

We shall now obtain a lower bound on the time complexity of ELEMENT UNIQUENESS in the algebraic decision-tree model.

A set of N real numbers $\{x_1, \ldots, x_N\}$ can be viewed as a point $(x_1, \ldots, x_N)$ in E^N. Using the terminology of Section 1.4, let $W \subseteq E^N$ be the membership set of ELEMENT UNIQUENESS on $\{x_1, \ldots, x_N\}$ (i.e., W contains all points, no two coordinates of which are identical). We claim that W contains $N!$ disjoint connected components. Indeed, any permutation π of $\{1, 2, \ldots, N\}$ corresponds to the set of points in E^N

$$W_\pi = \{(x_1, \ldots, x_N): x_{\pi(1)} < x_{\pi(2)} < \cdots < x_{\pi(N)}\}.$$

Clearly $W = \bigcup_{\text{all} \pi} W_\pi$, the W_π's are connected and disjoint, and $\#(W) = N!$. Therefore, as a consequence of Theorem 1.2 we have

Corollary 5.1. *In the algebraic decision tree model any algorithm that determines whether the members of a set of N real numbers are distinct requires $\Omega(N \log N)$ tests.*

We shall use this important result in the next section.

Remark. Three important low-complexity prototypes considered in computational geometry for the algebraic decision-tree model—sorting, extreme points, and element uniqueness—have complexity $\Omega(N \log N)$ but are not readily transformable to one another. However, they all share the fundamental trait that their complexity is derived from the cardinality of the set of permutations of N letters (the symmetric group S_N). Although it is tempting to look for a common "ancestor" of these prototypes—something like PERMUTATION IDENTIFICATION—we fail to see a natural way to formulate the latter.

5.3 Lower Bounds

As usual, before embarking in the study of algorithms for solving the problems of Section 5.1, we confront the question of their complexity. We begin with the search-type problems, NEAREST-NEIGHBOR SEARCH and k-NEAREST NEIGHBORS.

Here the computational prototype is BINARY SEARCH. We can easily

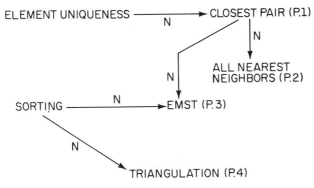

Figure 5.6   Relationship among computational prototypes and proximity problems.

show that

$$\text{BINARY SEARCH} \propto_{0(1)} \text{NEAREST NEIGHBOR.}$$

Suppose that N real numbers $x_1, x_2, \ldots, x_N$ are given. A binary search (with preprocessing allowed!) identifies the x_i that is the closest to a query number q. But we recognize that this very problem can be viewed in a geometric setting, whereby each x_i corresponds to the point $(x_i, 0)$ in the plane. Thus the "nearest-neighbor search" provides the answer sought by "binary search." Therefore we have by the standard information-theoretic argument:

Theorem 5.1. $\Omega(\log N)$ *comparisons are necessary to find a nearest neighbor of a point (in the worst case) in any dimension.*

In the decision-tree model, if we assume that q is equally likely to fall in any of the $N + 1$ intervals determined by the x_i, then Theorem 5.1 bounds the expected behavior of any nearest-neighbor search algorithm.

As regards k-NEAREST NEIGHBORS, the transformation NEAREST-NEIGHBOR SEARCH $\propto k$-NEAREST NEIGHBORS is immediate by setting $k = 1$. Thus, Theorem 5.1 applies to k-NEAREST NEIGHBOR as well.

Having rapidly disposed of the search type problems, we now turn our attention to the "one-shot" applications P.1–P.4. The problem transformations are shown diagrammatically (an arc replacing the usual symbol "$\propto$") in Figure 5.6.

We begin by showing the transformation ELEMENT UNIQUENESS $\propto_N$ CLOSEST PAIR. Given a set of real numbers $\{x_1, \ldots, x_N\}$, treat them as points on the $y = 0$ line and find the two closest. If the distance between them is nonzero, the points are distinct. Since a set in one dimension can always be embedded in k dimensions, the transformation generalizes.

The transformation CLOSEST PAIR $\propto_N$ ALL NEAREST NEIGHBORS is immediate, since one of the pairs obtained by the latter is a closest pair and ca be determined with $O(N)$ comparisons.

Figure 5.7   Illustration for the lower bound on TRIANGULATION.

We next consider Problem P.3, EMST. In the preceding section we have reviewed that the EMST contains a shortest edge of the Euclidean graph on the N given points, whence CLOSEST PAIR is trivially transformable in linear time to EMST. However we can also show

$$\text{SORTING} \propto_N \text{EMST}.$$

Indeed, consider a set of N real numbers $\{x_1, \ldots, x_N\}$. Each x_i is interpreted as a point $(x_i, 0)$ in the plane, and the resulting set of points possesses a unique EMST, namely, there is an edge from x_i to x_j if and only if they are consecutive in sorted order. A solution to the EMST problem consists of a list of $N-1$ pairs (i, j), giving the edges of the tree, and it is a simple exercise to transform this list into the sorted list of the x_i's in time $O(N)$.

Finally, we consider the problem TRIANGULATION (P.4), and show that

$$\text{SORTING} \propto_N \text{TRIANGULATION}.$$

Consider the set of N points $\{x_1, x_2, \ldots, x_N\}$ pictured in Figure 5.7, which consists of $N-1$ collinear points and another not on the same line. This set possesses only one triangulation, the one shown in the figure. The edge list produced by a triangulation algorithm can be used to sort the x_i in $O(N)$ additional operations, so $\Omega(N \log N)$ comparisons must have been made.[7]

The preceding analysis establishes the problem transformations illustrated in Figure 5.6. Since in the computation-tree model both ELEMENT UNIQUENESS and SORTING of N-element sets have a lower bound $\Omega(N \log N)$ we have the following theorem.

Theorem 5.2. *In the computation-tree model, any algorithm that solves any of the problems* CLOSEST PAIR, ALL NEAREST NEIGHBORS, EMST, *and* TRIANGULATION *requires* $\Omega(N \log N)$ *operations.*

We shall next undertake the study of algorithms for these problems, and the search for optimal algorithms.

_____

[7] An equivalent transformation—ORDERED HULL $\propto_N$ TRIANGULATION—is based on the fact that a triangulation of S is a planar graph (embedded in the plane), whose external boundary is the convex hull of S.

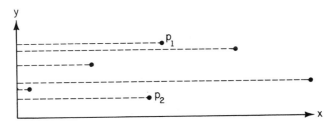

Figure 5.8   The failure of projection methods. Points p_1 and p_2 are the closest pair but are farthest in y-distance.

5.4 The Closest Pair Problem: A Divide-and-Conquer Approach

The lower bound of Theorem 5.2 challenges us to find a $\theta(N \log N)$ algorithm for CLOSEST PAIR. There seem to be two reasonable ways to attempt to achieve such behavior: a direct recourse to sorting or the application of the divide-and-conquer scheme. We can readily dispose of the former, since the environment where sorting is useful is a total ordering. The only reasonable way to obtain a total ordering seems to be to project all the points on a straight line: unfortunately projection destroys essential information, as illustrated informally in Figure 5.8: points p_1 and p_2 are the closest, but they are farthest when projected on the y-axis.

   A second way toward a $\theta(N \log N)$ performance is to split the problem into two subproblems whose solutions can be combined in linear time to give a solution to the entire problem [Bentley–Shamos (1976); Bentley (1980)]. In this case, the obvious way of applying divide-and-conquer does not lead to any improvement, and it is instructive to explore why it fails. We would like to split the sets into two subsets, S_1 and S_2, each having about $N/2$ points, and obtain a closest pair in each set recursively. The problem is how to make use of the information so obtained. The possibility still exists, though, that the closest pair in the set consists of one element of S_1 and one element of S_2, and there is no clear way to avoid making $N^2/4$ additional comparisons. Letting $P(N, 2)$ denote the running time of the algorithm to find the closest pair in two dimensions, the preceding observations lead to a recurrence of the form

$$P(N, 2) = 2P(N/2, 2) + O(N^2) \qquad \text{\footnotesize ⓐ}$$

whose solution is $P(N, 2) = O(N^2)$. Let us try to remedy the difficulty by retreating to one dimension.

   The only $\theta(N \log N)$ algorithm we know on the line is the one which sorts the points and performs a linear-time scan. Since, as noted above, sorting will not generalize to two dimensions, let us try to develop a one-dimensional

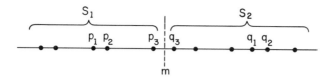

Figure 5.9   Divide-and-conquer in one dimension.

divide-and-conquer scheme that will. Suppose we partition a set of points on the line by some point m, into two sets S_1 and S_2 with the property that $p < q$ for all $p \in S_1$ and $q \in S_2$. Solving the closest pair problem recursively on S_1 and S_2 separately gives us two pairs of points $\{p_1, p_2\}$ and $\{q_1, q_2\}$, the closest pairs in S_1 and S_2, respectively. Let δ be the smallest separation found thus far (see Figure 5.9):

$$\delta = \min(|p_2 - p_1|, |q_2 - q_1|).$$

The closest pair in the whole set is either $\{p_1, p_2\}$, $\{q_1, q_2\}$ or some $\{p_3, q_3\}$, where $p_3 \in S_1$ and $q_3 \in S_2$. Notice, though, *and this is the key observation*, that both p_3 and q_3 must be within distance δ of m if $\{p_3, q_3\}$ is to have a separation smaller than δ. (It is clear that p_3 must be the rightmost point in S_1 and q_3 is the leftmost point in S_2, but this notion is not meaningful in higher dimensions so we wish to be somewhat more general.) How many points of S_1 can lie in the interval $(m - \delta, m]$? Since every semi-closed interval of length δ contains at most one point of S_1, $(m - \delta, m]$ contains at most one point. Similarly, $[m, m + \delta)$ contains at most one point. The number of pairwise comparisons that must be made between points in different subsets is thus at most *one*. We can certainly find all points in the intervals $(m - \delta, m]$ and $[m, m + \delta)$ in linear time, so an $O(N \log N)$ algorithm results (CPAIR1).

function CPAIR1(S)
Input: $X[1:N]$, N points of S in one dimension.
Output: δ, the distance between the two closest.
begin if $(|S| = 2)$ **then** $\delta := |X[2] - X[1]|$
    else if $(|S| = 1)$ **then** $\delta := \infty$
        else begin $m := \text{median}(S)$;
            Construct(S_1, S_2) $(*S_1 = \{p: p \le m\}, S_2 = \{p: p > m\}*)$;
            $\delta_1 := \text{CPAIR1}(S_1)$;
            $\delta_2 := \text{CPAIR1}(S_2)$;
            $p := \max(S_1)$;
            $q := \min(S_2)$;
            $\delta := \min(\delta_1, \delta_2, q - p)$
    end;
    return δ
end.

This algorithm, while apparently more complicated than the simple sort and scan, provides the necessary transition to two dimensions.

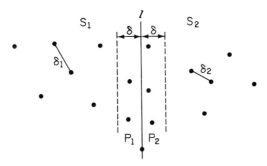

Figure 5.10   Divide-and-conquer in the plane.

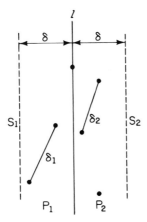

Figure 5.11   All points may lie within δ of l.

Generalizing as directly as possible, let us partition a two-dimensional set S into two subsets S_1 and S_2, such that every point of S_1 lies to the left of every point of S_2. That is, we cut the set by a vertical line l defined by the median x-coordinate of S. Solving the problem on S_1 and S_2 recursively, we obtain δ_1 and δ_2, the minimum separations in S_1 and S_2, respectively. Now let $\delta = \min(\delta_1, \delta_2)$. (See Figure 5.10.)

If the closest pair consists of some $p \in S_1$ and some $q \in S_2$, then surely p and q are both within distance δ of l. Thus, if we let P_1 and P_2 denote two vertical strips of width δ to the left and to the right of l, respectively, then $p \in P_1$ and $q \in P_2$. At this point complications arise that were not present in the one-dimensional case. On the line we found at most one candidate for p [8] and at most one for q. In two dimensions, every point can be a candidate because it is only necessary for a point to lie within distance δ of l. Figure 5.11 shows a set with this property. It again seems that $N^2/4$ distance comparisons will be

---

[8] In CPAIR1 there is *exactly* one candidate for p: $p = \max(S_1)$.

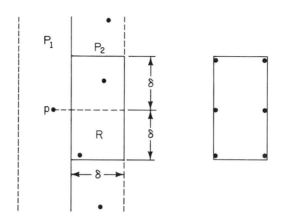

Figure 5.12   For every point in P_1 only a constant number of points in P_2 need to be examined (at most six).

required to find the closest pair, but we will now show that the points lying within the strips of width δ around l have special structure indeed.

Referring now to Figure 5.12, consider any point p in P_1. We must find all points q in P_2 that are within δ of p, but how many of these can there be? They must lie in the $\delta \times 2\delta$ rectangle R and we know that no two points in P_2 are closer together than δ.[9] The maximum number of points of separation at least δ that can be packed into such a rectangle is six, as shown in the figure. This means that for each point of P_1 we need only examine at most six points of P_2, not $N/2$ points. In other words, at most $6 \times N/2 = 3N$ distance comparisons will need to be made in the subproblem merge step instead of $N^2/4$.

We do not yet have an $O(N \log N)$ algorithm, however, because even though we know that only six points of P_2 have to be examined for every point of P_1, we do not know which six they are! To answer this question, suppose we project p and all the points of P_2 onto l. To find the points of P_2 that lie in rectangle R, we may restrict ourselves to consider the projection points within distance δ from the projection of p (at most six). If the points are sorted by y-coordinate, for *all* points in P_1 their nearest-neighbor "candidates" in P_2 can be found in a single pass through the sorted list. Here is a sketch of the algorithm as developed so far.

procedure CPAIR2(S)
1. Partition S into two subsets, S_1 and S_2, about the vertical median line l.
2. Find the closest pair separations δ_1 and δ_2 recursively.
3. $\delta := \min(\delta_1, \delta_2)$;
4. Let P_1 be the set of points of S_1 that are within δ of the dividing line l and let P_2 be the corresponding subset of S_2. Project P_1 and P_2 onto l and sort by y-coordinate, and let P_1^* and P_2^* be the two sorted sequences, respectively.

---

[9] This crucial observation is due to H. R. Strong (personal communication, 1974).

5. The "merge" may be carried out by scanning P_1^* and for each point in P_1^* by inspecting the points of P_2^* within distance δ. While a pointer *advances* on P_1^*, the pointer of P_2^* may oscillate within an interval of width 2δ. Let δ_l be the smallest distance of any pair thus examined.
6. $\delta_s \leftarrow \min(\delta, \delta_l)$.

If $T(N)$ denotes the running time of the algorithm on a set of N points, Steps 1 and 5 take $O(N)$ time, Steps 3 and 6 take constant time, and Step 2 takes $2T(N/2)$ time. Step 4 would take $O(N \log N)$ time, if sorting were to be performed at each execution of the step; however, we can resort to a standard device—called *presorting*—whereby we create once and for all a list of points sorted by y-coordinate, and, when executing Step 4, extract the points from the list *in sorted order* in only $O(N)$ time.[10] This trick enables us to write the recurrence for the running time $P(N, 2)$ of the closest-pair algorithm in two dimensions

$$P(N, 2) = 2P(N/2, 2) + O(N) = O(N \log N) \tag{5.1}$$

which gives the following theorem.

Theorem 5.3. *The shortest distance determined by N points in the plane can be found in $\theta(N \log N)$ time, and this is optimal.*

The central features of the described divide-and-conquer strategy, which enable its extension to higher-dimensional cases, are summarized as follows:

1. The step at which the subproblem solutions are combined takes place in one lower dimension (from the plane to the line).
2. The two point sets involved in the combination of the subproblem solutions have the property of *sparsity*, i.e., it is guaranteed that in each set the distance between any two points is lower-bounded by a known constant.

The emergence of sparsity in the combination step is really the crucial factor. More formally we say

Definition 5.1. Given real $\delta > 0$ and integer $c \geq 1$, a point set S in d-dimensional space has *sparsity c for given* δ, if in any hypercube of side 2δ [11] there are at most c points of S.

Here sparsity is defined as a (monotone nondecreasing) function of δ. Sparsity can be induced, as we shall see, by choosing δ as the minimum

---

[10] This algorithm was implemented nonrecursively by Hoey and Shamos and a curious phenomenon was observed: the number of distance computations made was always strictly less than N. That this is always the case will be shown later in this section. Of course, the behavior of the algorithm is still dominated by the sorting step.

[11] Such cube is also called a *box*.

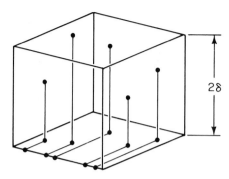

Figure 5.13   Sparsity is preserved through orthogonal projection (for $d \geq 2$).

distance between pairs of points. In the preceding example, the set of points within δ of the dividing line l is sparse, with sparsity $c = 12$, as can be inferred from Figure 5.12. (Note that $c = 12 = 4 \times 3$ for $d = 2$.)

For $d \geq 2$, the portion of E^d contained between two hyperplanes orthogonal to one of the coordinate axes and at distance 2δ from each other is called a δ-*slice*. Note that if a δ-slice has sparsity c for given δ, this sparsity is preserved through projection of the δ-slice on the hyperplane bisecting the δ-slice (a $(d - 1)$-dimensional variety). Indeed the d-dimensional cube of side 2δ projects to a $(d - 1)$-dimensional cube, also of side 2δ, containing exactly as many points as the original cube (Figure 5.13).

Suppose now to have a set S of N points in E^d. According to the previous pattern, we propose the following algorithm:

procedure CPAIR $d(S)$

1. Partition set S (in time $O(N)$) by means of a suitable hyperplane l (called *cut-plane*) orthogonal to a coordinate axis, into two subsets S_1 and S_2 of sizes αN and $(1 - \alpha)N$ respectively (where $0 < \alpha_0 \leq \alpha \leq 1 - \alpha_0$, to be later determined).
2. Apply recursively the algorithm to S_1 and S_2 and obtain the minimum distances δ_1 and δ_2 in S_1 and S_2, respectively, and let $\delta = \min(\delta_1, \delta_2)$.
3. Combine the results by processing a δ-slice of E^d of width 2δ bisected by l.

We now concentrate on Step 3. Assume, as a working hypothesis, that the set of points in the δ-slice, called S_{12}, has sparsity c for the given δ (we shall later satisfy ourselves that such c exists). Let S_{12}^* be the projection of S_{12} onto the hyperplane l (a $(d - 1)$-dimensional space). Clearly, a necessary condition for a pair of points of S_{12} to be the closest pair in S is that their projections in l have distance less than δ. Thus we have a sparse set, S_{12}^*, in which we must find all pairs of points at bounded distance ($<\delta$). This is an instance of the following problem, which is of interest in its own right.

PROBLEM P.7 (FIXED RADIUS NEAREST-NEIGHBOR REPORT IN SPARSE SET). Given real δ and a set of M points in E^d, with sparsity c for given δ, report all pairs at distance less than δ.

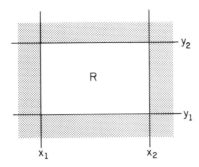

Figure 5.14   Definition of rectangle $R = [x_1, x_2] \times [y_1, y_2]$.

If we let $|S_{12}^*| = M$ and $F_{\delta,c}(M, d)$ denote the running time of Problem P.7, the preceding analysis leads to the recurrence

$$P(N, d) = P(\alpha N, d) + P((1 - \alpha)N, d) + O(N) + F_{\delta,c}(M, d - 1). \quad (5.2)$$

Our objective is the minimization of $F_{\delta,c}(M, d - 1)$. This, in turn, corresponds both to choosing a cut-plane minimizing M, and to developing an efficient algorithm for Problem P.7. Both of these tasks are realizable by virtue of the following interesting theorem.

Theorem 5.4 [Bentley–Shamos (1976)]. *Given a set S of N points in E^d, there exists a hyperplane l perpendicular to one of the coordinate axes with the following properties:*

(i) *both subsets S_1 and S_2 of S on either side of l contain at least $N/4d$ points;*
(ii) *there are at most $dbN^{1-1/d}$ points in a δ-slice of E^d around l, where $\delta = \min(\delta_1, \delta_2)$ and δ_i is the minimum interpoint distance in S_i $(i = 1, 2)$, and b is an upper bound to the number of points with least distance δ contained in a box of side 2δ.*

PROOF. We shall carry out the proof for $d = 2$, because of its more immediate intuitive appeal. Its d-dimensional generalization involves the same steps. We assume, with a negligible loss of generality, that N is a multiple of 8. On the x-axis we determine an interval $[x_1, x_2]$, such that there are $N/8$ points of S both to the left of x_1 and to the right of x_2; letting $C_x \subseteq S$ denote the set of points whose abscissae are between x_1 and x_2, we note that $|C_x| = 3N/4$. Similarly, we determine $[y_1, y_2]$ on the y-axis. The Cartesian product $[x_1, x_2] \times [y_1, y_2]$ is called R (Figure 5.14).

On the x-axis we determine the *largest* interval, totally contained in $[x_1, x_2]$ and containing the projections of $2bN^{1/2}$ points; we do likewise on the y-axis. Let γ be the maximum of the lengths of these two intervals. Without loss of generality, we assume that $[x_1', x_2']$ is the subinterval (of $[x_1, x_2]$) yielding γ. We claim that the line l of equation $x = (x_1' + x_2')/2$ (perpendicular bisector of the segment $\overline{x_1' x_2'}$) satisfies properties (i) and (ii) (see Figure 5.15).

To prove this claim, we begin by showing by contradiction that $\gamma < 2\delta$ is

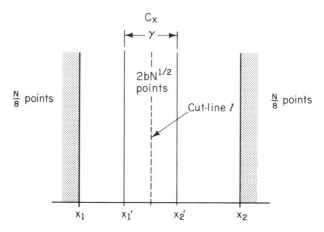

Figure 5.15   Relationship between $[x_1, x_2]$ and $[x'_2, x'_2]$.

impossible; we shall then see that the condition $\gamma \geq 2\delta$ ensures the desired result.

The condition $\gamma < 2\delta$ means that any horizontal and vertical strip of width 2δ projecting in the interior of $[y_1, y_2]$ and $[x_1, x_2]$, respectively, contains more than $2bN^{1/2}$ points.

If we now partition $[x_1, x_2]$ into subintervals of size 2δ, in each of the vertical strips of the plane projecting to these subintervals there are more than $2bN^{1/2}$ points. Since there are $\lfloor (x_2 - x_1)/2\delta \rfloor$ such subintervals, we have $\lfloor (x_2 - x_1)/2\delta \rfloor \cdot 2bN^{1/2} < |C_x| = 3N/4$, or $\lfloor (x_2 - x_1)/2\delta \rfloor < 3N^{1/2}/8b$. Arguing in the same way for the $[y_1, y_2]$ interval we obtain $\lfloor (y_2 - y_1)/2\delta \rfloor < 3N^{1/2}/8b$. Consider now the rectangle $R = [x_1, x_2] \times [y_1, y_2]$. This rectangle contains at most $\lceil (x_2 - x_1)/2\delta \rceil \cdot \lceil (y_2 - y_1)/2\delta \rceil$ squares of side 2δ. Each such square is exactly a box of side 2δ and so, by the hypothesis of the theorem, it contains at most b points. If follows that the number of points of S in R is at most

$$\left\lceil \frac{x_2 - x_1}{2\delta} \right\rceil \cdot \left\lceil \frac{y_2 - y_1}{2\delta} \right\rceil \cdot b \leq \left(\left\lfloor \frac{x_2 - x_1}{2\delta} \right\rfloor + 1 \right) \left(\left\lfloor \frac{y_2 - y_1}{2\delta} \right\rfloor + 1 \right) \cdot b$$

$$< \left(\frac{3N^{1/2}}{8b} + 1 \right)^2 \cdot b.$$

Since b is a small constant ≥ 1, the right side is maximized by $b = 1$, so that R contains at most $2(3/8)^2 N \cong 0.28N$ points for $N \geq 8$.

On the other hand, the vertical strips external to $[x_1, x_2]$ contain $2 \cdot N/8$ points, and so do the horizontal strips external to $[y_1, y_2]$ (see Figure 5.15). By the principle of inclusion–exclusion, the number of points external to R is at most $4N/8 = N/2$, so that R contains *no less than* $N/2$ points. A contradiction has been obtained whence the condition $\gamma < 2\delta$ is impossible.

Since $\gamma \geq 2\delta$, the result follows, for property (i) is ensured by the construction of $[x_1, x_2]$, and property (ii) holds since the δ-slice around l contains at most $2bN^{1/2}$ points. Therefore line l satisfies the specifications of the theorem.

$\square$

We use this theorem to show first that a cut-plane l can be chosen in Step 1 of CPAIR$d(S)$ so that $1/4d \le \alpha \le 1 - 1/4d$, and S_{12}, the set of points in the δ-slice around l, has cardinality at most $cdN^{1-1/d}$, where $c = 4 \cdot 3^{d-1}$. Indeed, if $\delta = \min(\delta_1, \delta_2)$, a box of side 2δ in the δ-slice contains at most $c = 4 \cdot 3^{d-1}$ points, as can be easily seen by induction on d. The induction starts with $d = 2$, for which we have already seen that at most 12 points are contained in a "box" of side 2δ. This proves that cut-plane l exists. To determine it, we presort S, once and for all, on *all* coordinates in time $O(dN \log N)$. By a simple $O(dN)$ scan of each of the resulting sorted sequences, we determine intervals so that there are $N/4d$ points on either sides of them. In each of these intervals, by another simple $O(N)$ scan we can determine the maximum of the widths of the windows (subintervals) containing exactly $\lfloor cdN^{1-1/d} \rfloor$ points. The maximum of these maxima yields a unique coordinate of the space and an interval whose bisector is the sought cut-plane. Note that, exclusive of presorting, the cut-plane can be determined in time linear in the size of S.

Thus, we have obtained a *sparse set* S_{12} *of cardinality* $\le cdN^{1-1/d}$ (with sparsity $c = 4 \cdot 3^{d-1}$ for the computed δ). We next solve Problem P.7 for S_{12}^*, the projection of S_{12} on the previously obtained cut-plane l. We shall use again the divide-and-conquer approach for this $(d-1)$-dimensional problem. Specifically, we find a cut-plane l', as guaranteed by Theorem 5.4, with given δ (as obtained in Step 2 of procedure CPAIRd), and c equal to the sparsity of S_{12}^*, just obtained. It follows that the point set in the δ-slice around the cut plane l' has size bounded by $(d-1) \cdot cM^{1-1/(d-1)}$, i.e., $o(M)$. In conclusion, the $(d-1)$-dimensional problem P.7 is solved by determining, in time $O(M)$, a cut-plane, followed by recursively solving the same problem on two subsets of sizes αM and $(1-\alpha)M$, and finally by solving a residual $(d-2)$-dimensional problem on a set size $o(M)$. This leads to the recurrence

$$F_{\delta,c}(M, d-1) = F_{\delta,c}\left(\frac{M}{4d}, d-1\right) + F_{\delta,c}\left(M\left(1 - \frac{1}{4d}\right), d-1\right) + O(M) \quad (5.3)$$

$$+ F_{\delta,c}(o(M), d-2).$$

This recurrence is readily solved as $F_{\delta,c}(M, d-1) = O(M \log M)$. This fact is interesting *per se* and is summarized in the following theorem.

Theorem 5.5. *All the pairs at distance less than δ in a sparse set of M points in E^d (with sparsity c for given δ) can be reported in time $O(M \log M)$.*

On the other hand, returning to the original closest-pair problem, M, the cardinality of S_{12}, is by construction $O(N^{1-1/d})$, i.e., it is $o(N/\log N)$. It follows that $F_{\delta,c}(M, d-1) = O(M \log M) = O(N)$. As a conclusion, recurrence relation (5.2) has solution $P(N, d) = O(N \log N)$. This computational work is of the same order as that of the initial presorting. So we have

Theorem 5.6. *The determination of a closest pair of points in a set of N points in E^d can be completed in time $\theta(N \log N)$, and this is optimal.*

5.5 The Locus Approach to Proximity Problems: The Voronoi Diagram

While the previous divide-and-conquer approach for the closest-pair problem is quite encouraging, it even fails to solve the ALL NEAREST NEIGHBORS problem, which would seem to be a simple extension. Indeed, if we try to set up the analogous recursion for ALL NEAREST NEIGHBORS, we find that the natural way of splitting the problem does not induce sparsity, and there is no apparent way of accomplishing the merge step in less than quadratic time. On the other hand, a valuable heuristic for designing geometric algorithms is to look at the defining loci and try to organize them into a data structure. In a two-dimensional formulation, we want to solve

PROBLEM P.8 (LOCI OF PROXIMITY). Given a set S of N points in the plane, for each point p_i in S what is the locus of points (x, y) in the plane that are closer to p_i than to any other point of S?

Note that, intuitively, the solution of the above problem is a partition of the plane into regions (each region being the locus of the points (x, y) closer to a point of S than to any other point of S). We also note that, if we know this partition, by searching it (i.e., by locating a query point q in a region of this partition), we could directly solve the NEAREST-NEIGHBOR SEARCH (Problem P.5). We shall now analyze the structure of this partition of the plane. Given two points, p_i and p_j, the set of points closer to p_i than to p_j is just the half-plane containing p_i that is defined by the perpendicular bisector of $\overline{p_i p_j}$. Let us denote this half-plane by $H(p_i, p_j)$. The locus of points closer to p_i than to any other point, which we denote by $V(i)$, is the *intersection of $N - 1$ half-planes*, and is a convex polygonal region (see Section 1.3.1) having no more than $N - 1$ sides, that is,

$$V(i) = \bigcap_{i \neq j} H(p_i, p_j).$$

$V(i)$ is called the *Voronoi polygon associated with p_i*. A Voronoi polygon is shown in Figure 5.16(a) [Rogers (1964)].[12]

---

[12] These polygons were first studied seriously by the emigré Russian mathematician G. Voronoi, who used them in a treatise on quadratic forms [Voronoi (1908)]. They are also called Dirichlet regions, Thiessen polygons, or Wigner–Seitz cells. Dan Hoey has suggested the more descriptive (and impartial) term "proximal polygons."

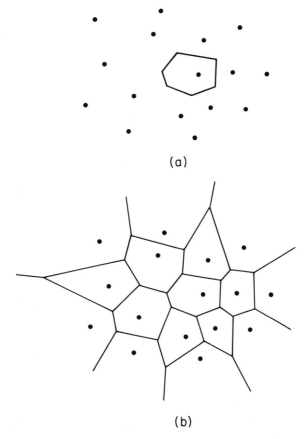

Figure 5.16 (a) A Voronoi polygon; (b) the Voronoi diagram.

These N regions partition the plane into a convex net which we shall refer to as the *Voronoi diagram*, denoted as Vor(S), which is shown in Figure 5.16(b). The vertices of the diagram are *Voronoi vertices*, and its line segments are *Voronoi edges*.

Each of the original N points belongs to a unique Voroni polygon; thus if $(x, y) \in V(i)$, then p_i is a nearest neighbor of (x, y). The Voronoi diagram contains, in a powerful sense, all of the proximity information defined by the given set.

5.5.1 A catalog of Voronoi properties

In this section we list a number of important properties of the Voronoi diagram. Although the Voronoi diagram can be defined for any number of dimensions, our review will refer to the planar case for a two-fold reason: first, to maintain an immediate link with intuitive evidence; second, to focus the

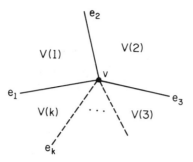

Figure 5.17   Voronoi edges incident on a Voronoi vertex.

treatment on the only known situation where efficient algorithmic techniques are available.

Throughout this section we make the following assumption.

Assumption A. *No four points of the original set S are cocircular.*

If this assumption is not true, inconsequential but lengthy details must be added to the statements and proofs of the theorems.

We begin by noting that every edge of the Voronoi diagram is a segment of the perpendicular bisector of a pair of points of S and is thus common to exactly two polygons. We then have

Theorem 5.7. *Every vertex of the Voronoi diagram is the common intersection of exactly three edges of the diagram.*

PROOF. Indeed, a vertex is the common intersection of a set of edges. Let e_1, e_2, $\ldots, e_k$ (for $k \geq 2$) be the clockwise sequence of edges incident on vertex v (refer to Figure 5.17). Edge e_i is common to polygons $V(i-1)$ and $V(i)$, for $i = 2$, $\ldots, k$, and e_1 is common to $V(k)$ and $V(1)$. Notice that v is equidistant from p_{i-1} and p_i since it belongs to e_i; on the other hand, by the same argument, it is equidistant from p_i and p_{i-1}, and so on. Thus v is equidistant from p_1, p_2, $\ldots, p_k$. But this means that $p_1, \ldots, p_k$ are cocircular, violating Assumption A if $k \geq 4$. Therefore $k \leq 3$. Suppose now that $k = 2$. Then e_1 is common to $V(2)$ and $V(1)$, and so is e_2: indeed they both belong to the perpendicular bisector of the segment $\overline{p_1 p_2}$, so that they do not intersect in v, another contradiction.   □

Equivalently, Theorem 5.7 says that the Voronoi vertices are the centers of circles defined by three points of the original set, and the Voronoi diagram is regular of degree three.[13] For a vertex v, we denote by $C(v)$ the above circle. These circles have the following interesting property.

---

[13] In graph-theoretical parlance, a graph is *regular* if all vertices have the same degree.

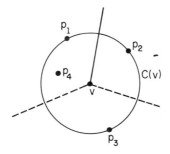

Figure 5.18   The circle $C(v)$ contains no other point of S.

Theorem 5.8. *For every vertex v of the Voronoi diagram of S, the circle $C(v)$ contains no other point of S.*

PROOF. By contradiction. Referring to Figure 5.18, let p_1, p_2, and p_3 be the three points of S determining circle $C(v)$. If $C(v)$ contains some other point, say p_4, then p_4 is closer to v than to any of p_1, p_2, or p_3, in which case v must lie in $V(4)$ and not in any of $V(1)$, $V(2)$, or $V(3)$, by the definition of a Voronoi polygon. But this is a contradiction since v is common to $V(1)$, $V(2)$, and $V(3)$. $\qquad\square$

Theorem 5.9. *Every nearest neighbor of p_i defines an edge of the Voronoi polygon $V(i)$.*

PROOF. Let p_j be a nearest neighbor of p_i and let v be the midpoint of their adjoining segment. Suppose that v does not lie on the boundary of $V(i)$. Then the line segment $\overline{p_i v}$ intersects some edge of $V(i)$, say the bisector of $\overline{p_i p_k}$ at u. (Figure 5.19). Then length$(\overline{p_i u}) <$ length$(\overline{p_i v})$, so length$(\overline{p_i p_k}) \leq$ 2 length$(\overline{p_i u}) <$ 2 length$(\overline{p_i v}) =$ length$(\overline{p_i p_j})$ and we would have p_k closer to p_i than p_j is, which is contradictory. $\qquad\square$

Theorem 5.10. *Polygon $V(i)$ is unbounded if and only if p_i is a point on the boundary of the convex hull of the set S.*

PROOF. If p_i is not on the convex hull of S, then it is internal to some triangle $p_1 p_2 p_3$ by Theorem 3.4. Consider the circles C_{12}, C_{13}, and C_{23} determined by p_i and each of the three pairs of vertices $\{p_1, p_2\}$, $\{p_1, p_3\}$, and $\{p_2, p_3\}$, respectively. (See Figure 5.20.) Each of these circles has finite radius. On circle C_{12} (and analogously for C_{13} and C_{23}) the external arc A_{12} is the circular arc between p_1 and p_2 not containing p_i; it is straightforward to show that any point of A_{12} is closer to p_1 or to p_2 than to p_i. Let C be a circle that encloses C_{12}, C_{13}, and C_{23}. We claim that any point x that lies outside C is closer to one of p_1, p_2, or p_3 than it is to p_i. Indeed, consider the segment $\overline{xp_i}$. By the Jordan curve theorem, $\overline{xp_i}$ intersects one of the sides of triangle $p_1 p_2 p_3$, say

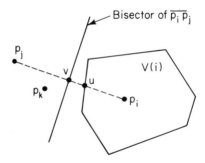

Figure 5.19   Every nearest neighbor of p_i defines an edge of $V(i)$.

$\overline{p_1 p_2}$; thus, it intersects also A_{12} in point u. But u is closer to either p_1 or p_2 than to p_i, whence the claim. Since x is closer to p_1, p_2, or p_3 than to p_i, $V(i)$ is contained entirely within C and hence is bounded.

   Conversely, assume that $V(i)$ is bounded and let $e_1, e_2, \ldots, e_k$ $(k \geq 3)$ be the sequence of its boundary edges. Each e_h $(h = 1, \ldots, k)$ belongs to the bisector of a segment $\overline{p_i p'_h}$, $p'_h \in S$. It is immediate to conclude that p_i is internal to the polygon $p'_1 p'_2 \cdots p'_k$, i.e., p_i is not on the convex hull of S.   □

   Since only unbounded polygons can have rays as edges, the rays of the Voronoi diagram correspond to pairs of adjacent points of S on the convex hull.

   We now consider the straight-line dual of the Voronoi diagram, i.e., the graph embedded in the plane obtained by adding a straight-line segment between each pair of points of S whose Voronoi polygons share an edge. The result is a graph on the original N points. (Figure 5.21.)

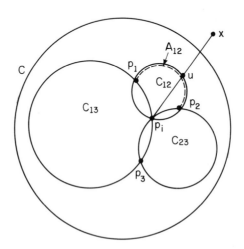

Figure 5.20   For the proof of Theorem 5.10.

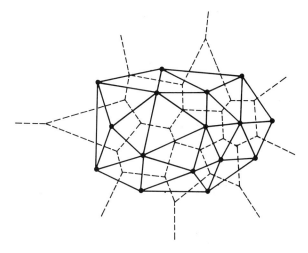

Figure 5.21   The straight-line dual of the Voronoi diagram.

The dual may appear to be unusual at first glance, since an edge and its dual may not even intersect (consider, e.g., the edges joining consecutive hull vertices.) Its importance is largely due to the following theorem of Delaunay (1934).

Theorem 5.11. *The straight-line dual of the Voronoi diagram is a triangulation of* S.[14]

(This implies that the Voronoi diagram can be used to solve the Triangulation problem, P.4, but the theorem has much more significant consequences.)

PROOF. To prove that the straight-line dual of the Voronoi diagram is a triangulation, we must show that the convex hull of S is partitioned into triangles determined by the points of S. To this end we shall show that a set of triangles $\mathscr{T} = \{T(v): v$ is a Voronoi vertex$\}$ can be constructed, so that no two intersect and each point in the interior of conv(S) belongs to one such triangle (and, therefore, to exactly one such triangle).

The triangles are constructed as follows. Let v be a Voronoi vertex shared by $V(1)$, $V(2)$, and $V(3)$ (see Theorem 5.7): $T(v)$ is defined as the triangle whose vertices are p_1, p_2 and p_3. We claim that if $T(v)$ intersects the interior of conv(S), then $T(v)$ is nondegenerate (i.e., p_1, p_2, and p_3 are *not* collinear). Indeed, assume, for a contradiction, that p_1, p_2, and p_3 are collinear: then the three segments $\overline{p_1 p_2}, \overline{p_1 p_3}$, and $\overline{p_2 p_3}$ all belong to the same straight line l, and their perpendicular bisectors are parallel. Since v is the common intersection of these three bisectors, v is a point at infinity, which implies that $V(1)$, $V(2)$,

---

[14] In this simple form the theorem fails if certain subsets of four or more points are cocircular. In this case, however, completing the triangulation will be straightforward.

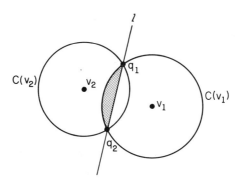

Figure 5.22   There is no point of $T(v_1)$ in the interior of the shaded region.

and $V(3)$ are unbounded. This means, by Theorem 5.10, that collinear p_1, p_2, and p_3 lie on the convex hull, contrary to the assumption that $T(v)$ intersects the *interior* of conv(S). This establishes the claim.

Next, consider the two triangles $T(v_1)$ and $T(v_2)$, for $v_1 \neq v_2$, and the corresponding circles $C(v_1)$ and $C(v_2)$ (notice that $C(v_i)$ is the circumcircle of $T(v_i)$). If $C(v_1)$ and $C(v_2)$ are disjoint, so are $T(v_1)$ and $T(v_2)$. So assume that $C(v_1)$ and $C(v_2)$ share internal points. Note that neither one can be entirely contained in the other, by Theorem 5.8: so $C(v_1)$ and $C(v_2)$ must intersect in two points q_1 and q_2, which define a straight line l (refer to Figure 5.22) separating v_1 from v_2. We claim that l also separates $T(v_1)$ and $T(v_2)$. Indeed, assume the contrary: say, there are points of $T(v_1)$ on the v_2-side of line l (i.e., in the shaded area in Figure 5.20, since $T(v_1) \subset C(v_1)$). This implies that there is a vertex of $T(v_1)$ in the shaded area, and hence in $C(v_2)$, contrary to Theorem 5.8. Thus, $T(v_1)$ and $T(v_2)$ do not share interior points.

Finally, consider an arbitrary point x in conv(S) and assume for a contradiction that $x \notin T(v)$, for any Voronoi vertex v. This implies that there is a small disc γ, centered in x, of points with the same property as x (refer to Figure 5.23). Let q be a point in $T(v)$, for an arbitrary v. We can choose a point $y \in \gamma$ so that the line l through q and y does not pass by any point of S. Line l intersects triangles of $\mathscr{T}$ in a set of closed intervals: let t be the endpoint of the interval closest to x, and let t be on the boundary edge $\overline{p_1 p_2}$ of $T(v_1)$, for some Voronoi vertex v_1. Next, let p_3 be the third vertex of $T(v_1)$ and consider the perpendicular l_1 from v_1 to $\overline{p_1 p_2}$. The initial portion of l_1 is the boundary edge of $V(1)$ and $V(2)$. Since x is assumed to be in conv(S), and x and p_3 are on opposite sides of $\overline{p_1 p_2}$, the segment $\overline{p_1 p_2}$ does not belong to the convex hull of S. This implies that the boundary edge on l_1 is *not* a ray (by Theorem 5.10), but a segment of which the other extreme is called v_2. Therefore, $T(v_2)$ has p_1 and p_2 as two of its vertices, and since $T(v_1) \cap T(v_2) = \varnothing$ (by a previous result), v_1 and v_2 lie on opposite sides of $\overline{p_1 p_2}$. This shows that t *is internal* to $T(v_1) \cup T(v_2)$, contrary to our (unfounded) assumption that the points of $\overline{ty}$ belong to no triangle of $\mathscr{T}$. $\square$

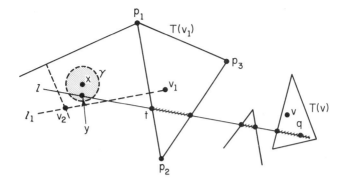

Figure 5.23   Each point in the convex hull of S belongs to some triangle.

An immediate consequence of the preceding theorem is

Corollary 5.2. *A Voronoi diagram on N points has at most $2N - 5$ vertices and $3N - 6$ edges.*

PROOF. Each edge in the straight-line dual corresponds to a unique Voronoi edge. Being a triangulation, the dual is a planar graph on N vertices, and thus by Euler's formula it has at most $3N - 6$ edges and $2N - 4$ faces. Therefore, the number of Voronoi edges is at most $3N - 6$; however, only the bounded faces (at most $2N - 5$) dualize to Voronoi vertices.   □

By the established duality, the Voronoi diagram can also be used to solve the TRIANGULATION problem (Problem P.4).

Since the Voronoi diagram is a planar graph, it can be stored in only *linear* space. This makes possible an extremely compact representation of the proximity data. Any given Voronoi polygon may have as many as $N - 1$ edges, but there are at most $3N - 6$ edges overall, each of which is shared by exactly two polygons. This means that the *average* number of edges in a Voronoi polygon does not exceed six.

In the next two sections we will use these properties to construct the Voronoi diagram quickly and employ it to solve the closest-point problems.

5.5.2 Constructing the Voronoi diagram

Even though we will be using it for other purposes, it is well to note that construction of Voronoi diagrams is an end in itself in a number of fields. In archaeology, Voronoi polygons are used to map the spread of the use of tools in ancient cultures and for studying the influence of rival centers of commerce [Hodder–Orton (1976)]. In ecology, the survival of an organism depends on the number of neighbors it must compete with for food and light, and the

Voronoi diagram of forest species and territorial animals is used to investigate the effect of overcrowding [Pielou (1977)]. The structure of a molecule is determined by the combined influence of electrical and short-range forces, which have been probed by constructing elaborate Voronoi diagrams.

By constructing the Voronoi diagram Vor(S) of a set of points S (hereafter referred to formally as Problem VORONOI DIAGRAM), we shall mean to produce a description of the diagram as a planar graph embedded in the plane, consisting of the following items (see Section 1.2.3.2):

1. The coordinates of the Voronoi vertices;
2. The set of edges (each as a pair of Voronoi vertices) and the two edges that are their counterclockwise successors at each extreme point (doubly-connected-edge-list, see Section 1.2.3.2). This implicitly provides the counterclockwise edge cycle at each vertex and the clockwise edge cycle around each face.

Next, according to a familiar pattern, we consider first the question of a lower bound to the time necessary for constructing the Voronoi diagram. The answer is provided by the following simple theorem.

Theorem 5.12. *Constructing a Voronoi diagram on N points in the plane must take $\Omega(N \log N)$ operations, in the worst case, in the algebraic computation-tree model.*

PROOF. We will see later that the closest-point problems are all linear-time transformable to VORONOI DIAGRAM, so many proofs of this theorem are possible but we content ourselves here with a very simple one. The Voronoi diagram of a set of points in *one* dimension consists of a sequence of $N - 1$ bisectors separating adjacent points on the line. From these consecutive pairs we can obtain a sorted list of the points in only linear time, whence SORTING can be transformed in linear time to VORONOI DIAGRAM.   □

A naive (almost brutal) approach to the construction of a Voronoi diagram is the construction of its polygons one at a time. Since each Voronoi polygon is the intersection of $N - 1$ half-planes, using a method to be described in Chapter 7, it can be constructed in time $O(N \log N)$, thereby resulting in an overall time $O(N^2 \log N)$. We will show next that the entire diagram can be obtained in optimal time $\theta(N \log N)$, which means that constructing the diagram is asympototically no more difficult than finding a single one of its polygons!

Indeed, in spite of its apparent complexity, VORONOI DIAGRAM is eminently suited to attack by divide-and-conquer. The method we employ depends for its success on various structural properties of the diagram that enable us to merge subproblems in linear time.

We can therefore attempt a rough sketch of the algorithm:

procedure VORONOI DIAGRAM (preliminary)

Step 1. Partition S into two subsets S_1 and S_2 of approximately equal sizes.

Step 2. Construct $\text{Vor}(S_1)$ and $\text{Vor}(S_2)$ recursively.

Step 3. "Merge" $\text{Vor}(S_1)$ and $\text{Vor}(S_2)$ to obtain $\text{Vor}(S)$.

We assume now, and later substantiate, that in any case Step 1 can be carried out in time $O(N)$. If we let $T(N)$ denote the overall running time of the algorithm, then Step 2 is completed in time approximately $2T(N/2)$. Thus, if $\text{Vor}(S_1)$ and $\text{Vor}(S_2)$ can be merged in linear time to form the Voronoi diagram $\text{Vor}(S)$ of the entire set, we will have a $\theta(N \log N)$ optimal algorithm. But before we tackle the algorithmic question, there is a deeper question to be settled: What reason is there to believe that $\text{Vor}(S_1)$ and $\text{Vor}(S_2)$ bear any relation to $\text{Vor}(S)$?

To answer this question, we begin by defining a geometric construct which is crucial to our approach.

Definition 5.2. Given a partition $\{S_1, S_2\}$ of S, let $\sigma(S_1, S_2)$ denote the set of Voronoi edges that are shared by pairs of polygons $V(i)$ and $V(j)$ of $\text{Vor}(S)$, for $p_i \in S_1$ and $p_j \in S_2$.

This collection $\sigma(S_1, S_2)$ of edges enjoys the following properties.

Theorem 5.13. $\sigma(S_1, S_2)$ *is the edge set of a subgraph of* $\text{Vor}(S)$ *with the properties*:

(i) $\sigma(S_1, S_2)$ *consists of edge-disjoint cycles and chains. If a chain has just one edge, this is a straight line; otherwise its two extreme edges are semi-infinite rays.*

(ii) *If S_1 and S_2 are linearly separated,*[15] *then, $\sigma(S_1, S_2)$ consists of a single monotone chain (see Section 2.2.2.2).*

PROOF. (i) If in $\text{Vor}(S)$ we imagine painting each of the polygons $\{V(i): p_i \in S_1\}$ in red and each of the polygons $\{V(j): p_j \in S_2\}$ in green, clearly $\text{Vor}(S)$ is a two-colorable map. Then it is well-known that the boundaries between polygons of different colors are edge-disjoint cycles and chains [Bollobás (1979)]. (Notice that two components of $\sigma(S_1, S_2)$ may share a vertex only if the degree of that vertex is at least four; so if all Voronoi vertices have degree three, then the components are also vertex-disjoint.) Each component of $\sigma(S_1, S_2)$ partitions the plane into two parts. Thus a chain either consists of a single straight line or it has rays as initial and final edges. This establishes (i).

(ii) Without loss of generality, assume that S_1 and S_2 are separated by a vertical line m, and let C be a component of $\sigma(S_1, S_2)$. Starting from a point q

---

[15] If more than one point belongs to the separating lines, all of these are assigned to the same set of the partition.

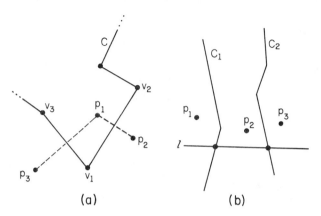

Figure 5.24 S_1 and S_2 are vertically separated. Then: (a) each component of $\sigma(S_1, S_2)$ is a vertically monotone chain; (b) $\sigma(S_1, S_2)$ consists of a single chain.

of an edge of C, we begin traversing C in the direction of decreasing y and continue until we reach a Voronoi vertex v_1 where C turns upward (see Figure 5.24(a)). Edge $\overline{v_1 v_2}$ is the bisector of $\overline{p_1 p_2}$, and $\overline{v_1 v_3}$ is the bisector of $\overline{p_1 p_3}$. Since $y(v_3) > y(v_1)$ and $y(v_2) > y(v_1)$, we have $x(p_3) \leq x(p_1) \leq x(p_2)$; however, by the shape of C, p_2 and p_3 belong to the same set, contrary to the assumption that S_1 and S_2 are separated by a vertical line. So, a vertex like v_1 is impossible and C is vertically monotone. This implies that C is a chain.

Suppose now that $\sigma(S_1, S_2)$ contains at least two (vertically monotone) chains, C_1 and C_2. A horizontal line l intersects each of C_1 and C_2 in a single point, due to monotonicity; assume also that the intersection with C_1 is to the left of that with C_2 (see Figure 5.24(b)). Then there are three points of S: p_1, p_2, and p_3, with $x(p_1) < x(p_2) < x(p_3)$, of which p_1 and p_3 belong to the same set of the partition $\{S_1, S_2\}$, contradicting the hypothesis of vertical separation of these two sets. So, $\sigma(S_1, S_2)$ consists of a single monotone chain. $\square$

When S_1 and S_2 are linearly separated, the single chain of $\sigma(S_1, S_2)$ will be referred to as σ. If the separating line m is chosen vertical, we can unambiguously say that σ cuts the plane into a *left* portion π_L and a right portion π_R. We then have the following decisive property.

Theorem 5.14. *If S_1 and S_2 are linearly separated by a vertical line with S_1 to the left of S_2, then the Voronoi diagram $\mathrm{Vor}(S)$ is the union of $\mathrm{Vor}(S_1) \cap \pi_L$ and $\mathrm{Vor}(S_2) \cap \pi_R$.*

Proof. All points of S to the right of σ are the points of S_2. Then all edges of $\mathrm{Vor}(S)$ to the right of σ separate polygons $V(i)$ and $V(j)$, where both p_i and p_j are in S_2. This implies that each edge of $\mathrm{Vor}(S)$ in π_R either coincides or is a portion of an edge of $\mathrm{Vor}(S_2)$. All analogous statement holds for π_L. $\square$

The preceding theorem answers our original question on how $\text{Vor}(S_1)$ and $\text{Vor}(S_2)$ relate to $\text{Vor}(S)$. Indeed, it states that when S_1 and S_2 are linearly separated (a situation which is entirely under our control and can be enforced in Step 1 of the algorithm), then it provides a method for merging $\text{Vor}(S_1)$ and $\text{Vor}(S_2)$. The algorithm is therefore revised as follows.

procedure VORONOI DIAGRAM

Step 1. Partition S into two subsets S_1 and S_2, of approximately equal sizes, by median x-coordinate.

Step 2. Construct $\text{Vor}(S_1)$ and $\text{Vor}(S_2)$ recursively.

Step 3'. Construct the polygonal chain σ, separating S_1 and S_2.

Step 3''. Discard all edges of $\text{Vor}(S_2)$ that lie to the left of σ and all edges of $\text{Vor}(S_1)$ that lie to the right of σ. The result is $\text{Vor}(S)$, the Voronoi diagram of the entire set.

Clearly, the success of this procedure depends on how rapidly we are able to

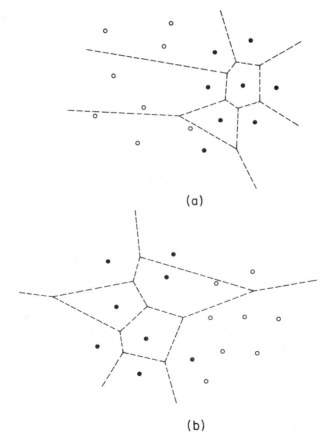

(a)

(b)

Figure 5.25   The Voronoi diagrams of the left set (a) and of the right set (b).

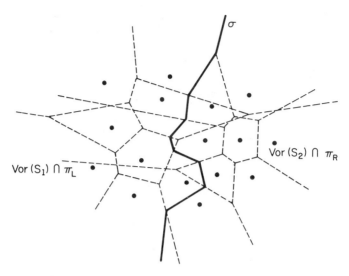

Figure 5.26 $\mathrm{Vor}(S_1)$, $\mathrm{Vor}(S_2)$, and σ superimposed.

construct σ, since Step 3″ poses no difficulties. For an illustration, the diagrams $\mathrm{Vor}(S_1)$ and $\mathrm{Vor}(S_2)$ are shown separately in Figure 5.25(a) and
5.25(b), respectively; $\mathrm{Vor}(S_1)$, $\mathrm{Vor}(S_2)$, and σ are shown superimposed in
Figure 5.26.

From a performance viewpoint, the initial partition of S according to the
median of the x-coordinates can be done in time $O(N)$ by the standard median
finding algorithms. Moreover, Step 3″ can be carried out in time $O(|S_1|$
$+ |S_2|) = O(N)$. What remains is to find an efficient way of constructing the
dividing chain σ. This will be our next task [Shamos–Hoey (1975); Lee (1978),
(1980a)].

5.5.2.1 Constructing the dividing chain

The first step in the construction of σ is to find its semi-infinite rays. We
observe that each ray of σ is the perpendicular bisector of a supporting
segment of $\mathrm{CH}(S_1)$ and $\mathrm{CH}(S_2)$. We also note that, since S_1 and S_2 are linearly
separated by hypothesis, there are just *two* supporting segments of $\mathrm{CH}(S_1)$
and $\mathrm{CH}(S_2)$ (thereby confirming that $\sigma(S_1, S_2)$ consists of just one chain σ). If
we now assume (inductively) that these two convex hulls are available, their
two supporting segments, denoted t_1 and t_2, are constructed in (at most) linear
time (see Section 3.3.5) and the rays of σ are readily determined (see Figure
5.27). Notice that as a byproduct of this activity we also obtain $\mathrm{CH}(S)$,
thereby providing the induction step for the availability of the convex hulls.

Once we have found a ray of σ, the construction continues, edge by edge,
until the other ray is reached. It is useful to refer to the example of Figure 5.28,
where, for simplicity, point p_j is shown by its index j. The upper ray of σ is the

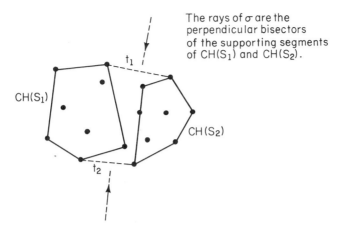

The rays of σ are the perpendicular bisectors of the supporting segments of CH(S_1) and CH(S_2).

Figure 5.27   Finding the rays of σ.

bisector of points 7 and 14. Imagine a point z on the ray, moving down from infinity. Initially z lies in polygons of Vor(S_1) and Vor(S_2). It will continue to do so until it crosses an edge of one of these polygons, when it will start moving in a different direction. In our example, z encounters an edge of Vor(S_2) before it hits any edge of Vor(S_1). This means that z is now closer to point 11 than it is to 14, so it must move off along the 7–11 bisector. It continues until the 6–7 bisector of Vor(S_1) is reached, and moves off along the 6–11 bisector. Eventually it hits the 10–11 bisector of $V(11)$ and proceeds via the 6–10 bisector. This jagged walk continues until the bottom ray of σ is reached.

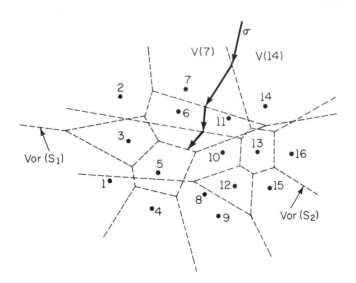

Figure 5.28   A zigzag walk to construct σ.

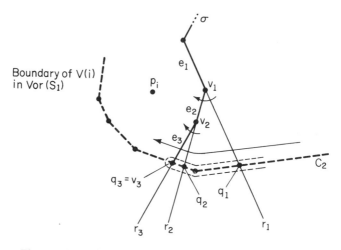

Figure 5.29 The intersections q_1, q_2, q_3 are ordered on C_2.

Assuming that σ is traversed in the direction of decreasing y, the advancing mechanism of the walk obtains from the current edge e and current vertex v (the upper extreme of e), the next edge e' and vertex v'. If e separates $V(i)$ and $V(j)$, for $p_i \in S_1$ and $p_j \in S_2$, then v' will be either the intersection of e with the boundary of $V(i)$ in $\text{Vor}(S_1)$ or the one with the boundary of $V(j)$ in $\text{Vor}(S_2)$, whichever is closer to v. So by scanning the boundaries of $V(i)$ in $\text{Vor}(S_1)$ and of $V(j)$ in $\text{Vor}(S_2)$ we can readily determine vertex v'. Unfortunately, σ may remain inside a given polygon $V(i)$, turning many times, before crossing an edge of $V(i)$. If we were to scan all of $V(i)$ each time, far too many edge examinations would be performed; in fact, this procedure could take as much as quadratic time. However, the particular structure of the Voronoi diagram affords a much more efficient scanning policy.

Indeed, assume that σ contains *a sequence of more than one edge* $e_1, e_2, \ldots,$ e_k of $V(i)$, where, say $p_i \in S_1$ (refer to Figure 5.29, where $k = 3$). Suppose to extend edge e_h from its endpoint v_h as a ray r_h and consider the intersection q_h of r_h with the boundary of $V(i)$ in $\text{Vor}(S_1)$. Consider the subchain $C_1 = e_1,$ $\ldots, e_k$ of σ and the portion C_2 of the boundary of $V(i)$ in $\text{Vor}(S_1)$ to the right of σ. C_2 is part of the boundary of a (possibly unbounded) convex polygon $V(i)$ of $\text{Vor}(S_1)$; C_1 is also convex and contained in $V(i)$. Since the angles from $\overline{v_h q_h}$ to $\overline{v_h q_{h+1}}$ ($h = 1, 2, \ldots, k - 1$) have all the same sign, the intersections q_1, $q_2, \ldots, q_k$ are *ordered* on C_2. This proves that C_2, i.e., the boundary of $V(i)$ in $\text{Vor}(S_1)$, needs to be scanned *clockwise, with no backtrack*, when determining $q_1, q_2, \ldots$. An analogous argument shows that the boundary of any $V(j)$ in $\text{Vor}(S_2)$ needs to be scanned *counterclockwise, with no backtrack*.

Specifically, we assume that both $\text{Vor}(S_1)$ and $\text{Vor}(S_2)$ be given as DCELs (see Section 1.2.3.2), where the face cycles are directed clockwise in $\text{Vor}(S_1)$ and counterclockwise in $\text{Vor}(S_2)$. In the DCEL of $\text{Vor}(S_1)$ we have a pointer

$\text{NEXT}_1[\ ]$, which is used for traversing the boundary of a face clockwise; similarly we define $\text{NEXT}_2[\ ]$ for $\text{Vor}(S_2)$. The algorithm maintains three edges: two edges e_L and e_R, in $\text{Vor}(S_1)$ and $\text{Vor}(S_2)$ respectively, and the "current edge" e of σ (actually, the straight line containing edge e). Of edge e, it also maintains its initial point v (for the initial ray e^*, v^* is a point of conveniently large ordinate on e^*). Finally, it maintains two points of S, $p_L \in S_1$ and $p_R \in S_2$, so that the current edge e is the perpendicular bisector of $\overline{p_L p_R}$. By $I(e, e')$ we denote the intersection of edges e and e'; $I(e, e') = \Lambda$ means that e and e' do not intersect. Again, t_1 and t_2 are the two supporting segments, with $t_1 = \overline{pq}$. Thus the implementation of Step 3' runs as follows:

1. **begin** $p_L := p$;
2. $p_R := q$;
3. $e := e^*$;
4. $v := v^*$;
5. $e_L :=$ first edge on (open) boundary of $V(p_L)$;
6. $e_R :=$ first edge on (open) boundary of $V(p_R)$;
7.   **repeat**
8.   **while** $(I(e, e_L) = \Lambda)$ **do** $e_L := \text{NEXT}_1[e_L]$ (*scan boundary of $V(p_L)$*);
9.   **while** $(I(e, e_R) = \Lambda)$ **do** $e_R := \text{NEXT}_2[e_R]$ (*scan boundary of $V(p_R)$*);
10.   **if** $(I(e, e_L)$ is closer to v than $I(e, e_R))$ **then**
11.   **begin** $v := I(e, e_L)$;
12. $p_L :=$ point of S on other side of e_L;
13. $e :=$ bisector of $\overline{p_L p_R}$;
14. $e_L :=$ reverse of e_L (*the new e_L is an edge of $V(p_L)$*)
            end
15.   **else begin** $v := I(e, e_R)$;
16. $p_R :=$ point of S on other side of e_R;
17. $e :=$ bisector of $\overline{p_L p_R}$;
18. $e_R :=$ reverse of e_R
            end
19.   **until** $(\overline{p_L p_R} = t_2)$
    end.

After the initialization (lines 1–6) we begin the walk on σ (lines 7–19). The actual advancing mechanism is expressed by lines 8–18: here the actual scanning of the boundaries of $V(p_L)$ and $V(p_R)$ is done by lines 8 and 9, respectively, without backtracking. Lines 11–14, or 15–18, have the task to update the relevant parameters among $\{v, p_L, p_R, e_L, e_R\}$ in the two alternative cases, and each runs in constant time. Since there are no more than $3N - 6$ edges in $\text{Vor}(S_1)$ and $\text{Vor}(S_2)$ together, and $O(N)$ vertices in σ, then the entire construction of σ takes only linear time. A detailed implementation of this procedure appears in Lee (1978).

Recall that to form the final Voronoi diagram we must discard all edges of

Vor(S_1) that lie to the right of σ and all edges of Vor(S_2) that lie to the left. This is done as part of the clockwise and counterclockwise scans performed during the construction of σ. It follows that the process of merging Vor(S_1) and Vor(S_2) to form Vor(S) takes only linear time. We conclude the section with the following theorem.

Theorem 5.15. *The Voronoi diagram of a set of N points in the plane can be constructed in $O(N \log N)$ time, and this is optimal.*

PROOF. The time required by the recursive merge procedure is described by the recurrence relation $T(N) = 2T(N/2) + O(N) = O(N \log N)$. Optimality was shown in Theorem 5.12.   □

5.6 Proximity Problems Solved by the Voronoi Diagram

As we alluded in Section 5.1, all of the proximity problems there described can be solved efficiently by means of the Voronoi diagram. This section is devoted to a detailed illustration of this claim for problems P.1, P.2, P.4, and P.5.

Beginning with problem P.2, ALL NEAREST NEIGHBORS, we have

Theorem 5.16. *The ALL NEAREST NEIGHBORS problem is linear-time transformable to VORONOI DIAGRAM and thus can be solved in $O(N \log N)$ time, which is optimal.*

PROOF. By Theorem 5.9, every nearest neighbor of a point p_i defines an edge of $V(i)$. To find a nearest neighbor of p_i it is only necessary to scan each edge of $V(i)$. Since every edge belongs to two Voronoi polygons, no edge will be examined more than twice. Thus, given the Voronoi diagram, all nearest neighbors can be found in linear time.   □

Obviously, since CLOSEST PAIR (Problem P.1) is transformed in linear time to ALL NEAREST NEIGHBORS, then the Voronoi diagram can be used also optimally to solve Problem P.1.

In nearest-neighbor searching, (Problem P.5), we are given a set of points, and we wish to preprocess them so that given a query point q, its nearest neighbor can be found quickly. However, finding the nearest neighbor of q is equivalent to finding the Voronoi polygon in which it lies. The preprocessing just consists of creating the Voronoi diagram! Since the diagram is a planar straight-line graph, it can be searched using any of the methods given in Section 2.2.2. We then have

Theorem 5.17. *Nearest-neighbor search can be performed in $O(\log N)$ time, using $O(N)$ storage and $O(N \log N)$ preprocessing time, which is optimal.*

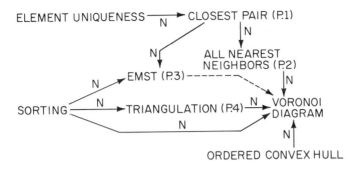

Figure 5.30 Relationship among computational prototypes and proximity problems. (Figure 5.6 revisited.)

PROOF. $O(N \log N)$ time is used to construct the Voronoi diagram, then Theorem 2.7 applies.   □

Since we have already seen that the dual of the Voronoi diagram is a triangulation (the Delaunay triangulation, Theorem 5.11), it follows that

Theorem 5.18. *A triangulation with the property that the circumcircle of every triangle is empty can be found in* $\theta(N \log N)$ *time, and this is optimal for any triangulation.*

Although the Voronoi diagram can be used to obtain, in optimal time, a triangulation of a set of points, and thereby to solve Problem P.4, the general problem of planar triangulations is a broader one and we shall return to it in Section 6.2. Moreover, in a later section we will examine the interesting relationship between EUCLIDEAN MINIMUM SPANNING TREE and VORONOI DIAGRAM.

At this point, it is interesting to revisit the diagram in Figure 5.6, which illustrates the relationship among proximity problems. This diagram, with the additions reflecting the results of this section, is repeated in Figure 5.30. This new diagram also shows the known fact that SORTING is linear-time transformable to VORONOI DIAGRAM (Theorem 5.12) and the new transformation from CONVEX HULL to VORONOI DIAGRAM, established by the following theorem.

Theorem 5.19. *Given the Voronoi diagram on N points in the plane, their convex hull can be found in linear time.*

PROOF. The Voronoi diagram is given as a DCEL with, say, counterclockwise orientation of the edge cycles of its faces. We then examine the Voronoi edges until a ray r is found (refer to Figure 5.31). If we direct r toward its endpoint at finite, let $V(i)$ be the polygon to its right: then p_i is a hull vertex (Theorem 5.10). Next we scan the edges of $V(i)$ until another ray r' is found. By reversing

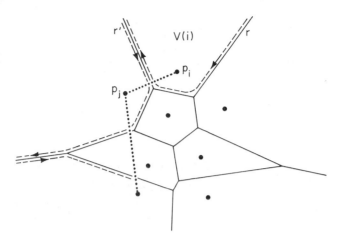

Figure 5.31   Construction of the convex hull from the Voronoi diagram.

the direction of r', we obtain another hull point p_j, and we now scan $V(j)$, etc., until we return to $V(i)$. An edge is examined only when one of the polygons containing it is scanned. Since each edge occurs in exactly two polygons, no edge is examined more than twice, and linear time suffices.   □

It is indeed remarkable that such a diverse collection of problems can be solved by a single unifying structure. However, this is not the whole story: from other applications, to be examined later, we shall not cease being surprised at the versatility of the Voronoi diagram.

5.7 Notes and Comments

The results discussed in the preceding sections of this chapter refer to proximity problems in the Euclidean metric. The Euclidean metric can be readily generalized to the notion of L_p-metric (Minkowski metric) in the following way.

Definition 5.3. In the Euclidean space E^d of coordinates $x_1, \ldots, x_d$, for any real $1 \le p \le \infty$ the L_p-distance of two points q_1 and q_2 is given by the norm

$$d_p(q_1, q_2) = \left(\sum_{j=1}^{d} \left| x_j(q_1) - x_j(q_2) \right|^p \right)^{1/p}. \tag{5.4}$$

Thus, the conventional Euclidean metric coincides with the L_2-metric. All of the problems discussed earlier can be considered in the L_p-metric for arbitrary p, and such research has been undertaken, among others, by Lee and Wong (1980) and Hwang (1979). Beside the L_2-distance, of particular interest are the L_1-distance (also called *Manhattan distance*), given by the length of a shortest path each edge of which is parallel to a coordinate axis, and the L_∞-distance, given by the largest of the differences in the coordinates. Such metrics are relevant to various applications, such as modelling of arm movements in disc transport mechanisms and in integrated circuit layout.

It is reasonable to expect that for the specialization of a general problem to a restricted class of data one may find a more efficient algorithm. One such restricted class occurs when the N-point set S is the vertex set of a convex polygon. Indeed, convexity enables us to solve the corresponding all-nearest-neighbors problem in optimal time $\theta(N)$ [Lee–Preparata (1978)]. On the other hand, although the solution of this problem provides N edges of the Delaunay triangulation (i.e., N edges of the Voronoi diagram), no algorithm is known to compute the Voronoi diagram of a convex polygon in time $o(N \log N)$. Indeed, this is one of the outstanding open problems in computational geometry.

In this chapter the Delaunay triangulation has been presented as a byproduct of the construction of the Voronoi diagram. It is possible, however, to define the Delaunay triangulation as the unique triangulation such that the circumcircle of each triangle does not contain any other point in its interior. It is then possible to construct directly the Delaunay triangulation by means of a recursive procedure analogous to the procedure presented for the Voronoi diagram and running within the optimal time bound, as has been shown by Lee and Schachter (1980). Indeed, this is the approach suggested for the L_1-metric, since in this metric the Voronoi diagram is not unique, thereby destroying the close tie between Voronoi diagrams and Delaunay triangulation.

Another interesting class of problems concerns the distance between sets of points. The distance between two sets A and B of points is normally defined as the minimum distance between a point in A and a point in B, namely, $d(A, B) = \min_a \min_b d(a, b)$, where $a \in A$, $b \in B$, and d is the Euclidean distance. With this definition $d(A, B) = d(B, A)$. Another measure that is of interest is the *Hausdorff* distance [Grünbaum (1967)], which is not symmetric. The *Hausdorff* distance from A to B is $\max_a \min_b d(a, b)$, and the Hausdorff distance between two sets A and B is equal to $\max\{d(A, B), d(B, A)\}$. For finite sets A and B, the CLOSEST PAIR BETWEEN SETS problem is solved with the aid of the Voronoi diagram in optimal $\theta(N \log N)$ time. When the points are the vertices either of a simple polygon or of a convex polygon, faster solutions are expected. Atallah (1983) discusses the Hausdorff distance between two convex polygons and gives an $O(N)$ time algorithm. Schwartz (1981) considers the problem of finding the closest pair of points between two convex polygons (two nonfinite sets), and gives an $O(\log^2 N)$ time algorithm; this result was later improved to $O(\log N)$ [Edelsbrunner (1982); Chin–Wang (1983)]. To find the closest pair of vertices of two convex polygons, $O(N)$ time is both sufficient and necessary [Chin–Wang (1984); Toussaint (1983a)].

5.8 Exercises

1. *Lee.* Given two sets A and B of points in the plane, each containing N elements, find the two closest points, one in A and the other in B. Show that this problem requires $\Omega(N \log N)$ operations. What if these two sets A and B are linearly separable?

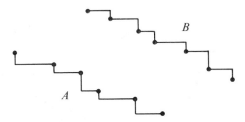

2. *Lee.* Given two sets A and B of points in the plane, each arranged as a staircase (that is, each set coincides with its set of maxima in the dominance relation—Section 4.1.3):
   (a) Find the pair (p_i, p_j), $p_i \in A$ and $p_j \in B$, closest in the L_1-metric.
   (b) Is linear time achievable?

3. *Inverse of Voronoi diagram.* Given an N-vertex planar map with valence 3 (each vertex has degree 3), develop an efficient algorithm to test if it is the Voronoi diagram of a finite set S of points. In the affirmative, your algorithm should also explicitly construct S.

4. *Voronoi diagram on the sphere.* Let S be a set of N points on the 3-dimensional sphere S^3. In the Euclidean metric, consider the Voronoi diagram S^3-$\mathrm{Vor}(S)$ of S on S^3:
   (a) Show that each edge of S^3-$\mathrm{Vor}(S)$ is an arc of a great circle of S^3.
   (b) Suppose that all points of S lie in a hemisphere of S^3 (all on one side of a plane through the center of S^3). Show how the algorithm developed for the planar Voronoi diagram (Section 5.5) can be used to construct S^3-$\mathrm{Vor}(S)$.

5. (a) In the plane, characterize the Voronoi diagram of a set of N points in the L_1-metric.
   (b) Solve the same problem for the L_∞-metric.
   (c) What is the relationship between the Voronoi diagram in the L_1-metric and that in the L_∞-metric?

6. In the L_2-metric the points of a set S whose Voronoi polygons are unbounded form the convex hull of S (Theorem 5.10). Does the same hold in the L_1-metric?

7. Formulate and prove the analog of Theorem 5.8 in the L_1-metric.

8. *Lee.* Consider the following definition for the Delaunay triangulation of a set of N points in the plane, no four of which are cocircular. Two points p_i and p_j determine an edge of the Delaunay triangulation if and only if there exists a circle passing by these two points that does not contain any other point in its interior. Show that this definition leads to a triangulation that satisfies the circumcircle property, i.e., the circumcircle of each triangle does not contain any other point in its interior. Therefore, it is the same as the dual graph of the Voronoi diagram of the set of points.

9. *Lee.* Show that the above definition for Delaunay triangulation gives a unique triangulation (assuming no four points are cocircular).

10. Given two sets of points A and B in the plane, use the Voronoi diagram to compute

$$\min_{a \in A} \min_{b \in B} \mathrm{dist}(a, b)$$

(in the Euclidean metric) in time $O(N \log N)$, where $N = |A| + |B|$.

CHAPTER 6
Proximity: Variants and Generalizations

The main conclusion derived from the preceding chapter is that the Voronoi diagram is both an extremely versatile tool for the solution of some fundamental proximity problems and an exceptionally attractive mathematical object in its own right. Indeed, these two facets—the instrumental and the aesthetic—have been the inspiration of a considerable amount of research on the topic. This chapter is devoted to the illustration of these extensions. Specifically, we shall at first discuss two very important applications: the already mentioned Euclidean Minimum Spanning Tree problem (and its ramifications) and the general problem of plane triangulations. We shall then show how the locus concept can be generalized in a number of directions. We shall then close the chapter with the analysis of "gaps and covers," which will afford us a chance to appreciate the power of different computation models.

6.1 Euclidean Minimum Spanning Trees

In the preceding chapter we considered the Euclidean Minimum Spanning Tree problem (EMST) (Problem P.3, Section 5.1) and showed that SORTING is transformable to it in linear time. This establishes an $\Omega(N \log N)$ lower bound to the time for finding the EMST on a set of N points. In this section we shall see that this bound is algorithmically attainable.

The algorithm to be described, as any known minimum spanning tree algorithm on general graphs, is based on the following straightforward lemma.

Lemma 6.1 [Prim (1957)]. *Let $G = (V, E)$ be a graph with weighted edges and let $\{V_1, V_2\}$ be a partition of the set V. There is a minimum spanning tree of G which contains the shortest among the edges with one extreme in V_1 and the other in V_2.*

In our case the vertex set V is the point set S, and the length of an edge is the Euclidean distance between its two endpoints. The EMST algorithm will handle, at each step, a forest of trees (which will turn out to be subtrees of the final EMST). The initial forest is the collection of the points (i.e., each point is an individual edgeless tree). The general step of the algorithm runs as follows (here $d(u, v)$ is the length of the segment $\overline{uv}$):

(i)   Select a tree T in the forest;
(ii) Find an edge (u', v') such that $d(u', v') = \min\{d(u, v): u \text{ in } T, v \text{ in } S - T\}$
(iii) If T' is the tree containing v', merge T and T' by binding them with edge (u', v').

The algorithm terminates when the forest consists of a single tree, the EMST of the given point set.

Although activities (i) and (iii) require careful implementation, it is clear that activity (ii) is the crux of the method. Indeed one may instinctively fear that it may use work proportional to $|T| \times (N - |T|)$, corresponding to examining the distances from each vertex in T to each vertex outside of T. Fortunately, the machinery of the Voronoi diagram comes to our rescue with the following lemma.

Lemma 6.2. *Let S be a set of points in the plane, and let $\Delta(p)$ denote the set of points adjacent to $p \in S$ in the Delaunay triangulation of S. For any partition $\{S_1, S_2\}$ of S, if $\overline{qp}$ is the shortest segment between points of S_1 and points of S_2, then q belongs to $\Delta(p)$.*

PROOF. Let $\overline{pq}$ realize the minimum distance between points of S_1 and points of S_2, with $p \in S_1$ and $q \in S_2$ (see Figure 6.1). We claim that $q \in \Delta(p)$. For, if $q \notin \Delta(p)$, the perpendicular bisector of segment $\overline{pq}$ does not contain a segment of the boundary of $V(p)$, the Voronoi polygon of point p (Theorem 5.9). This implies that $V(p)$ intersects $\overline{pq}$ in a point u between p and the midpoint M of $\overline{pq}$. The Voronoi edge l containing u is the perpendicular bisector of a segment $\overline{pp'}$, where $p' \in S$. For all possible choices of u and l, p' is confined to the interior of the disk C with center M and diameter $\overline{qp}$. Now we have two cases: (i) $p' \in S_1$. In this case $d(q, p') < d(q, p)$ whence p', and not p, is closest to q (a contradiction); (ii) $p' \in S_2$. In this case $d(p, p') < d(q, p)$, whence p', and not q, is closest to p (again a contradiction).   □

In other words, this lemma tells us that we need only examine the edges of the Delaunay triangulation, that is, we must compute the minimum spanning tree of a planar graph [Shamos (1978)].

We now return to the implementation of the general step of the EMST

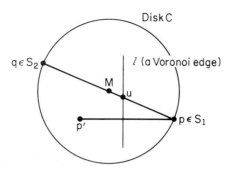

Figure 6.1 Illustration for the proof that $q \in \Delta(p)$.

algorithm. Cheriton and Tarjan (1976) propose, among other techniques, the following simple strategy. For the selection of the tree T (to be merged with another tree of the forest) they suggest the following *uniform selection rule*, where initially all single-vertex trees are placed in a queue:

1. Pick tree T at the front of the queue.
2. If T'' is the tree obtained by combining T with some other T', delete T and T' from the queue, and enter T'' at the back of the queue.

Since two trees are combined each time and there are initially N trees in the queue, the general step is executed exactly $(N - 1)$ times.

Suppose now that for each T in the queue we define an integer called *stage*(T), where *stage*$(T) = 0$ if $|T| = 1$ and *stage*$(T) = \min($*stage*(T'), *stage*$(T'')) + 1$ if T results from the combination (merging) of trees T' and T''. An interesting invariant of the queue of trees is that at any time during the execution of the algorithm the stage numbers of its members form a nondecreasing sequence from front to back. We shall then say that *stage j* has been completed when T is deleted from the queue, *stage*$(T) = j$ and no other T' in the queue has *stage*$(T') = j$. Note then that at the completion of *stage j* the queue contains at most $N/2^j$ members, and that there are at most $\lceil \log_2 N \rceil$ stages. We could then proceed as follows: each time we access T (where *stage*$(T) = j$) at the front of the queue, we select the shortest edge among the edges connecting vertices of T to vertices outside of T. In this manner, at each stage each edge is examined at most twice (except those already in some tree T). Thus each stage can be completed in time proportional to the number of edges, which is $O(N)$ due to the planarity of the graph. Since—as we noted earlier—there are at most $\lceil \log_2 N \rceil$ stages we have obtained an $O(N \log N)$ algorithm. The reader may observe at this point that we could content ourselves with this result, since $O(N \log N)$ time is used to compute the Delaunay Triangulation, and we have an optimal algorithm anyway. However, it is interesting to seek a technique that enables us to obtain the EMST from the Voronoi diagram in *linear time*. This achievement is made possible by the "clean-up" refinement proposed by Cheriton and Tarjan.

The *clean-up activity* has the objective of shrinking the original graph G (in our case, the Delaunay Triangulation) to a reduced graph G^*, which, at each point in the execution of the algorithm, contains *exactly the relevant information*. This means that each tree T in the forest F is shrunk to a single vertex of G^* (i.e., all unselected (chordal) edges connecting vertices of T are deleted), and all but the shortest of the unselected edges bridging different trees T' and T'' are also deleted. It is important to note that if G is planar, so is G^*.

The clean-up work is carried out as a clean-up step to be executed immediately after the completion of a stage (see above). It is rather straightforward to devise an implementation which runs in time proportional to the number of edges of the graph to be cleaned-up.

Less informally, we can summarize the preceding description as follows, where with each tree T we associate a list $\mathcal{E}(T)$ of the unselected edges incident on vertices of T.

```
    procedure EMST
1.  begin F := ∅;
2.      for i := 1 to N do
3.          begin stage(p_i) := 0;
                F ⟸ p_i
            end; (*the queue is initialized*)
4.      j := 1; (*the stage number is initialized to the next value*)
5.      while (F contains more than one number) do
6.          begin T ⟸ F;
7.              if (stage(T) = j) then begin clean-up;
8.                                          j := j + 1
                                      end;
9.              (u, v) := shortest unselected edge incident on T (with u in
                    T);
10.             T' := tree in F containing v;
11.             T'' := merge(T, T');
12.             delete T' from F;
13.             stage(T'') := min(stage(T), stage(T')) + 1;
14.             F ⟸ T'' (*T'' is entered at the back of the queue*)
        end
    end.
```

We now have all the items to be combined for the analysis of performance. At the completion of stage $(j - 1)$ there are at most $N/2^{j-1}$ trees in the queue, whence the corresponding G^* has at most $N/2^{j-1}$ vertices and fewer than $3N/2^{j-1}$ edges, being a planar graph. Stage j is completed in time proportional to the number of edges (each edge being examined at most twice by line 9 during the stage) and so is the clean-up work at the completion of stage j (lines 7–8, just before entering stage $(j + 1)$). Thus loop 6–14 uses time $O(N/2^{j-1})$. Recalling that there are at most $\lceil \log_2 N \rceil$ stages the total running time of loop

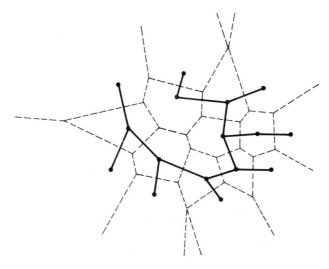

Figure 6.2   A set of points, its Voronoi diagram, and its EMST.

6–14 is upper bounded by

$$\sum_{j=1}^{\lceil \log_2 N \rceil} K_1 \frac{N}{2^{j-1}} < 2K_1 N_1$$

for some constant K_1. This proves the following theorem.

Theorem 6.1. *An EMST of a set S of N points in the plane can be computed from the Delaunay Triangulation of S in optimal time $\theta(N)$.*

Combining this result with the fact that the Delaunay triangulation is computable in time $\theta(N \log N)$ we have

Corollary 6.1. *An EMST of a set S of N points in the plane can be computed in optimal time $\theta(N \log N)$.*

In Figure 6.2 we illustrate a set of points, its Voronoi diagram, and an EMST of this set.

Next we shall consider an interesting application of the EMST of a finite set of points.

6.1.1   Euclidean traveling salesman

PROBLEM P.9 (EUCLIDEAN TRAVELING SALESMAN). Find a shortest closed path through N given points in the plane.

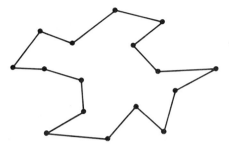

Figure 6.3   A traveling salesman tour.

A shortest tour is shown in Figure 6.3. (A reader familiar with the complexity of this problem may wonder how we were able to obtain the solution to a 16-point example. This was done by applying the Christofides heuristic described below.)

This problem differs from the ordinary traveling salesman problem in the same way that Euclidean minimum spanning tree differs from the MST problem in graphs: the interpoint distances are not arbitrary, but are inherited from the Euclidean metric. The general traveling salesman problem (TSP) is known to be NP-hard.[1] One may be tempted to suspect (or to hope) that properties of the Euclidean metric could be used to produce a polynomial-time algorithm in the plane. However, Garey, Graham, and Johnson (1976), Papadimitriou and Steiglitz (1976), and Papadimitriou (1977) have succeeded in proving that the ETSP is NP-hard, among a number of NP-completeness results for other geometric problems. We will therefore not attempt an efficient worst-case ETSP algorithm, but will concentrate on the relationship between ETSP and other closest-point problems, with a view toward developing good approximation methods.

We begin by considering the simpler (and less effective) of two known methods, expressed by the following theorem.

Theorem 6.2 [Rosenkrantz–Stearns–Lewis (1974)]. *A minimum spanning tree can be used to obtain an approximate* TSP *tour whose length is less than twice the length of a shortest tour.*

PROOF. Let T^* denote a Euclidean minimum spanning tree of the given set S of points, and let Φ denote the optimal traveling salesman tour. We begin by doubling each edge of T^*, thereby obtaining a graph W, each vertex of which has even degree (see Figure 6.4). Graph W is a connected Euler graph, i.e., its edges can be numbered so that the resulting sequence is a circuit, or, in other words, a tour that visits all vertices (some more than once). Therefore,

---

[1] For definitions relating to NP-complete problems and a proof of the NP-hardness of the TSP, see [Karp (1972); Garey–Johnson (1979)].

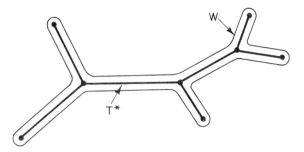

Figure 6.4   Doubling the EMST results in an Euler tour of all vertices of the set.

length(W) = 2 length(T^*). Observe now that if we remove an edge e from Φ we obtain a path called Φ_{path}, which is also a spanning tree of S. Clearly length(T^*) $\leq$ length(Φ_{path}) (by the definition of T) and length(Φ_{path}) < length(Φ) (trivially). Thus, we obtain length(W) < 2 length(Φ).   □

Notice that the approximate tour shown in Figure 6.4 could be further shortened by bypassing all unnecessary stops, i.e., by never revisiting a point that has already been visited. Specifically, we choose an arbitrary direction on the circuit W and start from an arbitrary vertex (which we leave unmarked). Each visited vertex is marked, and the next vertex on the final tour is obtained by proceeding to the first unmarked vertex on the directed W. With this refinement the example of Figure 6.4 gives the tour shown in Figure 6.5. (Notice that this refinement cannot increase the total length of the tour, since the distances obey the triangle inequality.)

The next approximate result makes use of a *minimum weighted matching* on a set of points.

PROBLEM P.10 (MINIMUM EUCLIDEAN MATCHING). Given $2N$ points in the plane, join them in pairs by line segments whose total length is a minimum.

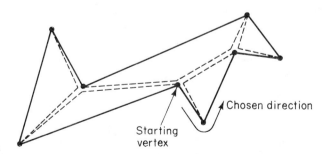

Figure 6.5   "Short-cuts" on the Euler tour ensure that each vertex is visited exactly once.

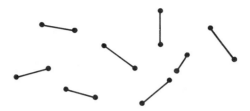

Figure 6.6   A minimum Euclidean matching.

Such a matching is shown in Figure 6.6.

Edmonds (1965) has shown that a minimum weight matching in an arbitrary graph can be obtained in polynomial time, and an $O(N^3)$ implementation is given in Gabow (1972). The following result, which relates minimum spanning trees, matchings, and the traveling salesman problem, is due to Christofides (1976).

Theorem 6.3. *An approximation to the traveling salesman problem whose length is within $3/2$ of optimal can be obtained in $O(N^3)$ time if the interpoint distances obey the triangle inequality.*

PROOF. The following algorithm achieves the desired result on the given set S:

1. Find a minimum spanning tree T^* of S.
2. Find a minimum Euclidean matching M^* on the set $X \subseteq S$ of vertices of *odd* degree in T^*. (X has always even cardinality in any graph.)
3. The graph $T^* \cup M^*$ is an Eulerian graph, since all of its vertices have even degree. Let Φ_e be an Eulerian circuit of it.
4. Traverse Φ_e edge by edge and bypass each previously visited vertex. Φ_{appr} is the resulting tour.

Denoting as usual by Φ the optimal tour, we note: (i) length$(T^*) <$ length(Φ); (ii) length$(M^*) \leq \frac{1}{2}$ length(Φ). Indeed if we select every other edge from Φ we obtain two matchings on S (the selected edges and the remaining ones), the length of the shorter of these two matchings is no more than $(1/2)$ length(Φ) and certainly no less than length(M^*), since M^* is minimal. We have shown earlier that length$(T^*) <$ length(Φ). Finally since the distances obey the triangle inequality, length$(\Phi_{appr}) \leq$ length(Φ_e). Combining these results we have

$$\text{length}(\Phi_{appr}) \leq \text{length}(\Phi_e)$$
$$= \text{length}(T^*) + \text{length}(M^*)$$
$$< \text{length}(\Phi) + \tfrac{1}{2}\,\text{length}(\Phi)$$
$$= \tfrac{3}{2}\,\text{length}(\Phi).$$

The time bound follows from Gabow's result.   □

Remark. By employing the Euclidean minimum matching one is able to improve the approximation from twice to $3/2$ of the optimal value, while the computational cost grows from $O(N \log N)$ to $O(N^3)$. For several years since Christofides' result no one has succeeded either in improving the approximation below $3/2$ of the optimal or in reducing the computation time. It is a remarkable fact in its own right that the method used to obtain the Euclidean minimum matching makes no use of geometric properties. Indeed the only known method consists in transforming the given problem into a maximum weighted matching problem (by replacing the length l of each edge by $M - l$ where $M = \max_{\text{all edges}} l$) and by applying to the transformed problem a technique that was developed for general graphs.

6.2 Planar Triangulations

We have noted in Section 5.1 the importance of planar triangulations for a multitude of practical applications in surface interpolation, both for the purposes of a graphical display and for calculations in numerical analysis. In addition to those very significant applications, the ability to perform planar triangulations is a very useful tool in its own right for its use in other problems in computational geometry. Suffice it to mention two instances: (i) Kirkpatrick's geometric searching technique (Section 2.2.2.3) assumes that the planar subdivision is a triangulation; (ii) a polyhedron intersection algorithm (see Section 7.3.1) requires that the polyhedral surface be pretriangulated.

We have already shown in the preceding chapter that a triangulation—the Delaunay triangulation—of a set of points can be found in optimal time. However, this does not preempt the problem, since, although the Delaunay triangulation is a remarkable object enjoying very attractive properties, there may be applications whose requirements are not satisfied by the Delaunay triangulation. For example, one may wish to minimize the total length of the triangulation edges (Minimum-Weight Triangulation). It had been conjectured that the Delaunay Triangulation is also a minimum-weight triangulation. Another conjecture stated that a triangulation method to be described in Section 6.2.1 (the "greedy triangulation" method) also yields a minimum-weight triangulation. Both conjectures were disproved by Lloyd (1977), and as of now the status of the problem (i.e., its NP-hardness or the existence of a polynomial-time algorithm) remains open.[2]

Other criteria do not involve the edge lengths, but refer to the sizes of the internal angles of the triangles. Indeed, in many applications it is desirable to

---

[2] However, the minimum-weight triangulation of the interior of the single polygon with N vertices can be computed on time $O(N^3)$ [Gilbert (1979)].

have triangles as "regular" as possible, that is, triangles that are "more or less equilateral." The Delaunay triangulation is very attractive from this viewpoint, since the minimum angle of its triangles is maximum over all triangulations. (Indeed, this is equivalent to the fact that the circumcircle of a Delaunay triangle does not contain any point of the given set in its interior [Lawson (1977); Lee (1978)].)

We refer the reader to Section 5.6 for a detailed discussion of the Delaunay triangulation. In what follows we shall consider some other triangulation methods, either for sets of points or for composite sets of points and segments. As a general reference we recall from Section 5.3 that $\Omega(N \log N)$ operations are required for any algorithm that triangulates a set of N points in the plane.

6.2.1 The greedy triangulation

A "greedy" method—as is well known—is one that never undoes what it did earlier. Thus, a greedy triangulation method adds one edge of the triangulation at a time and terminates after the required number of edges (entirely determined by the size of the point set and of its convex hull) has been generated. If one's objective is the minimization of the total edge length all that he can do in a greedy method is to adopt the *local* criterion to add at each stage the shortest possible edge that is compatible with the previously generated edges, i.e., it does not intersect any of them.

This is the essence of the technique. A straightforward implementation of this idea runs as follows (as usual S is the point set and N is its size). All the $\binom{N}{2}$ edges between points of S are generated and ordered by increasing lengths (pool of edges) and the triangulation is initialized as empty. At the general step, we pick and remove from the pool the shortest edge: if this edge does not intersect any of the current triangulation edges, we add it to the triangulation, otherwise we discard it. The process correctly terminates either when the triangulation is complete (by tracking the number of added edges) or when the pool of edges is empty [Düppe–Gottschalk (1970)].

This outlined method is not only very simple to implement but also very simple to analyze. The initial sorting of the edge lengths uses $O(N^2 \log N)$ operations. Next, there are $\binom{N}{2}$ selections of edges from the pool and each selected edge is matched against the edges currently in the triangulation in some time $\varphi(N)$: the form of φ depends upon the technique used to carry out this test. In conclusion the running time of the method is $O(N^2 \log N + N^2 \varphi(N))$.

The most naive way to carry out the test is to check whether the selected edge intersects each of the k edges currently in the triangulation. Since each of these intersection tests runs in constant time, φ is of the form $\varphi(k) = C_1 k$ (for some constant C_1), thereby resulting in an overall running time $O(N^3)$.

A more efficient approach was proposed by Gilbert (1979). The objective is here to balance the works of the decision task (whether the selected edge

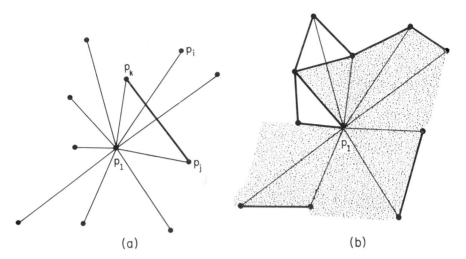

Figure 6.7 "Star of spokes" at p_1. Edges in the triangulation are shown as heavy segments.

belongs to the triangulation) and of the selection task (the scanning of the edges by increasing length). The latter is clearly $O(N^2 \log N)$, whether through presorting or through the management of a heap [Aho–Hopcroft–Ullman (1974)]. So, the goal is that the handling of an edge should cost no more than $O(\log N)$ operations. This can be achieved as follows. Referring to Figure 6.7(a), suppose that the currently selected edge is $\overline{p_1 p_i}$. Consider the set of edges connecting p_1 to every other point in S: we refer to it as $\text{STAR}(p_1)$, the "star of spokes" of p_1. These "spokes" are ordered, say, counterclockwise, and they subdivide the 2π-angle at p_1 into $(N - 1)$ sectors, or angular intervals. If an edge $\overline{p_k p_j}$ has been placed into the triangulation, it spans a set of *consecutive* sectors in the star of spokes of p_1, and, for that matter, of any p_l for $l \neq k, j$. Notice also that no two edges currently in the triangulation intersect (by definition), so that the edges present in each angular interval of $\text{STAR}(p_l)$, $l = 1, \ldots, N$ are *ordered*. Assume that in $\text{STAR}(p_1)$ point p_i falls in the sector $p_j p_1 p_k$. To decide whether $\overline{p_1 p_i}$ is to be added to the triangulation, we must check whether it intersects the triangulation edges that cut this sector, and, specifically, the one that is *closest to* p_1 in the sector. Thus, the decision has been reduced to a special case of planar point location, where in each sector of $\text{STAR}(p_l)$ (for all l) we must maintain one triangulation edge (called *spanning edge*). A typical situation is illustrated in Figure 6.7(b), where the shaded area corresponds to the acceptance of triangulation edges.

The search data structure designed to support the acceptance test is of a dynamic type, since it must be updated each time an edge is accepted in the triangulation. Recalling however that an update involves a *sequence of consecutive sectors* in each star of spokes, it is natural to resort to segment trees (see Section 1.2.3.1). Specifically, the sectors of a generic $\text{STAR}(p_l)$ are organized as the $(N-1)$ leaves of a segment tree T_l. Each node v of T_l is associated with an edge $e(v)$, which, among those allocated to v by insertions into the segment tree, is closest to p_l. (Obviously $e(v)$ may be the empty edge.) To test for $\overline{p_i p_j}$ we search T_i by tracing the path identified by p_j; at each node v in this path we test p_j against $e(v)$, and eventually accept $\overline{p_i p_j}$ only if it does not intersect any of the edges so encountered. This search completes the task when the currently selected edge is to be rejected. In the other case—i.e., when edge $\overline{p_i p_j}$ is accepted—each $\text{STAR}(p_l)$, for $l \neq i, j$ must still be updated: this is simply done by inserting $\overline{p_i p_j}$ in $\text{STAR}(p_l)$ and, updating, as required, each node v in the insertion paths. (The reader is referred back to Section 1.2.3.1 for a detailed illustration of the mechanics of segment trees.) Clearly, each $\text{STAR}(p_l)$ is updated in time $O(\log N)$. Since there are $(N-2)$ stars to be updated at each edge acceptance and there are $O(N)$ acceptances in the triangulation, the entire updating task is completed in time $O(N^2 \log N)$.

In conclusion, since the management of the edge selection, the edge acceptance tests, and the updates of the data structures can each be performed in time $O(N^2 \log N)$, we have

Theorem 6.4. *The greedy triangulation of a set of N points can be constructed in time $O(N^2 \log N)$ and space $O(N^2)$.*

6.2.2 Constrained triangulations

In many cases the triangulation problem may be of a constrained nature, that is, a set of triangulation edges may be prespecified in the problem statement. Typically, this is the case when we are asked to triangulate the interior of a simple polygon.

Of the previously described techniques, the greedy method will succeed for such constrained problems, but it has the drawback that its performance is substantially far from optimal. On the other hand, there is no immediate way to adapt to this case the Delaunay triangulation method. (See Sections 5.6 and 6.5.) So, a new technique has to be found.

As usual, we are given a set S of N points; in addition, we have a set E of M nonintersecting edges on S (the constraints), which must appear in the triangulation. We slightly, and insignificantly, modify our previous convention by stipulating that the region to be triangulated is the smallest rectangle, $\text{rect}(S)$, with sides parallel to the coordinate axes, which inscribes S. (Notice that in $\text{rect}(S)$ there are at least two vertices with largest y, and at least two with smallest y.)

The objective is to efficiently decompose $\text{rect}(S)$ into simpler polygonal regions that can be easily triangulated. These polygons are characterized as follows [Garey–Johnson–Preparata–Tarjan (1978)].

Definition 6.1. A polygon P is said to be *monotone* with respect to a straight line *l* if it is simple and its boundary is the union of two chains monotone with respect to *l* (see Section 2.2.2.2).[3]

In the following we assume that the line *l* is the *y*-axis and that no two points of *S* have the same ordinate (as usual, this assumption is in no way crucial, but simplifies the presentation). We shall present later a simple algorithm for the triangulation of a monotone polygon; now we illustrate how to partition rect(*S*).

The technique is of the plane-sweep type and is essentially the same as the "regularization" procedure of a planar graph described in Section 2.2.2.2. Indeed the sets *S* and *E* define a planar graph embedded in the plane. The regularization procedure (which is not described again here, to avoid unnecessary repetitions) in time $O(N \log N)$ adds edges so that the following properties hold:

(i)   No two edges intersect (except at vertices);
(ii) Each vertex (except the ones with largest *y*-coordinate) is joined directly to at least one vertex with a larger *y*-coordinate;
(iii) Each vertex (except the ones with smallest *y*-coordinate) is joined directly to at least one vertex with a smaller *y*-coordinate.

We claim that each of the regions of the planar graph resulting from the regularization procedure is a monotone polygon. The proof of the claim is based on the notion of interior cusp. A vertex *v* of a simple polygon is an *interior cusp* if the internal angle at *v* exceeds π and its adjacent vertices both have not larger *y*-coordinates than *v* or both have not smaller *y*-coordinates than *v*. It follows from properties (ii) and (iii) above that no vertex of the regularized graph can be an interior cusp. The following lemma is the key to our claim:

Lemma 6.3. *If P is a simple polygon with no interior cusp, then P is monotone with respect to the y-axis.*

PROOF. Let $v_1, v_2, \ldots, v_m$ be the clockwise sequence of the vertices of *P*, and let v_1 and v_s be the vertices with the largest and smallest *y*, respectively (see Figure 6.8). If *P* is not monotone, then at least one of the two chains from v_1 to v_s formed by the boundary edges of *P* is not strictly decreasing by *y*-coordinate. Consider the case in which the chain passing through v_2 fails to be strictly decreasing (the other case is symmetric). Choose v_i, $1 < i < s$, to be the first vertex on this path such that the *y*-coordinate of v_{i+1} exceeds that of v_i.

We first observe that the three-vertex sequence $(v_{i-1} v_i v_{i+1})$ must form a right turn, for otherwise v_i would be an interior cusp of *P* (see Figure 6.8(a)). Now consider the line through v_i and v_s (see Figure 6.8(b)), and let $r \neq v_i$ be the

---

[3] It has been shown that a simple polygon can be tested for monotonicity is linear time [Preparata–Supowit (1981)].

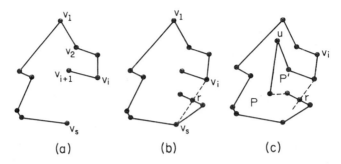

Figure 6.8   (a) v_i is the first vertex in $v_1 v_2 \ldots$ for which $y(v_{i+1}) > y(v_i)$. (b) The line from v_i to v_s first intersects P in point r. (c) The vertex u with highest y in P' is an interior cusp of P.

first point on the boundary of P encountered when traveling from v_i to v_s along this line (r might be v_s). Then the line segment joining v_i to r divides the exterior of P into two parts, one of which is a finite polygon P', as shown in Figure 6.8(c). Except for r, the vertices of P' are all vertices of P. Among all the vertices of P', let u be the one with largest y. Then u is also a vertex of P, which must also be an interior cusp of P (a contradiction).   □

Lemma 6.3 and the fact that no polygon obtained by the regularization procedure has an interior cusp, prove the claim that each polygon is monotone. We can now turn our attention to the triangulation of a monotone polygon [Garey–Johnson–Preparata–Tarjan (1978)].

6.2.2.1 Triangulating a monotone polygon

Since P is monotone with respect to the y-axis, by merging the two monotone chains forming the boundary of P we can sort in time $O(N)$ the vertices of P in order of decreasing y-coordinate. Let $u_1, u_2, \ldots, u_N$ be the resulting sequence. Clearly monotonicity means that for each u_i ($1 \leq i \leq N - 1$) there is a u_j, with $i < j$, so that $\overline{u_i u_j}$ is an edge of P.

The triangulation algorithm processes one vertex at a time in order of decreasing y. In this process *diagonals* of P are generated. Each diagonal bounds a triangle and leaves a polygon, with one less side, still to be triangulated. The algorithm makes use of a stack containing vertices that have been visited but not yet reached by a diagonal. Both the top and the bottom of the stack are accessible by the algorithm (the bottom only for inspection).

The stack content, v_1 (= bottom of STACK), $v_2, \ldots, v_i$, form a chain on the boundary of P and $y(v_1) > y(v_2) > \cdots > y(v_i)$, and if $i \geq 3$ angle $(v_j v_{j+1} v_{j+2}) \geq 180°$ for $j = 1, \ldots, i - 2$ (see Figure 6.9(a)–(c)). The algorithm runs as follows.

Initial step. Vertices u_1 and u_2 are placed in the stack.
General step. Let u be the current vertex.

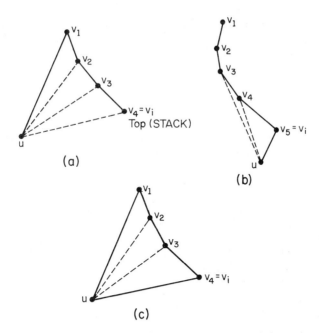

(a)

(b)

(c)

Figure 6.9   The three cases of the general step. Dashed lines are the added diagonals.

(i) if u is adjacent to v_1 but not v_i (top(STACK)), then add diagonals $\overline{uv_2}$, $\overline{uv_3}, \ldots, \overline{uv_i}$. Replace stack content by v_i, u. (See Figure 6.9(a)).

(ii) if u is adjacent to v_i but not to v_1, then while $i > 1$ and angle $(uv_iv_{i-1}) <$ $180°$, add diagonal $\overline{uv_{i-1}}$ and pop v_i from stack. When the two-fold condition no longer holds (or if it did not hold to begin with) add u to stack (see Figure 6.9(b)).

(iii) otherwise (u adjacent to both v_1 and v_i), add diagonals $\overline{uv_2}, \overline{uv_3}, \ldots, \overline{uv_{i-1}}$. (In this case processing is completed—see Figure 6.9(c).)

The correctness of the algorithm depends upon the fact that the added diagonals lie completely inside the polygon. Consider for example the diagonal $(\overline{uv_2})$ constructed in case (i). None of the vertices $v_3, \ldots, v_i$ can lie inside or on the boundary of the triangle (u, v_1, v_2) because the internal angles of at least $180°$ at $v_2, v_3, \ldots, v_{i-1}$ force u and $v_3, v_4, \ldots, v_i$ to lie on opposite sides of the line containing $\overline{v_1 v_2}$. No other vertex of the polygon lies inside or on the boundary of this triangle because all such vertices have smaller y-coordinates than u. No point inside the triangle can be external to the polygon, because the polygon would then have to pass through the interior of the triangle, and at least one of its vertices would be in the interior of the triangle. Thus, the diagonal $\overline{uv_2}$ lies completely within the polygon. The proofs for cases (ii) and (iii) are analogous, and it is also quite simple to verify that the stack properties are an invariant of the algorithm.

From a performance viewpoint, $O(N)$ time is taken by the initial merge.

In the triangulation process, each vertex is added to the stack and visited just once, except when, in case (ii), the condition "while $i > 1$ and angle $(uv_iv_{i-1}) < 180°$" fails. If we charge the corresponding work, as well as the access to top(STACK) and bottom(STACK), to the current vertex u, it is clear that this task is also completed in time $O(N)$. In conclusion we have the following theorem.

Theorem 6.5. *A monotone polygon with N sides can be triangulated in optimal time $O(N)$.*

Combining this result with the previous result that rect(S) can be decomposed into monotone polygons in time $O(N \log N)$, and further recalling the lower bound for the triangulation problem, we obtain

Theorem 6.6. *The constrained triangulation on a set of N points can be algorithmically constructed in optimal time $\theta(N \log N)$.*

6.3 Generalizations of the Voronoi Diagram

We recall from the preceding chapter the definition of the Voronoi diagram: "The Voronoi diagram of a finite *set S of points* in the *plane* is a partition of the plane so that each region of the partition is the locus of points which are closer to *one member* of S than to any other member." In this definition we have italicized three items (set of points, plane, one member), which are exactly the ones that are susceptible of generalization. Indeed, generalizations have been attempted and obtained with varying degrees of success, in each of these three directions.

First of all, while still remaining in the plane (i.e., in two dimensions), the given set may be extended to contain other geometric objects besides points, such as segments, circles and the like. This direction, however, will not be treated in detail in this book (see Notes and Comments at the end of the chapter).

Next, to seek a data structure that efficiently supports k-nearest-neighbor searches (see Problem P.6, Section 5.1), one may wish to define loci of points closer to a given subset of k members of S than to any other subset of the same size. It is interesting that, if one lets k become $N - 1$, one obtains the farthest point Voronoi diagram.

Finally, although the definition of the standard Voronoi diagram in any number of dimensions is straightforward, its construction is plagued with considerable algorithmic difficulties. Indeed, it is readily shown that for given N points the number of items required to describe the Voronoi diagram grows exponentially with the dimension. This is a phenomenon we already observed for the convex hull problem: indeed, the connection between Voronoi dia-

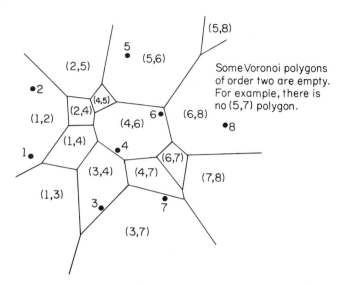

Figure 6.10   A Voronoi diagram of order two. Points are shown by their indices.

grams and convex hulls goes well beyond the exponential growths of their respective descriptions.

6.3.1 Higher-order Voronoi diagrams (in the plane)

The Voronoi diagram, while very powerful, has no means of dealing with farthest points, k closest points and other distance relationships. As such, it is unable to deal with some of the problems we have posed. The difficulty is that we have been working with the Voronoi polygon associated with a single point, but such a restriction is not necessary and it will be useful to speak of the *generalized Voronoi polygon* $V(T)$ of a subset T of points [Shamos–Hoey (1975)], defined by

$$V(T) = \{p: \forall v \in T \forall w \in S - T, d(p,v) < d(p,w)\}.$$

That is, $V(T)$ is the locus of points p such that each point of T is nearer to p than is any point not in T. An equivalent definition is

$$V(T) = \cap H(p_i, p_j), p_i \in T, p_j \in S - T,$$

where $H(p_i, p_j)$ is, as usual, the half-plane containing p_i that is defined by the perpendicular bisector of $\overline{p_i p_j}$. This shows that a generalized Voronoi polygon is still convex. It may, of course, happen that $V(T)$ is empty. In Figure 6.10 for example, there is *no* point with the property that its two nearest neighbors are p_5 and p_7. A set S with N points has 2^N subsets. How many of these can possess nonempty Voronoi polygons? If the number is not large, there will be some hope of performing k-nearest-neighbor searching without excessive storage.

Let us define the *Voronoi diagram of order k*, denoted $\text{Vor}_k(S)$, as the collection of all generalized Voronoi polygons of k-subsets of S, so

$$\text{Vor}_k(S) = \cup\, V(T), \qquad T \subset S, |T| = k.$$

In this notation, the ordinary Voronoi diagram is just $\text{Vor}_1(S)$. It is proper to speak of $\text{Vor}_k(S)$ as a "diagram" because its polygons partition the plane. Given $\text{Vor}_k(S)$, the k points closest to a new given point q can be determined by finding the polygon in which q lies. Figure 6.10 shows a Voronoi diagram of order two, the set of loci of nearest pairs of points.

Before tackling computational problems, it is necessary to investigate the structure of generalized Voronoi diagrams. The next subsection is devoted to this task.

6.3.1.1 Elements of inversive geometry

We begin with a brief digression on Inversive Geometry, a tool that is particularly suited to our objective. We have

Definition 6.2. *Inversion* in E^d is a point-to-point transformation of E^d which maps a vector $\mathbf{v}$ applied to the origin to the vector $\mathbf{v}' = \mathbf{v} \cdot 1/|\mathbf{v}|^2$ applied to the origin.

We notice at first that inversion is involutory, since the inner product of $\mathbf{v}$ and of its image $\mathbf{v}'$ is

$$\mathbf{v} \cdot \mathbf{v}'^T = \mathbf{v} \cdot \mathbf{v}^T \frac{1}{|\mathbf{v}|^2} = 1$$

(the involutory property derives immediately from $\mathbf{v} \cdot \mathbf{v}'^T = 1$, which shows that $\mathbf{v}$ and $\mathbf{v}'$ are the inversive images of each other). Next, we will show the characteristic property of inversion, as expressed by the following theorem.

Theorem 6.7. *Inversion of E^d maps hyperspheres to hyperspheres.*

PROOF. A hypersphere σ of center $\mathbf{c}$ (a vector applied at the origin) and radius r is the set of all vectors $\mathbf{v}$ applied at the origin satisfying $|\mathbf{v} - \mathbf{c}|^2 = r^2$. From this we readily obtain $2\mathbf{v} \cdot \mathbf{c}^T = |\mathbf{v}|^2 + |\mathbf{c}|^2 - r^2$. Now consider the quantity $|\mathbf{v}' - \mathbf{c}/(|\mathbf{c}|^2 - r^2)|^2$ where $\mathbf{v}'$ is the inversive image of $\mathbf{v}$, i.e., $\mathbf{v}' = 1/|\mathbf{v}|^2 \cdot \mathbf{v}$. We now obtain

$$\left| \mathbf{v}' - \frac{\mathbf{c}}{|\mathbf{c}|^2 - r^2} \right|^2 = \frac{1}{|\mathbf{v}|^2} + \frac{|\mathbf{c}|^2}{(|\mathbf{c}|^2 - r^2)^2} - \frac{2\mathbf{v} \cdot \mathbf{c}^T}{|\mathbf{v}|^2 \cdot (|\mathbf{c}|^2 - r^2)}$$

$$= \frac{(|\mathbf{c}|^2 - r^2)^2 + |\mathbf{v}|^2|\mathbf{c}|^2 - |\mathbf{v}|^2(|\mathbf{c}|^2 - r^2) - (|\mathbf{c}|^2 - r^2)^2}{|\mathbf{v}^2| \cdot (|\mathbf{c}|^2 - r^2)^2}$$

$$= \frac{r^2}{(|\mathbf{c}|^2 - r^2)^2}.$$

Since the latter quantity is a constant (indeed it depends on constants $|\mathbf{c}|^2$ and r) we have that the inversive image of σ is a hypersphere of center $\mathbf{c}/(|\mathbf{c}|^2 - r^2)$ and radius $r/(|\mathbf{c}|^2 - r^2)$. □

The only singularity of the inversive mapping is that the origin of E^d is the image of the hyperplane at infinity. This leads to the consequence that the interior of a hypersphere σ maps to the interior of its inversive image σ' if and only if σ (and therefore σ') does not contain the origin in its interior; in the opposite case, the interior of σ maps to the exterior of σ'.

Of particular interest for our purposes are the hyperspheres passing by the origin, since they represent the limiting case of the two classes mentioned above. Specifically, a hypersphere through the origin maps to a hyperplane, and the interior of the hypersphere maps to the half-space (determined by the image hyperplane) *not* containing the origin. In other words, the family of hyperplanes of E^d maps to the family of hyperspheres through the origin.

To provide both intuitive support and illustration, from now on we shall refer to the two- and three-dimensional cases. Moreover, the three-dimensional space will provide the setting for the discussion of higher-order Voronoi diagrams in the plane; however, the figures—for simplicity—will illustrate two-dimensional instances. Therefore, the preceding discussion yields that inversion in the plane maps circles to circles, whereas in the space it maps spheres to spheres.

A unique role in E^3 is played by the *unit sphere* (sphere of radius one centered at the origin) because its points are the *fixed points* of the inversion. For example, the transformation is illustrated in Figure 6.11 for two dimensions, where line l_i and circle C_i, for $i = 1, 2, 3$, are inversion pairs.

6.3.1.2 The structure of higher-order Voronoi diagrams

We now have all the necessary premises. Let S be a set of N points in a plane. We immerge this plane in E^3 of coordinates x, y, and z and identify it with the plane $z = 1$ (notice however that any other choice would be equally suitable). We construct the inversive image of this plane, i.e., the sphere C of radius $\frac{1}{2}$ centered at $(0, 0, \frac{1}{2})$. (See Figure 6.12 for an analogy in one less dimension.) We then map each point of the set S to its image on C, by a straightforward projection from the origin of E^3.[4] Thus we obtain a set S' of N points on the sphere C.

Consider now a plane π determined by three points p'_i, p'_j, and p'_l of S' (respectively the inversive images of p_i, p_j, and p_l of S.) (Refer to Figure 6.12.) This plane determines a spherical cap on C not containing the origin, and let $S'(\pi) \subseteq S'$ be the subset of S' contained in the interior of this spherical cap. Note that $S'(\pi)$ is a set of points in the interior of the half-space $H(\pi)$ determined by π and not containing the origin. Next consider the inversive

---

[4] This projection is known as *stereographic*.

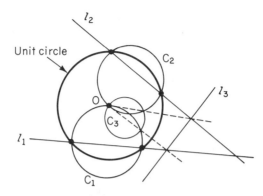

Figure 6.11   Illustration of inversion in the plane, with restriction to straight lines and circles by the origin.

image $C(\pi)$ of π: we know that $C(\pi)$ is a sphere by the origin and passing by points p_i, p_j, and p_l (indeed, the sphere determined by these four points). Since the half-space $H(\pi)$ maps to the interior of sphere $C(\pi)$, the inversive images of the points of $S'(\pi)$ lie in the interior of $C(\pi)$. In particular, they lie in the interior of the circle $C(p_i, p_j, p_l)$ passing by p_i, p_j, and p_l, since this circle is the intersection of the sphere $C(\pi)$ with the plane $z = 1$. Let v be the center of circle $C(p_i, p_j, p_l)$ and let $R \subset S$ be the "set of points in the interior" of $C(p_i, p_j, p_l)$ (obviously R is the image of $S'(\pi)$). We claim that v is a vertex in some higher-order Voronoi diagram. Indeed, let $|R| = k$, and assume, for a moment, that $0 < k < N - 3$. Clearly, R is the set of points closer to v than any of $\{p_i, p_j, p_l\}$. Now, v appears in the order-$(k + 1)$ Voronoi diagram as the common point of $V(R \cup \{p_i\})$, $V(R \cup \{p_j\})$, and $V(R \cup \{p_l\})$ and in the order-$(k + 2)$ diagram

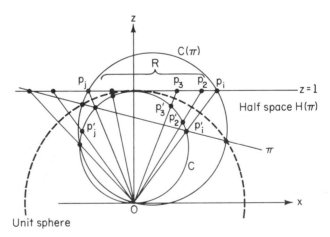

Figure 6.12   Illustration of the inversion between a point set on the plane $z = 1$ and its image on the sphere C.

as the common point of $V(R \cup \{p_i, p_j\})$ $V(R \cup \{p_j, p_l\})$, and $V(R \cup \{p_i, p_l\})$. If $R = \varnothing$, that is, $k = 0$, then v is a vertex of just $\text{Vor}_1(S)$, whereas for $|R| = N - 3$, v is a vertex of just $\text{Vor}_{N-1}(S)$. We now remark that plane π can be chosen in $\binom{N}{3}$ ways, and each choice determines a Voronoi vertex (most of them appearing in two higher-order Voronoi diagrams). Since each Voronoi diagram is a planar graph and each Voronoi vertex has degree at least three (and barring degeneracies, exactly three), in each $\text{Vor}_k(S)$, $k = 1, \ldots, N - 1$, the number of polygons is proportional to the number of vertices. Combining these facts we obtain the following interesting result [Brown (1979a), (1979b)].

Theorem 6.8. *The number of Voronoi polygons of all orders is* $O(N^3)$.

To gain more insight into the structure of the family of higher-order Voronoi diagrams we follow an approach of the type recently proposed by Edelsbrunner and Seidel (1984). Consider the stereographic projection which describes the restriction of the inversion on E^3 to the plane $z = 1$ (or equivalently, to the sphere C, inversive image of this plane). It is convenient to explicitly refer to the equation of C, that is,

$$x^2 + y^2 + \left(z - \frac{1}{2}\right)^2 = \frac{1}{4}. \tag{6.1}$$

Suppose that we now apply a suitable projective transformation φ to E^3 and let ξ, η, and ζ be the coordinates in the transformed space. Specifically, we seek a transformation that maps the plane $z = 0$ to the plane at infinity. We carry out the transformation as follows: first we introduce homogeneous coordinates x_1, x_2, x_3, and x_4 in E^3 (see Section 1.3.2 for a discussion of the correspondence between homogeneous and inhomogeneous coordinates); then we consider these as the coordinates of E^4, and finally we apply a rotation to E^4 expressed by the following relation (where ξ_1, ξ_2, ξ_3, and ξ_4 are the homogeneous coordinates in the transformed space):

$$(\xi_1, \xi_2, \xi_3, \xi_4)^T = \begin{bmatrix} 1 & 0 & 0 & 0 \\ 0 & 1 & 0 & 0 \\ 0 & 0 & 0 & 1 \\ 0 & 0 & 1 & 0 \end{bmatrix} \cdot (x_1, x_2, x_3, x_4)^T. \tag{6.2}$$

By rewriting (6.1) in homogeneous coordinates and applying transformation (6.2), we obtain the homogeneous-coordinate equation of $\varphi(C)$:

$$\xi_1^2 + \xi_2^2 + \xi_4^2 = \xi_3 \xi_4. \tag{6.3}$$

Restoring now the inhomogeneous coordinates $\xi = \xi_1/\xi_4$, $\eta = \xi_2/\xi_4$, and $\zeta = \xi_3/\xi_4$, we obtain the inhomogeneous-coordinate equation of $\varphi(C)$:

$$\zeta = \xi^2 + \eta^2 + 1, \tag{6.4}$$

which describes a (rotation) paraboloid $\mathscr{P}$.

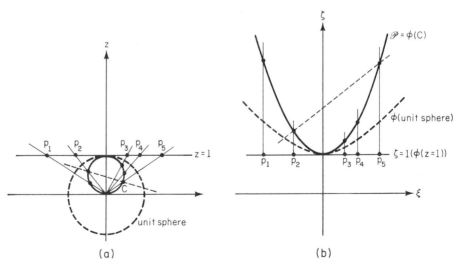

Figure 6.13   Transformation φ maps diagram (a) to diagram (b).

In addition, φ maps plane $z = 1$ to plane $\zeta = 1$, the unit sphere to the hyperboloid of equation $\xi^2 + \eta^2 - \zeta^2 + 1 = 0$, and the origin to the point at infinity of the ζ-axis. In summary, the diagram of Figure 6.12, repeated for convenience in Figure 6.13(a), is mapped under φ to the diagram of Figure 6.13(b). Consider now the images, under φ, of pairs of points, respectively in C and $z = 1$, which are mutual inversive images. These new points, respectively in $\mathscr{P}$ and $\zeta = 1$, are now obtained by a simple projection parallel to the ζ-axis (vertical). It is useful to denote by γ the vertical projection $\gamma:(\zeta = 1) \to \mathscr{P}$.

Again, let S be a finite point set in the plane $\zeta = 1$ and let $S' = \{\gamma(p): p \in S\}$. Given any two points p_1 and p_2 in S, consider the planes $\pi(p_1)$ and $\pi(p_2)$ tangent to $\mathscr{P}$ at the points $\gamma(p_1)$ and $\gamma(p_2)$, respectively. A very important property of these two planes is provided by the following lemma, whose proof is left as an exercise.

Lemma 6.4. *The projection of the intersection line of $\pi(p_1)$ and $\pi(p_2)$ onto the plane $\zeta = 1$ is the perpendicular bisector of the segment $\overline{p_1 p_2}$.*

Suppose now we have constructed the set of planes $\{\pi(p): p \in S\}$. These planes partition the space into cells $\{D_1, D_2, \ldots, D_M\}$. We establish the connection between cells and Voronoi polygons (of some $\mathrm{Vor}_k(S)$) on the plane $\zeta = 1$. For each $p \in S$, we define as $\mathrm{HS}(p)$ the half-space determined by $\pi(p)$ (again, $\pi(p)$ is the plane tangent to $\mathscr{P}$ at $\gamma(p)$) and external to the paraboloid $\mathscr{P}$. Let us now consider a generic cell D of the partition of the space determined by the set of planes $\{\pi(p): p \in S\}$. With D we associate a unique set $T(D) \subseteq S$

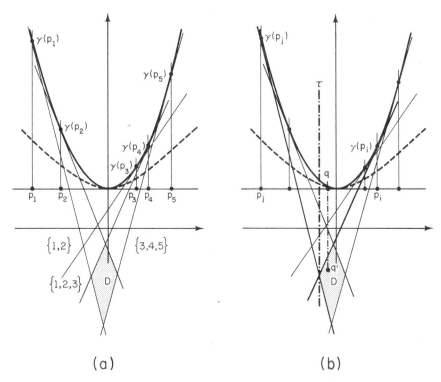

Figure 6.14 (a) Two-dimensional analog of the space partition determined by tangent planes. (b) Illustration for the proof of Theorem 6.10.

defined as the unique subset of S such that

$$\text{for each } p \in T(D), D \subseteq HS(p).$$

(See Figure 6.14 for a two-dimensional analog.) Let q' be a generic point in D and q its vertical projection on the plane $\zeta = 1$; moreover, let $p_i \in T(D)$ and $p_j \in S - T(D)$ (refer to Figure 6.14(b)). By the definition of $T(D)$, we have that $q' \in HS(p_i)$ and $q' \notin HS(p_j)$. The vertical plane τ passing by the intersection of $\pi(p_i)$ and $\pi(p_j)$ determines two half-spaces; the condition $q' \in HS(p_i)$ and $q' \notin HS(p_j)$ implies that q' lies in the half-space containing p_i. Recalling Lemma 6.4, we have that q belongs to the half-plane $H(p_i, p_j)$, and, due to the arbitrariness of p_i in $T(D)$ and of $p_j \in S - T(D)$, we reach the following conclusion.

Theorem 6.9. *The projection on $\zeta = 1$ of a cell D of the space partition determined by $\{\pi(p): p \in S\}$ is a Voronoi polygon $V(T)$, where T is the largest subset of S so that $D \subseteq HS(p)$ for each $p \in T$.*

This shows that the above partition of the space provides complete infor-

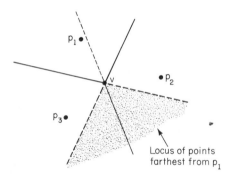

Figure 6.15   Closest and farthest point Voronoi diagrams for a three-point set.

mation about the family of higher-order Voronoi diagrams. In the next section we shall address the question of their algorithmic construction.

6.3.1.3 Construction of the higher-order Voronoi diagrams

To develop the necessary algorithmic mechanism, we shall now explore the relationship between $Vor_k(S)$ and $Vor_{k+1}(S)$. (Note, incidentally, that $Vor_{N-1}(S)$ is appropriately called the farthest point Voronoi diagram, since each of its polygons is the locus of the points of the plane closer to any member of $V - \{p_i\}$ (for some i) than to p_i, i.e., the locus of the points for which p_i is the farthest point.)

We begin by considering a set S of three points: p_1, p_2, and p_3 (refer to Figure 6.15). In this case there is a single Voronoi vertex v, where the three bisectors meet. The ordinary Voronoi diagram is shown with solid lines. If we now take the complementary ray on each bisector (shown as a broken line), we partition the plane into three regions, such that each point in any given region is closer to two of the given points than to the third one: we have obtained the farthest point Voronoi diagram of $S = \{p_1, p_2, p_3\}$.

Consider now a polygon $V(T)$ of the diagram $Vor_k(S)$, where $T = \{p_1, \ldots, p_k\}$. We know that $V(T)$ is a convex polygon and let v be one of its vertices. Vertex v is common to three polygons of $Vor_k(S)$: $V(T)$, $V(T_1)$, and $V(T_2)$. Now we distinguish two cases[5]:

(i) $|T \oplus T_1 \oplus T_2| = k + 2$,   (void if $k = N - 1$)
(ii) $|T \oplus T_1 \oplus T_2| = k - 2$,   (void if $k = 1$)

Vertex v is called of the *close-type* in case (i), and of the *far-type* in case (ii). To gain some intuition on this classification, note that in case (i) we have, for example:

(a) $T = R_1 \cup \{p_k\}$, $T_1 = R_1 \cup \{p_{k+1}\}$, $T_2 = R_1 \cup \{p_{k+2}\}$

---

[5] Here $\oplus$ denotes the symmetric difference.

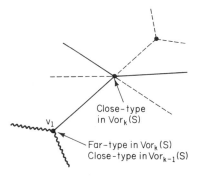

Figure 6.16   Illustration of the "life-cycle" of a Voronoi vertex.

where $R_1 = \{p_1, \ldots, p_{k-1}\}$. So, if we remove R_1 from S, v and its incident edges (now becoming infinite rays) from the *closest* point Voronoi diagram of $\{p_k, p_{k+1}, p_{k+2}\}$. By contrast, in case (ii) we typically have:

(b) $T = R_2 \cup \{p_{k-1}, p_k\},$ $T_1 = R_2 \cup \{p_{k-1}, p_{k+1}\},$
            $T_2 = R_2 \cup \{p_k, p_{k+1}\},$

where $R_2 = \{p_1, \ldots, p_{k-2}\}$. If, again, we remove R_2 from S, v and its incident edges form the *farthest* point Voronoi diagram of $\{p_{k-1}, p_k, p_{k+1}\}$.[6] It is interesting to point out that the Delaunay circle centered at v contains $k-1$ points of S in its interior if v is of the close type, and $(k-2)$ points in the other case.

Moreover, suppose that v is a vertex of the close-type of $\mathrm{Vor}_k(S)$, and let it be the common point of $V(T)$, $V(T_1)$ and $V(T_2)$ as given in (a) above. If in a suitable neighborhood of v we replace each of the three Voronoi edges incident on v with its extension beyond v, vertex v becomes the common vertex of three new polygons, which are readily recognized as $V(T \cup T_1)$, $V(T \cup T_2)$, and $V(T_1 \cup T_2)$. But $|T \cup T_1| = |T \cup T_2| = |T_1 \cup T_2| = k+1$. Thus, by extending the Voronoi edges incident on a close-type vertex of $\mathrm{Vor}_k(S)$, this vertex has become a far-type vertex of $\mathrm{Vor}_{k+1}(S)$! This is, in a nutshell, the key idea for the construction of the sequence $\mathrm{Vor}_2(S)$, $\mathrm{Vor}_3(S)$, $\ldots$, $\mathrm{Vor}_{N-1}(S)$.

To analyze the mechanism further, suppose that v is a close-type vertex of $\mathrm{Vor}_k(S)$ (refer to Figure 6.16) and suppose it is adjacent to a far-type vertex v_1. This means that v has been generated in $\mathrm{Vor}_k(S)$ by extending, in the sense described earlier, the edges incident on close-type vertices of $\mathrm{Vor}_{k-1}(S)$ (one such vertex is v_1). In the construction of $\mathrm{Vor}_{k+1}(S)$, in turn, we shall extend the edges incident on v. In the process, however, we shall delete edge $\overline{v_1 v}$, so that v_1 disappears from $\mathrm{Vor}_{k+1}(S)$. This illustrates, so to speak, the life cycle of a Voronoi vertex: it appears in two Voronoi diagrams of consecutive orders, first as a close-type vertex and then as a far-type vertex.

---

[6] Note that the $\mathrm{Vor}_1(S)$ contains only close-type vertices, while $\mathrm{Vor}_{N-1}(S)$ contains only far-type vertices.

The construction of $\mathrm{Vor}_{k+1}(S)$ is essentially the partitioning of each polygon of $\mathrm{Vor}_k(S)$ into portions of polygons of $\mathrm{Vor}_{k+1}(S)$. This is equivalent to obtaining the intersection of each polygon $V(T)$ of $\mathrm{Vor}_k(S)$ with the ordinary Voronoi diagram $\mathrm{Vor}_1(S - T)$. However, we do not have to compute the entire $\mathrm{Vor}_1(S - T)$, but only that portion of it that is internal to $V(T)$. Specifically, let e be an edge of $V(T)$ incident on a close-type vertex v; then e is the common boundary of $V(T)$ and $V(T')$, where $|T \oplus T'| = 2$, and will become internal to $V(T \cup T')$, a polygon in $V_{k+1}(S)$. It follows that denoting by s the number of close-type vertices on the boundary of $V(T)$, $V(T)$ will be decomposed into s portions if bounded and into $(s + 1)$ portions if unbounded. Since the close-type vertices identify the subset of $S - T$ affecting the partition of $V(T)$, the latter can be obtained either by computing an ordinary Voronoi diagram in time $O(s \log s)$ or by using a specific technique due to Lee (1976, 1982), which also runs time in $O(s \log s)$. In all cases, if $V_k^{(1)}$ is the number of close-type vertices in $\mathrm{Vor}_k(S)$, $\mathrm{Vor}_{k+1}(S)$ can be obtained in time at most $O(V_k^{(1)} \log V_k^{(1)})$.

Lee (1982) has produced a detailed analysis of the various parameters of $\mathrm{Vor}_k(S)$. In particular we have (stated here without proof)

Lemma 6.5. *The number $V_k^{(1)}$ of close-type vertices of $\mathrm{Vor}_k(S)$ is upper-bounded by*

$$2k(N - 1) - k(k - 1) - \sum_{i=1}^{k} v_i,$$

where v_i is the number of unbounded regions in $\mathrm{Vor}_i(S)$.

Clearly $V_k^{(1)} = O(kN)$. It follows that $\mathrm{Vor}_{k+1}(S)$ can be obtained from $\mathrm{Vor}_k(S)$ in time $O(kN \log N)$, and, globally, $\mathrm{Vor}_{k+1}(S)$ can be obtained from S in time

$$\sum_{i=1}^{k-1} O(iN \log N) = O(k^2 N \log N).$$

Thus we summarize as follows

Theorem 6.10. *The order-k Voronoi diagram of an N point set is obtained in time $O(k^2 N \log N)$.*

This approach can be iterated to the construction of *all* $\mathrm{Vor}_k(S)$ for $k = 1, \ldots, N - 1$, and would exhibit a running time $O(N^3 \log N)$. However, a more recent approach [Edelsbrunner–Seidel (1984)], based on the theory outlined in the preceding section 6.3.1.2, constructs the space partition determined by the family of planes $\{\pi(p) : p \in S\}$ tangent to the paraboloid $\mathscr{P}$, and runs in time proportional to the intersections of triplets of planes, that is, in time $O(N^3)$. Thus we have

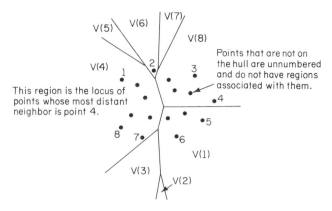

Figure 6.17 The farthest-point Voronoi diagram.

Theorem 6.11. *The family of all higher-order Voronoi diagrams for a given point set S in the plane can be constructed in time $O(N^3)$.*

We close this section with two observations. The first is that Problem P.6, k-NEAREST NEIGHBORS search reduces to a point location problem in $\text{Vor}_k(S)$. This task uses a searching time—location in a region of $\text{Vor}_k(S)$—and a report time. The latter is trivially $O(k)$, while the former is proportional to $\log V_k$, where V_k is the number of vertices of $\text{Vor}_k(S)$. Recalling that $V_k = V_k^{(1)} + V_{k-1}^{(1)}$ and that $V_k = O(kN)$, we have that the k-NEAREST NEIGHBORS search is carried out in time $O(\log kN + k) = O(\log N + k)$. We summarize this discussion as a theorem (refer to Section 2.2 for the pertinent results on searching):

Theorem 6.12. *The k nearest out of N neighbors at a point in the plane can be found in time $O(\log N + k)$ with $O(k^2 N \log N)$ preprocessing.*

The second observation is that the notion of generalized Voronoi diagram unifies closest- and farthest-point problems, since the locus of points whose k nearest neighbors are the set T is also the locus of points whose $N - k$ farthest neighbors are the set $S - T$. Thus, the order-k closest-point diagram is exactly the order-$(N - k)$ farthest-point diagram. Let us examine one of these more closely, the order-$(N - 1)$ closest-point diagram, or equivalently, the order-1 farthest-point diagram (Figure 6.17).

Associated with each point p_i is a convex polygonal region $V_{N-1}(p_i)$ such that p_i is the farthest neighbor of every point in the region. This diagram is determined only by points on the convex hull, so there are no bounded regions. Of course, $\text{Vor}_{N-1}(S)$ can be constructed by a straightforward application of the general method just described. However, the task would be completed in $O(N^3)$ (Theorem 6.12). It must then be pointed out that there is a

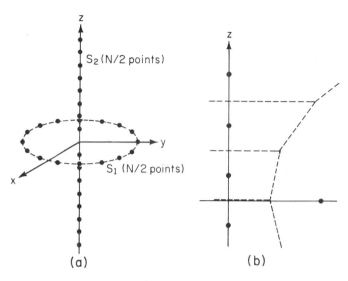

Figure 6.18   Illustration of a three-dimensional N-point set with $\theta(N^2)$ Voronoi edges.

direct method, based on the divide-and-conquer scheme and analogous to the algorithm for the closest-point diagram, which achieves the result in optimal $\theta(N \log N)$ time [Shamos (1978); Lee (1980b)]. Having found the farthest-point diagrams of the left and right halves of the set, the polygonal dividing line σ has the same properties as in the closest-point case. This time, however, we discard all segments of $\mathrm{Vor}_{N-1}(S_1)$ that lie to the left of σ, and also remove those segments of $\mathrm{Vor}_{N-1}(S_2)$ that lie to the right. We shall have the opportunity to further discuss the intimate relationship existing between $\mathrm{Vor}_1(S)$ and $\mathrm{Vor}_{N-1}(S)$ in the next section.

6.3.2 Multidimensional closest-point and farthest-point Voronoi diagrams

As we saw earlier, the generalized planar Voronoi diagrams (of all orders) have the topology of planar graphs. This implies that the numbers of vertices, faces, and edges of each of them are of the same order, and also allows the development of optimal time algorithms for the construction of the closest-point and the farthest-point Voronoi diagrams.

Can the same result be obtained in dimensions higher than two? Any hope is readily demolished by observing [Preparata (1977)] that the closest-point Voronoi diagram on N points in three dimensions may have $O(N^2)$ edges. An instance exhibiting this property is easily found. (Refer to Figure 6.18.) Consider a set S_1 of $N/2$ points uniformly placed on a circle in the (x, y)-plane and centered at the origin. Next consider a set S_2 of $N/2$ points uniformly

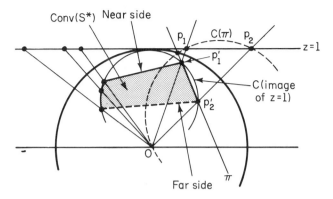

Figure 6.19   Illustration of the relation between convex hulls sand Voronoi diagram. The near side of conv(S^*) is in solid lines, the far side in broken line.

placed on the z-axis with median at the origin, and let $S = S_1 \cup S_2$. We claim that any segment joining a point of S_1 and a point of S_2 is an edge of the Delaunay graph, that is, it dualizes to a polygon of the Voronoi diagram (a facet of a Voronoi polytope). This is readily seen by considering a section with a plane containing the z-axis and a point of S_1 (Figure 6.18(b)). (The details are left to the reader.) Therefore, although the Voronoi diagram has $O(N)$ regions, it may have $O(N^2)$ vertices and edges. This finding was later extended to arbitrary dimension d by V. Klee (1980), and is summarized as follows.

Theorem 6.13. *If $M(d, N)$ is the maximum number of vertices of the Voronoi diagram on N points in the d-dimensional space, then*

$$\lceil d/2\rceil! n^{\lceil d/2\rceil} \leq M(d, N) \leq 2(\lceil d/2\rceil! n^{\lceil d/2\rceil}), \quad \text{for } d \text{ even}$$

and

$$\frac{(\lceil d/2\rceil - 1)!}{e} \cdot n^{\lceil d/2\rceil} \leq M(d, N) < \lceil d/2\rceil! n^{\lceil d/2\rceil}, \quad \text{for } d \text{ odd}$$

This exponential growth in the complexity of the Voronoi diagram reveals an intriguing similarity with the behavior of higher-dimensional convex hulls. This similarity has deep roots, which are brought to full evidence in the context of the inversive transformation described in Section 6.3.1.1., as first noted by K. Brown (1979b).

We examine, without loss of generality, the Voronoi diagrams in the plane. Let $S = \{p_1, \ldots, p_N\}$ be a set of N points in the plane $z = 1$ of E^3, let C be the inversive image of this plane in E^3, and let $S' = \{p'_1, \ldots, p'_N\}$ be the inversive image (on C) of S (here p_i and p'_i are reciprocal inversive images). Consider now the convex hull CH(S'), of S', and partition it into two parts, the *far side* and the *near side*, depending upon whether its points are "visible" or not from the origin (refer to Figure 6.19 for a two-dimensional illustration).

Let π be a plane containing a facet F of $CH(S')$, determined by three points $p_1', p_2',$ and p_3' (here we assume that $CH(S')$ is simplicial; i.e., each facet is a triangle). The inversive image of π—a sphere $C(\pi)$—passes by $O, p_1, p_2,$ and p_3, and intersects $z = 1$ in the circle determined by $p_1, p_2,$ and p_3. We distinguish two cases, depending upon whether F belongs to the near side or the far side of $CH(S')$:

(i) *F belongs to the near side of* $CH(S')$. In this case the half-space determined by π and not containing the origin contains *no* point of S' in its interior. Since this half-space maps to the interior of $C(\pi)$, the circle by $p_1, p_2,$ and p_3 is a Delaunay circle, whose center is a vertex of $Vor_1(S)$. Thus each facet of the near-side of $CH(S')$ determines a triangle of the Delaunay triangulation of S (and, dually, a vertex of $Vor_1(S)$).

(ii) *F belongs to the far side of* $CH(S')$. Again, the half-space determined by π and not containing the origin maps to the interior of $C(\pi)$. In this case, however, the half-space contains $N - 3$ points of S' in its interior, so that the circle by $p_1, p_2,$ and p_3 contains $N - 3$ points of S in its interior. It follows that the center of this circle is a vertex of the farthest-point Voronoi diagram of S.[7]

This observation not only establishes a closer link between $Vor_1(S)$ and $Vor_{N-1}(S)$, but also connects the nearest-point Voronoi diagram on N points in d dimensions to the convex hull on N points (lying on a spherical surface) in $(d + 1)$ dimensions. Therefore, one can resort to the existing machinery on multi-dimensional convex hulls (Section 3.3) to obtain multi-dimensional Voronoi diagrams. Work on these lines has been done by Avis and Bhattacharya (1983) and Seidel (1982).

6.4 Gaps and Covers

We shall now study a few additional very significant proximity problems, which can be characterized under the succinct heading of "gaps and covers." *Gaps* are the "balls" in space, (in the plane, the disks) that are void of given points, while *covers* are sets of balls whose union contains all the given points. We begin with

PROBLEM P.11 (SMALLEST ENCLOSING CIRCLE).[8] Given N points in the plane, find the smallest circle that encloses them.

---

[7] With the usual hypothesis that no four points of S are cocircular, the circle intersection of $C(\pi)$ with $z = 1$ passes by exactly three points of S. Correspondingly, F is determined by their three images and no other point of S', that is, F is a triangle ($CH(S')$ is a simplicial polytope).

[8] This problem is also frequently referred to as MINIMUM SPANNING CIRCLE.

This is a classical problem with an immense literature, the search for an efficient algorithm having apparently begun in 1860 [Sylvester (1860)]. The smallest enclosing circle is unique and is either the circumcircle of some three points of the set or defined by two of them as a diameter [Rademacher–Toeplitz (1957), Ch. 16]. Thus there exists a finite algorithm which examines all pairs and triples of points, and chooses the smallest circle determined by them which still encloses the set. The obvious implementation of this procedure would run in $O(N^4)$ time and an improvement to $O(N^2)$ time was proposed by Elzinga and Hearn (1972a, 1972b).

The enclosing circle problem is familiar in Operations Research as a minimax *facilities location problem*, in which we seek a point $p_0 = (x_0, y_0)$ (the center of the circle) whose greatest distance to any point of the set is a minimum. We may characterize p_0 by

$$\min_{p_0} \max_i (x_i - x_0)^2 + (y_i - y_0)^2. \tag{6.5}$$

The minimax criterion is used in siting emergency facilities, such as police stations and hospitals, to minimize worst-case response time [Toregas *et al.* (1971)]. It has also been used to optimize the location of a radio transmitter serving N discrete receivers so as to minimize the RF power required [Nair–Chandrasekaran (1971)]. Some authors have apparently been misled by Equation (6.5) into treating the smallest enclosing circle as a *continuous* optimization problem because there seem to be no restrictions on the location of p_0. This approach has given rise to a number of iterative algorithms [Lawson (1965); Zhukhovitsky–Avdeyeva (1966)], even though the problem is known to be discrete. This illustrates an important point: *Just because a problem $\mathscr{A}$ can be formulated as a special case of $\mathscr{B}$ is no reason for believing that a general method for solving $\mathscr{B}$ is an efficient way of solving $\mathscr{A}$.* We saw another instance of the validity of this maxim in connection with the Euclidean minimum spanning tree problem (Section 6.1): even though it can be embedded in the complete graph over the given set of points, the search for a method that capitalizes on the specific properties of the problem was rewarded with an optimal algorithm.

We now turn our attention to

PROBLEM P.12 (LARGEST EMPTY CIRCLE). Given N points in the plane, find a largest circle that contains no points of the set and whose center is internal to the convex hull of those points (see Figure 6.20).

The restriction on the center is necessary, for otherwise the problem would be unconstrained and would not possess a bounded solution. This problem is dual to P.11 in that it is *maximin*. In other words, we want p_0 as defined by

$$\max_{p_0 \in \text{Hull}(S)} \min_i (x_i - x_0)^2 + (y_i - y_0)^2. \tag{6.6}$$

Obviously, there may be more than one p_0 satisfying (6.6). The largest

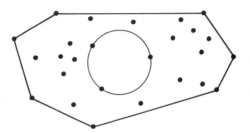

Figure 6.20   A largest empty circle whose center is internal to the hull.

empty circle is another facilities location problem, but one in which we would like to position a new facility so that it is as far as possible from any of N existing ones. The new site may be a source of pollution which should be placed as to minimize its effect on the nearest residential neighborhood, or it may be a new business that does not wish to compete for territory with established outlets. Such problems arise frequently in industrial engineering [Francis–White (1974)]. An early solution of the present problem is an algorithm whose worst-case running time is $O(N^3)$ [Dasarathy–White (1975)].

It is a quite puzzling question that $O(N^3)$ is achievable for problem P.12 while we seem to obtain a better result ($O(N^2)$) for its "dual." Is there a deep difference between the two? We shall now see that the contrast between $O(N^3)$ and $O(N^2)$ reflects an imperfect analysis of the problems, since both can certainly be solved in $O(N \log N)$ time with the aid of the Voronoi machinery; whereas time $\theta(N \log N)$ is optimal for P.12, we will see that an optimal $\theta(N)$ solution is attainable for P.11.

We begin by considering how P.11 can be solved by means of the Voronoi diagram. We know that the smallest enclosing circle C is determined either by the diameter of the given set S or by three of its points. Recall that the circle centered at a vertex of the farthest point Voronoi diagram of S ($\text{Vor}_{N-1}(S)$) and passing through its three determiners (points of S) encloses all the points of S. Moreover, the circle centered at any point of an edge e of $\text{Vor}_{N-1}(S)$ and passing through the two points of which e is the perpendicular bisector, also encloses all the points of S. Thus if C is determined by three points of S, then its center lies at a vertex of $\text{Vor}_{N-1}(S)$. If, on the other hand, C is determined by two points, then its center lies on an edge of $\text{Vor}_{N-1}(S)$. A possible approach [Shamos (1978)] consists of first finding the diameter of S in $O(N \log N)$ time (for example, with the method of Section 4.2.3) and of determining whether it encloses the set. If so, we are done. Otherwise, the center c of C is a vertex of $\text{Vor}_{N-1}(S)$. This diagram contains only $O(N)$ points and the circumradius associated with each vertex is the distance from it to any of the three points of whose polygons it is the intersection. The minimum over all vertices of this distance is the radius of the circle C. This task can clearly be carried out in total time $O(N \log N)$, since $O(N \log N)$ time is used by both the determination of the diameter of S and by the construction of $\text{Vor}_{N-1}(S)$;

additional $O(N)$ time is used by the inspection of the vertices of $\text{Vor}_{N-1}(S)$.[9]
We can summarize the discussion as follows.

Theorem 6.14. *A smallest enclosing circle of a set of N points can be obtained in time* $O(N \log N)$.

Is this result optimal? The determination of the minimum enclosing circle does not require in reality complete knowledge of the farthest point Voronoi diagram, but rather only of a portion of it in a suitable neighborhood of the center of the circle. It is indeed this idea that enabled Megiddo to obtain an optimal $\theta(N)$ algorithm [Megiddo (1983)], by viewing the original formulation (6.5) as an instance of convex programming problems. This new elegant technique will be described in Section 7.2.5.

We now turn our attention to the dual of the problem just discussed, i.e., the largest empty circle. In two or more dimensions, we will now show that the problem can be solved with the aid of the Voronoi diagram. For concreteness and simplicity, we shall as usual refer to the planar case, although the multidimensional generalization is rather straightforward. We have the following theorem.

Theorem 6.15. *The largest empty circle for an N-point set in the plane can be constructed in time* $O(N \log N)$.

PROOF. Given a set S of N points in the plane consider the function $f(x, y)$, the distance of point $p = (x, y)$ to the nearest point of S. Since we have constrained the center of the largest empty circle to lie within the convex hull of S, we shall consider the intersection of $\text{Vor}(S)$ with $\text{conv}(S)$, which is a collection of convex polygons (each of these polygons is the intersection of $\text{conv}(S)$ with a Voronoi polygon). Within a Voronoi polygon, $f(x, y)$ is a downward-convex function of both x and y, and the same applies for each polygon of the described partition. Thus, $f(x, y)$ attains its maximum at a vertex of one such polygon. This vertex is either a Voronoi vertex (in which case the largest empty circle is just a Delaunay circle) or the intersection of a Voronoi edge with a hull edge. All of the Voronoi points can be found in $O(N \log N)$ time, so it only remains to show that the intersection of the convex hull $CH(S)$ and of the Voronoi diagram $\text{Vor}(S)$ can be found quickly. We begin by noting two facts:

Property 1. *A Voronoi edge intersects at most two edges of* $CH(S)$. *Indeed, by convexity of* $\text{conv}(S)$, *the intersection of any straight line with* $\text{conv}(S)$ *consists of a (possibly empty) single segment.*

Property 2. *Each edge of* $CH(S)$ *intersects at least one Voronoi edge. Indeed,*

---

[9] An alternative approach, achieving the same time bound, is discussed in [Preparata (1977)].

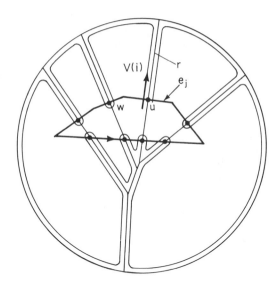

Figure 6.21   Illustration of the tour of Vor(S) to compute the intersections of Vor(S) with CH(S).

any edge of CH(S) *joins two distinct points of S, which belong to two different Voronoi polygons.*

Next, we conventionally "close" each unbounded Voronoi polygon by means of a "segment" on the line at infinity of the plane (see Figure 6.21).

Let $e_1, e_2, \ldots, e_n$ be the counterclockwise sequence of the edges of CH(S), and let u be the intersection of an arbitrarily chosen edge e_j with an edge r of the Voronoi diagram (by Property 2 above, u always exists). If from u we direct r toward the exterior of conv(S), let $V(i)$ be the Voronoi polygon lying to the left of r. If we traverse the boundary of $V(i)$ counterclockwise (in the exterior of conv(S)), we shall intersect again CH(S) in a point w for the first time: due to convexity of $V(i)$, and by Property 2, w belongs either to e_j or to e_{j+1}. Thus each edge of $V(i)$ has to be tested for intersection with just two edges of CH(S), i.e., in constant time. Once w has been found, the process is repeated with w playing the role of u, and so on until we reach u again. In this manner all intersections of Vor(S) with CH(S) can be found.

In this tour, each visited segment of Vor(S) is traversed at most twice; since each edge has at most two such segments, by Property 1, we conclude that each edge of Vor(S) is inspected at most four times, and the intersection of Vor(S) and CH(S) is computed in time $O(N)$.[10]   □

At this point we may wonder about the optimality of the preceding result, and, as usual, it is convenient to choose the one-dimensional case as the setting

---

[10] The approach described is related to an analogous result due to Toussaint (1983b).

for our argument. In this case, the problem reduces to finding a pair of consecutive points on a line that are farthest apart (MAXIMUM GAP), since a "circle" in one dimension is just a segment. Unfortunately, for the general algebraic computation tree model, no proof has been obtained so far that $\Omega(N \log N)$ operations are necessary.[11] (Contrast this situation with that of CLOSEST PAIR, discussed in Sections 5.2 and 5.3). However, Gonzalez (1975) has obtained the most surprising result that, if we modify the computation model, the problem can actually be solved in linear time. The modification consists of adding the (nonanalytic) *floor function* "$\lfloor \quad \rfloor$" to the usual repertoire. Here is Gonzalez's remarkable algorithm:

procedure MAX GAP
    Input: N real numbers $X[1:N]$ (unsorted)
    Output: MAXGAP, the length of the largest gap between consecutive
                  numbers in sorted order.
begin MIN := min $X[i]$;
    MAX := max $X[i]$;
    (*create $N - 1$ buckets by dividing the interval from MIN to MAX
    with $N - 2$ equally-spaced points. In each bucket we will retain
    HIGH[i] and LOW[i], the largest and smallest values in bucket i*)
    for $i := 1$ **until** $N - 1$ **do**
        begin COUNT[i] := 0;
                LOW[i] := HIGH[i] := Λ
        end; (*the buckets are set up*)
    (*hash into buckets*)
    for $i := 1$ **until** $N - 1$ **do**
        begin BUCKET := $1 + (N - 1) \times \lfloor (X[i] - \text{MIN})/$
                (MAX − MIN) $\rfloor$;
                COUNT[BUCKET] := COUNT[BUCKET] + 1;
                LOW[BUCKET] := min $(X[i], \text{LOW[BUCKET]})$;[12]
                HIGH[BUCKET] := max $(X[i], \text{HIGH[BUCKET]})$[12]
    end;
    (*Note that $N - 2$ points have been placed in $N - 1$ buckets, so by
    the pigeonhole principle some bucket must be empty. This means that
    the largest gap cannot occur between two points in the same bucket.
    Now we make a single pass through the buckets*)
    MAXGAP := 0;
    LEFT := HIGH[1];
    for $i := 2$ **until** $N - 1$ **do**
        if (COUNT[i] $\neq 0$) **then**
            begin THISGAP := LOW[i]-LEFT;

---

[11] An $\Omega(N \log N)$ lower bound, however, has been established in the linear decision-tree model by Manber and Tompa (1982).

[12] Here, by convention, $\min(x, \Lambda) = \max(x, \Lambda) = x$.

$$\text{MAXGAP} := \max(\text{THISGAP}, \text{MAXGAP});$$
$$\text{LEFT} := \text{HIGH}[i]$$

      end

end.

This algorithm sheds some light on the computational power of the "floor" function. In view of the apparent similarity between MAXGAP and CLOSEST PAIR in one dimension, it is remarkable that a linear algorithm is possible. Unfortunately, no generalization to two dimensions seems to be possible.

6.5 Notes and Comments

Although the Euclidean Minimum Spanning Tree of a planar set can be computed optimally, the construction of the EMST in higher dimensions is by and large an open problem. Exploiting the geometric nature of the problem Yao (1982) has developed an algorithm to construct the EMST of N points in d dimensions in time $T(N, d) = O(n^{2-a(d)}(\log n)^{1-a(d)})$, with $a(d) = 2^{-(d+1)}$, with a time bound of $O(N \log n)^{1.8})$ for $d = 3$ (this technique is closely related to Yao's algorithm for point-set diameter, Section 4.2.1.3).

A problem closely related to the Voronoi diagram is the construction of the skeleton of a polygon, a construct frequently studied in pattern recognition because it is highly suggestive of the "shape" of the polygon. The *skeleton* of a simple polygon P is the locus σ of its internal points such that each $p \in \sigma$ is equidistant from at least two points on the boundary of P; for this reason, the skeleton is also known under the name of *medial axis* [Duda–Hart (1973)]. One may also imagine to construct the skeleton by applying fire simultaneously to all sides of P: in the hypothesis that the fire propagates at constant speed ("prairie fire"), the skeleton is the locus of points where fire waves meet. In another interpretation, the skeleton is a subgraph of the planar map formed by the loci of proximity of the sides of P. This offers an extension of the notion of Voronoi diagram to a particular set of line segments (the sides of a simple polygon).

A more powerful generalization of the Voronoi diagram—mentioned in Section 6.3—consists of extending the notion to a finite collection of points and open line segments in arbitrary positions. In this case the edges of the ensuing Voronoi diagram are not just straight-line segments, but also arcs of parabola, since they separate the loci of proximity of object pairs of the types (point, point), (line, line), (point, line); the last of these pairs give rise to arcs of parabola. Earlier suboptimal solutions to this problem were offered by Drysdale and Lee [Drysdale (1979); Lee (1978); Lee–Drysdale (1981)] (the latter having $O(N \log^2 N)$ running time for a mixed collection of N objects); an optimal $\theta(N \log N)$ solution was later presented by Kirkpatrick (1979). Kirkpatrick first solves the independently significant problem of merging in linear time the Voronoi diagrams of two sets of points that are not linearly separated. He then makes use of this construction, and of a linear time algorithm to obtain the minimum spanning tree of a planar graph (See Section 6.1), to solve the general problem. Note that this technique implicitly solves the *medial axis* problem.

The machinery of the Voronoi diagram provides an elegant method to obtain a rather efficient solution of the circular range search problem. The problem is stated as follows: given a set S of N points in the plane and a circle C of center q and radius r (the query), report the subset of S contained in C. The target query-time performance is, as is usual for range-search problems, $O(f(N) + K)$, where K is the size of the retrieved set. If this is attainable, then we characterize an algorithm by the pair $(M(N), f(N))$, where $M(N)$ is the storage requirements. In Chapter 2 we have noted the inadequacy of the rectangular-range-search approaches to tackle this problem. The original idea to resort to Voronoi diagrams is to be credited to Bentley and Maurer (1979). They proposed to construct the family of Voronoi diagrams $\{\text{Vor}_{2^i}(S): i = 0, 1, \ldots, \lfloor \log_2 N \rfloor\}$, and to search this collection in succession for $i = 0, 1, 2, \ldots,$. The search consists of locating q in a region of $\text{Vor}_{2^i}(S)$—and examining the associated neighbor list; the search stops as soon as the currently examined neighbor list contains a point at distance larger than r from q. A straightforward analysis shows that this algorithm has an $(N^3, \log N \cdot \log \log N)$ performance. This idea has been recently refined in several respects [Chazelle–Cole–Preparata–Yap (1984)], within the general approach of "filtering search," discussed in Section 2.4, resulting in an $(N(\log N \cdot \log \log N)^2, \log N)$-algorithm.

In Section 6.1 we noted that the minimum spanning tree (MST) of a planar set S of N points is a subgraph of the Delaunay triangulation (DT). There are other interesting geometric structures similarly related to the DT, which have found applications in pattern recognition and cluster analysis. They are the Gabriel graph and the relative neighborhood graph. The *Gabriel graph* (GG) of S is defined as follows [Gabriel–Sokal (1969)]: Let disk(p_i, p_j) be the disk having $\overline{p_i p_j}$ as a diameter; the GG of S has an edge between p_i and p_j in S if and only if disk(p_i, p_j) contains no point of S in its interior. An efficient algorithm to construct the GG of S in time $O(N \log N)$ removes from the DT each edge not intersecting its dual Voronoi edge [Matula–Sokal (1980)]. The *relative neighborhood graph* (RNG) of S is defined as follows [Toussaint (1980b)]: there is an edge between p_i and p_j in S if and only if

$$\text{dist}(p_i, p_j) \leq \min_{k \neq i,j} \max(\text{dist}(p_i, p_k), \text{dist}(p_j, p_k)).$$

This definition means that edge (p_i, p_j) exists if and only if lune(p_i, p_j), obtained by intersecting the disks of radius length$(\overline{p_i p_j})$ centered respectively at p_i and p_j, contains no point of S in its interior. The construction of the RNG is considerably more complex than that of the GG: an $O(N \log N)$-time algorithm has been developed by Supowit (1983). There is an interesting relation among the four mentioned graphs

$$\text{MST} \subseteq \text{RNG} \subseteq \text{GG} \subseteq \text{DT}. \tag{6.7}$$

The construction of these four graphs is well-understood in the planar case, but little is known in higher dimension. Particularly, an outstanding open problem is the analysis and the design of worst-case algorithms for the MST problem in three or more dimensions.

Finally, we mention the class of decomposition problems, whose goal is to partition a given geometric object into a collection of simpler "primitive" geometric objects. Frequently such primitives are convex polygons, star-shaped polygons, etc. One example of these problem is the triangulation of a polygon which partitions the interior of a given simple polygon into a collection of triangles.

Basically there are two types of decompositions: *partition*, which disallows overlapping of component parts, and *covering*, which does allow overlapping parts. Sometimes additional vertices, called *Steiner points*, may be introduced to obtain decompositions with the minimum number of parts. A recent survey by Keil and Sack (1985) discusses minimal decompositions in great detail.

6.6 Exercises

1. Develop in detail an algorithm that constructs the minimum spanning tree of a set S of N points in the plane by means of a heap containing the edges of the Delaunay triangulation of S. An edge extracted from the heap is rejected or accepted depending upon whether it forms a cycle with the previously accepted edges. Show that this algorithm runs in time $\theta(N \log N)$.

2. Give a counterexample to the conjecture that the Delaunay triangulation is a minimum-weight triangulation.

3. Give a counterexample to the conjecture that the greedy triangulation is a minimum-weight triangulation.

4. *Lee.* Show that in a minimum Euclidean matching (Problem P.10) the line segments determined by matched pairs of points do not intersect.

5. *Seidel.* Let $S = \{x_1 , \ldots, x_N\} \subset \mathbb{R}$. Show that one can construct in *linear time* set $S' = \{p_1 , \ldots, p_N\} \subset \mathbb{R}^2$ along with a triangulation of S', where $x(p_i) = x_i$ for $1 \le i \le N$.

6. The *k-th nearest neighbor diagram* of an N-point set S is a partition of the plane into (not necessarily *internally* connected) regions such that each region is the locus R_j of the points of the plane for which a given $p_j \in S$ is the k-th neighbor (that is, for any $q \in R_j$, length$(\overline{qp_j})$ is the k-th term in the ordered sequence of the lengths of $\overline{qp_i}$, $i = 1, \ldots, N$.). Show how the k-th nearest neighbor diagram is related to $\text{Vor}_{k-1}(S)$ and $\text{Vor}_k(S)$, for $k > 1$.

7. Show that in the L_1-metric, the farthest point Voronoi diagram of an N-point set $S(N \ge 4)$ consists of at most *four* regions, in all cases.

8. Consider the inversion of the plane that maps $\mathbf{v}$ to $\mathbf{v}' = \mathbf{v}(1/|\mathbf{v}|^2)$.
   (a) Given a line of equation $ax + by + c = 0$, obtain the equation of its inversive image (a circle by the origin).
   (b) Prove formally that if a circle C contains the origin in its interior, the interior of C maps to the exterior of its inversive image.

9. Prove formally Lemma 6.4.

10. Given an N-point set S, let l be the perpendicular bisector of $\overline{p_i p_j}$, for $p_i, p_j \in S$. Suppose that l contains a sequence $v_1, v_2, \ldots, v_s$ of Voronoi vertices, ordered from one end to the other. (Each v_i is the circumcenter of some triplet of points of S.) Show that if the ray terminating at v_1 appears as an edge in $\text{Vor}_h(S)$, then the ray terminating at v_s appears as an edge in $\text{Vor}_{N-h}(S)$.

11. Prove Lemma 6.5.

12. *Seidel.* Let $S = \{(x_i, y_i): 1 \le i \le N\} \subset \mathbb{R}^2$ be a set of N points in the plane, S^1 be the z-projection of S onto the paraboloid of rotation $z = x^2 + y^2$, i.e., $S^1 = \{(x_i, y_i, x_i^2 + y_i^2): 1 \le i \le N\} \subset \mathbb{R}^3$. Consider P, the convex hull of S^1.
    (a) Show as directly as possible that the orthogonal projection on the (x, y)-plane

of the faces of P that are "below" P (near-side) coincides with the Delaunay triangulation of S.

(b) Generalize these results to higher dimensions.

13. *Seidel.* Consider the following generalization of Voronoi diagrams (known as Dirichlet Cell Complexes, Power Diagram, or Laguerre diagram). Let S be a set of N points in E^2. For each $p \in S$ let r_p be a real number associated with p. For each $p \in S$ let

$$GV_p(S) = \{x \in \mathbb{R}^2 : d(p, x)^2 + r_p \le d(q, x)^2 + r_q, \quad \forall q \in S\}$$

be the generalized Voronoi region of p (with respect to S).

Call $GVor(S) = \{GV_p(S) : p \in S\}$ the generalized Voronoi Diagram of S.

(a) Show that for each $p \in S$, $GV_p(S)$ is convex and has straight line edges.

(b) Show that in the case $r_p = 0$ for all $p \in S$, the generalized Voronoi Diagram turns out to be the ordinary Voronoi Diagram of S.

(c) Show that for some choice of the r_ps, $GV_p(S)$ can be empty for some $p \in S$.

(d) Give a $\theta(N \log N)$ algorithm to construct the generalized Voronoi Diagram of S.

(*Hint*: Use the ideas presented in Section 6.3.1 (higher-order Voronoi diagrams).)

14. The SMALLEST BOMB problem requires the determination of the smallest circle that encloses at least k of N given points in the plane. Give a polynomial time algorithm to solve this problem.

15. Prove that the Gabriel Graph of an N-point set S (see Section 6.5 for a definition) is constructed by removing from the Delaunay triangulation of S each edge that does not cross its dual Voronoi edge.

16. Prove relation (6.7) in Section 6.5.

17. *Lee.* Show that the Relative Neighborhood Graph of N points in the plane under L_1-metric remains a subgraph of the corresponding Delaunay Triangulation and give an algorithm to compute it.

18. Give an $O(N \log N)$-time algorithm to construct the medial axis of an N-edge convex polygon (see Section 6.5 for a definition of medial axis).

19. *Research problem.* Find a suitable problem transformation to prove by means of Ben-Or's approach that the MAXIMUM GAP problem (Section 6.4) on N values requires time $\Omega(N \log N)$ in the algebraic computation-tree model.

CHAPTER 7

Intersections

Much of the motivation for studying intersection problems stems from the simple fact that two objects cannot occupy the same place at the same time. An architectural design program must take care not to place doors where they cannot be opened or have corridors that pass through elevator shafts. In computer graphics, an object to be displayed obscures another if their projections on the viewing plane intersect. A pattern can be cut from a single piece of stock only if it can be laid out so that no two pieces overlap. The importance of developing efficient algorithms for detecting intersection is becoming apparent as industrial applications grow increasingly more ambitious: a complicated graphic image may involve one hundred thousand vectors, an architectural database often contains upwards of a million elements, and a single integrated circuit may contain millions of components. In such cases even quadratic-time algorithms are unacceptable.

Another reason for delving into the complexity of intersection algorithms is that they shed light on the inherent complexity of geometric problems and permit us to address some fundamental questions. For example, how difficult is it to tell whether a polygon is simple? Although one would be justified in investigating such a topic even if it had no practical applications, we will find no shortage of uses for the algorithms of this chapter. Because two figures intersect only if one contains a point of the other[1] it is natural that intersection algorithms should involve testing for inclusion. We may thus consider intersection problems to be natural extensions of the inclusion problems treated in the context of Geometric Searching in Chapter 2.

To gain more insight into this area, we shall now consider some salient applications in more detail.

---

[1] Depending on how boundary intersections are defined.

7.1 A Sample of Applications

7.1.1 The hidden-line and hidden-surface problems

A pivotal problem in computer graphics and one that has absorbed the energy of many researchers [Desens (1969); Freeman–Loutrel (1967); Galimberti–Montanari (1969); Loutrel (1970); Matsushita (1969); Newman–Sproull (1973); Sutherland (1966); Warnock (1969); Watkins (1970)], are the *hidden-line* and *hidden-surface problems*. A two-dimensional image of a three-dimensional scene is necessarily a projection. We may not, however, merely project each object onto the plane of the observer, for some objects may be partially or totally obscured from view. In order to produce a faithful display, those lines which a real observer cannot see must be eliminated from the picture. Figure 7.1 shows a scene before and after hidden lines have been removed.

One object obscures another if their projections intersect, so detecting and forming intersections is at the heart of the hidden-line problem. A considerable investment has been made in developing hardware to perform this task, which is particularly difficult in practice because of the real-time requirements of graphic display systems and the fact that objects are usually in motion.

In view of the effort that has gone into graphic hardware development, it is surprising that the complexity of the hidden-line problem has received so little study, for it is here that the potential gains are the greatest. Building a black box with a program implemented in microcode cannot in general achieve, even with the exploitation of parallelism, effects comparable to those achievable with innovative algorithmic design (where the speed-up improves with increasing problem size).

In many cases, particularly for vector graphic devices, scene components are represented as polygons. If the projections of two objects are the polygons P_1 and P_2 and P_1 lies nearer to the viewer than P_2, what must be displayed is P_1 and $P_2 \cap \bar{P_1}$ (obviously $P_2 \cap \bar{P_1}$ is the intersection of P_2 and the complement of P_1).

The basic computational problem in hidden line removal is thus to *form the intersection of two polygons*. In practice we must do more than this, since the image will consist of many separate polygons, all of which must be displayed,

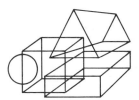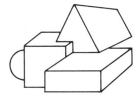

Figure 7.1   Elimination of hidden lines.

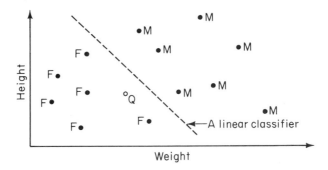

Figure 7.2   A two-variable classification problem.

but one may expect that the primitive remains pairwise intersection. Without an optimal algorithm for polygon intersection we cannot hope to perform hidden-line elimination efficiently. In this chapter we will obtain tight bounds for intersecting convex, star-shaped, and general polygons. The polygon problem is an example of the first type of intersection problem we will consider. To keep a desirable degree of generality, we shall refer to the generic category of geometric objects, which may be polygons, segments, polyhedra, etc., as the specific application demands. So we have

PROBLEM TYPE I.1 (CONSTRUCT INTERSECTION). Given two geometric objects, form their intersection.

7.1.2 Pattern recognition

One of the major techniques of pattern recognition is classification by *supervised learning*.[2] Given *N* points, each of which is identified as belonging to one of *m* samples, we wish to preprocess them so that a new point (*query point*) can be correctly classified. Figure 7.2 is a two-dimensional example in which the axes represent the weights and heights of a group of people of the same age. Males are designated by "*M*," females by "*F*," The point "*Q*" represents a person whose weight and height are known. Can we classify *Q* based on these quantities alone? What sort of decision rule should be used?

It is desirable, if possible, to obtain a *linear classifier*, that is, a linear function *f* such that a single computation (evaluation of a linear function and a comparison) will suffice to determine the sample to which *Q* belongs:

$$\textbf{if} f(x_Q, y_Q) > T \quad \textbf{then } Q \in M \quad \textbf{else } Q \in F.$$

In the above expression *T* is a threshold value. In *k* dimensions the locus

---

[2] Details on a number of geometric problems arising in pattern recognition can be found in Andrews (1972), Duda–Hart (1973), and Meisel (1972).

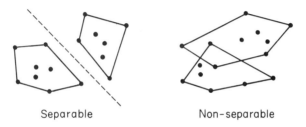

Separable   Non-separable

Figure 7.3   Two sets are separable if and only if their convex hulls are disjoint.

$f(x_1, \ldots, x_k) = T$ is a hyperplane; in two dimensions it is a straight line. A linear classifier performs well if it *separates* the two samples such that all points of M lie on one side and all points of F lie on the other.

Definition 7.1. Two sets are said to be *linearly separable* if and only if there exists a hyperplane H that separates them.

Determining the existence of a linear classifier is thus a matter of deciding whether the training samples are separable.
    Separability is a classical question in combinatorial geometry. A crucial criterion for linear separability is provided by the following theorem.

Theorem 7.1 [Stoer–Witzgall (1970), Theorem 3.3.9.]. *Two sets of points are linearly separable if and only if their convex hulls do not intersect.*

This theorem is illustrated in Figure 7.3 in the plane. Since we know that the convex hull of a finite point set is a convex polytope, linear separability is established by testing whether two convex polytopes intersect. The latter problem is an instance of the following problem type.

PROBLEM TYPE I.2 (INTERSECTION TEST). Given two geometric objects, do they intersect?

We shall see later that detection of intersection is frequently easier than the corresponding construction problem.

7.1.3 Wire and component layout

With circuit microminiaturization proceeding at a fantastic pace, the number of components on chips, conductors on boards, and wires in circuitry has grown to the point that such hardware cannot be designed without the aid of machines. The number of elements in a single integrated circuit may easily exceed the million and each must be placed by the designer subject to a variety of electronic and physical constraints. The programs that assist in this process are largely heuristic and often produce solutions that are not feasible because two components overlap or two conductors cross (see [Akers (1972); Hanan–Kurtzberg (1972); and Hanan (1975)]). Heuristic methods are used because some of the problems involved in component placement are *NP*-complete [Garey–Johnson–Stockmeyer (1976)]. The designs must therefore be subjected to exhaustive verification that involves pairwise comparisons of all items on the chip, an expensive and time-consuming operation. This motivates the following theoretically significant class of problems.

PROBLEM TYPE I.3 (PAIRWISE INTERSECTION). Given N geometric objects, determine whether any two intersect.

We will, of course, be looking for an algorithm that avoids testing each object against every other. The solution to a problem of this type, which we develop in Section 7.2.3 has extensive practical and theoretical applications (see also Chapter 8).

7.1.4 Linear programming and common intersection of half-spaces.

Linear programming can be viewed as a fourth type of intersection problem. The feasible region of a linear programming problem is the intersection of the half-spaces determined by its constraint set. The objective function is maximized at some vertex of this convex polyhedral region, which is specified in a different way from that studied in Chapter 3. Here we are not given a set of points among which the hull vertices are to be found, but rather a collection of half-spaces that bound the hull, and we are asked to find the vertices. Clearly, by constructing the common intersection of these N objects, we obtain a solution of the linear programming problem. However, all we need is the identification of one vertex for which the objective function is extremized (either maximized or minimized) and there is no reason to think that—even for small dimensionality—the *complete* construction of the polyhedral region is necessary.

In one dimension linear programming is trivial. It may be formulated as

$$\text{Maximize } ax + b \quad \text{subject to } a_i x + b_i \le 0, \quad i = 1, \ldots, N. \qquad (7.1)$$

The feasible region is either empty, an interval, or a half-line because it is an intersection of half-lines that extend either to $-\infty$ or $+\infty$.

Let L be the leftmost point of the positively extending half-lines and let R be the rightmost point of the negatively extending ones, If $L > R$, the feasible region is empty. If $L \le R$ it is the interval $[L, R]$. Clearly L and R can be found in linear time, so linear programming in one dimension is an $O(N)$ process. In higher dimensions, a linear programming problem can be solved by constructing the common intersection of half-spaces. However, the two problems are not equivalent as we shall see in more depth in Section 7.2.5.

We shall now discuss some basic algorithms for the solution of the problems outlined above. Most of the algorithms refer to problems of the
CONSTRUCT INTERSECTION type (Problem Type I.1), although, wher-
ever appropriate we shall contrast them with the corresponding "detection"
type problem. We begin with planar applications and shall later proceed to
three-dimensional instances. Needless to say, the state of knowledge on this
topic in relation to the number of dimensions mirrors a pattern by now
familiar in the whole of computational geometry.

7.2 Planar Applications

7.2.1 Intersection of convex polygons

In this section, by "polygon" we mean its boundary and its interior; the edge
cycle of the polygon will be referred to explicitly as the polygon boundary. The
problem is stated as follows.

PROBLEM I.1.1 (INTERSECTION OF CONVEX POLYGONS). Given two
convex polygons, P with L vertices and Q with M vertices, form their
intersection.

We assume without loss of generality that $L \leq M$.

Theorem 7.2. *The intersection of a convex L-gon and a convex M-gon is a convex*
polygon having at most $L + M$ vertices.

PROOF. The intersection of P and Q is the intersection of the $L + M$ interior
half planes determined by the two polygons.   □

The boundary of $P \cap Q$ consists of alternating chains of vertices of the two
polygons, interspersed with points at which the boundaries intersect. (See
Figure 7.4.)
The obvious method for forming the intersection is to proceed around P,
edge by edge, finding by rote all of the boundary intersection points involving
Q (of which there are at most two along any edge) and keeping a list of the
intersection points and vertices along the way. This will take $O(LM)$ time
because each edge of P will be checked against every edge of Q to see if they
intersect. Of course, it is natural to attempt to exploit the convexity of the
polygons to reduce the computational work.
One approach consists in subdividing the plane into regions, in each of
which the intersection of the two polygons can be easily computed. The
creation of new geometric objects to solve a given problem is not infrequent in

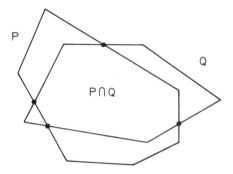

Figure 7.4   Intersection of convex polygons.

computational geometry; the reader may recall its crucial use in the various geometric searching techniques illustrated in Chapter 2. In our case, the simplest partition of the plane—the one induced by a star of rays—is adequate [Shamos–Hoey (1976)]. Specifically, we choose an arbitrary point O in the plane. (This point O may be chosen at infinity; indeed the original method of Shamos–Hoey is based on choosing O as the point at infinity on the y-axis of the plane.) From O we trace all the rays to the vertices of the polygons P and Q. These rays partition the plane into sections. Our goal is to sort the collection of these rays in angular order around O. Fortunately, this is readily done. Indeed, if O lies inside a polygon, say P, then the rays to the vertices of P are obviously sorted around O as the vertices, whose order is given. If, on the other hand, O lies outside P, then the vertices of P form two chains—both comprised between the two vertices of P reached by the supporting lines from O—and for each chain the corresponding rays from O are sorted in angular order. Thus, in all cases the sectors determined by the vertices of P are sorted in angular order in time linear in the number of vertices. Once these two sequences (one for P and one for Q) are available, they can be merged in linear time.

The key observation is that the intersection of each sector with a convex polygon is a *quadrilateral* (a *trapezoid* when O is a point at infinity). Thus, within any single sector, the intersection of P and Q is an intersection of two quadrilaterals, which can be found in constant time. The resulting pieces can be fitted together in a single linear-time sweep over the sectors. A clean-up pass is then executed to remove spurious vertices that occur at boundaries between sectors.

Theorem 7.3. *The intersection of a convex L-gon and a convex M-gon can be found in* $\theta(L + M)$ *time.*

In Figure 7.5 we illustrate the method, where the choice of O is the one originally proposed by Shamos–Hoey. In this case the angular sectors are

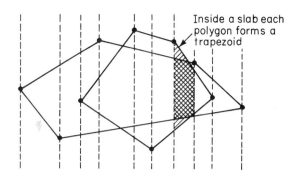

Figure 7.5   Slabs defined by the vertices of two convex polygons.

appropriately called slabs and the polar angle at O is appropriately replaced by the abscissa x.

Another approach is, surprisingly, an elegant refinement [O'Rourke–Chien–Olson–Naddor (1982)] of the crude method readily disposed of at the beginning of this section. The basic idea is informally described as follows. Suppose that $P \cap Q \neq \emptyset$, and consider the polygon $P^* \triangleq P \cap Q$. The boundary of P^* is an alternating sequence of fragments of the boundaries of P and of Q. If one such fragment is part of the boundary of, say, P, then a fragment of the boundary of Q wraps around it in the exterior of P^* (see Figure 7.6). It is descriptively appropriate, due to the convexity of both fragments, to call each pair of facing fragments a "sickle," of which the *internal* and *external chains* are naturally defined. A sickle is comprised between two intersection points, called *initial* and *terminal*, according to the common orientation of P and Q. A vertex (of P or Q) is said to belong to a sickle either (i) if it lies between the initial and terminal intersection points (on either chain), or (ii) if it is the terminus of the edge of the internal chain containing the terminal point of the sickle (see Figure 7.7). Note that there are vertices which belong to two sickles.

To develop the advancing mechanism let $(p_1, p_2, \ldots, p_L)$ and $(q_1,$

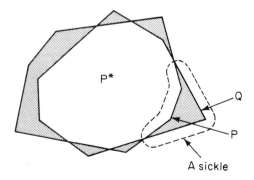

Figure 7.6   Two intersecting polygons. The "sickles" are shown shaded.

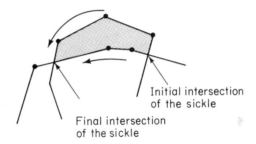

Figure 7.7   Sickle nomenclature. Solid vertices belong to the shaded sickle.

$q_2, \ldots, q_M$) be the counterclockwise cycles of the vertices of P and Q, respectively (so that a polygon is conventionally "to the left" of its boundary). Suppose that we are advancing on both boundaries and that p_i and q_j are the current vertices on the two polygons; moreover, the *current edges* are those terminating at p_i and q_j; there are four distinct situations to consider, illustrated in Figure 7.8. (Other situations are reducible to these four, by exchanging the roles of P and Q.) Let $h(p_i)$ denote the half-plane determined by $\overline{p_{i-1}p_i}$ and containing P ($h(q_j)$ is analogously defined). Obviously P^* is contained in the intersection of $h(p_i)$ and $h(q_j)$, that is, in the shaded regions in Figure 7.8.

The idea is *not to advance on the boundary* (either of P or of Q) *whose current*

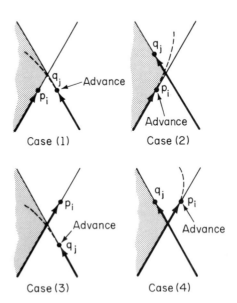

Figure 7.8   Illustration of the advancing mechanism. We advance on one boundary or the other depending upon the relative positions of the current vertices p_i and q_j (current edges are shown as solid lines).

edge may contain a yet to be found intersection. Thus, in Case (2) we shall advance on P, since the current edge $\overline{q_{j-1}q_j}$ of Q may contain a yet to be found intersection; by an analogous reasoning, in Case (3) we shall advance on the boundary of Q. In Case (4), all intersections on the current edge of P have already been found, while the current edge of Q may still contain an undiscovered intersection; thus we advance on P. Finally, in Case (1) the choice is arbitrary, and we elect to advance on Q. The handling of these four cases completely specifies the advancing mechanism of the method, embodied by a subroutine ADVANCE. The sentence "ADVANCE implements Case (j)" (for $j = 1, 2, 3, 4$) means that the advancing action corresponding to Case (j) is correctly executed.

Neglecting, for clarity and simplicity, the degenerate cases (when the boundary of P contains a vertex of Q, and *vice versa*) which require special care,[3] we can now formalize the algorithm. Note that p_L and q_M immediately precede p_1 and q_1, respectively (this means that, conventionally $p_0 = p_L$, and $q_0 = q_M$).

procedure CONVEX POLYGON INTERSECTION
begin $i := j := k := 1$;
    repeat
        begin if $(\overline{p_{i-1}p_i}$ and $\overline{q_{j-1}q_j}$ intersect$)$ **then** print intersection;
               ADVANCE (*either i or j is incremented*);
               $k := k + 1$
        end
    until $k = 2(L + M)$;
    if (no intersection has been found) **then**
        begin if $p_i \in Q$ **then** $P \subseteq Q$
               else if $q_j \in P$ **then** $Q \subseteq P$
                   else $P \cap Q = \varnothing$
        end
end.

We now establish the correctness of the algorithm. If p_i and q_j belong to the same sickle, the rules of ADVANCE guarantee that the terminal intersection of the sickle will be reached (as the *next* intersection), since we never advance on a boundary whose current edge may contain the sought intersection point. This, in turn, guarantees that once both current vertices belong to the same sickle, all the sickles will be constructed in succession. To complete the proof, we must show that the algorithm finds an intersection if P and Q intersect. Note that intersections occur in pairs, so there are at least two intersection points. Let $\overline{p_{i-1}p_i}$, the current edge of P, be an edge containing an intersection point v with $\overline{q_{r-1}q_r}$, so that $q_r \in h(p_i)$ (see Figure 7.9). If w is the next intersection point in the sickle beginning at v, there are two significant vertices of Q:

---

[3] Details can be found in the original paper by [O'Rourke *et al.* (1982)].

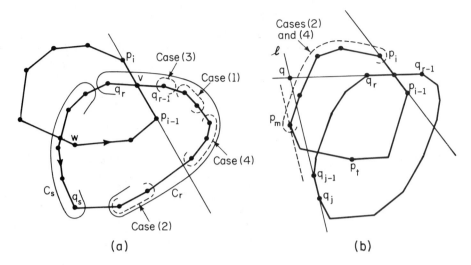

Figure 7.9 Illustration for the proof that the algorithm finds an intersection point if P and Q intersect.

vertex q_r, already defined, and vertex q_s, which is the farthest in $h(p_i)$ from the line determined by $\overline{p_{i-1}p_i}$. The boundary of Q is partitioned by these two vertices into two directed chains C_r and C_s, which respectively have q_r and q_s as their terminal vertices. Letting, as usual, $\overline{q_{j-1}q_j}$ denote the current edge of Q, we distinguish two cases:

(i) $q_j \in C_r$. ADVANCE successively implements sequences (all possibly empty) of Cases (2), (4), (1) and (3) as illustrated in Figure 7.9(a). The march takes place on Q while $\overline{p_{i-1}p_i}$ remains stationary, until v is found.

(ii) $q_j \in C_s$ (Figure 7.9(b)). Let l be the line determined by $\overline{q_{j-1}q_j}$. There are two supporting lines of P parallel to l. Let p_m be a vertex belonging to one such supporting line, which is first reached when tracing the boundary of P from p_i. Clearly, ADVANCE (Cases (2) and (4)) marches from p_i to p_m while $\overline{q_{j-1}q_j}$ remains stationary. Then if $p_m \notin h(q_j)$, ADVANCE further proceeds on to the first vertex $p_t \in h(q_j)$. In either case ADVANCE currently points to q_j and to a vertex p^* of P (either $p^* = p_m$ or $p^* = p_t$) lying inside $h(q_j)$. Note that the current situation is not different from that where a single vertex q replaces the subchain $(q_r, q_{r+1}, \ldots, q_{j-1})$: q is the intersection of the lines determined by $\overline{q_{r-1}q_r}$ and $\overline{q_{j-1}q_j}$, respectively. Thus it is as if q_j and p^* belong to the same sickle, in which case ADVANCE reaches the terminal point of this sickle.

If an edge like $\overline{p_{i-1}p_i}$ exists (i.e., the boundaries of P and Q intersect) after $(L + M)$ ADVANCE steps it has certainly been reached; indeed, after $(L + M)$ steps the boundary of at least one polygon has been completely traversed and each polygon contains at least one edge like $\overline{p_{i-1}p_i}$. It is then easy to

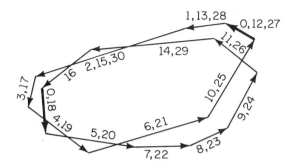

Figure 7.10   Illustration of the execution of the algorithm of O'Rourke *et al.* An edge bears the label *j* if the ADVANCE loop reaches it at its *j*-th iteration. Initial edges are shown in heavy line.

realize that at most $(L + M)$ additional ADVANCE steps are sufficient to obtain the boundary of the intersection of P and Q. Thus we conclude that, if after $2(L + M)$ ADVANCE steps the algorithm fails to find an intersection, the boundaries of the two polygons do not intersect at all. From a performance viewpoint, the work done by ADVANCE is completed in time $O(L + M)$; additional $O(L + M)$ steps are sufficient to decide, if necessary, among the alternatives $P \subseteq Q$, $Q \subseteq P$, or $P \cap Q = \emptyset$. Thus, this method provides an alternative proof of Theorem 7.3.

An illustration of the action of the algorithm is given in Figure 7.10. There an edge receives the label *j* if the *repeat* loop (ADVANCE step) reaches it at its *j*-th iteration. The two initial edges (one of P, the other of Q shown in heavy line) are correctly labelled with the integer zero.

7.2.2 Intersection of star-shaped polygons

Because star polygons are angularly simple, a property they share with convex polygons, we might suspect that their intersection can also be found quickly. This is not the case, as Figure 7.11 shows.

The intersection of P and Q is not a polygon itself, but is a union of many polygons. P and Q both have N vertices, and every edge of P intersects every edge of Q, so the intersection has on the order of N^2 vertices. This gives a (trivial) lower bound:

Theorem 7.4. *Finding the intersection of two star-shaped polygons requires* $\Omega(N^2)$ *time in the worst case.*

This means that the hidden line problem for arbitrary polygons must also take quadratic time in the worst case, since merely drawing the intersection requires that $\Omega(N^2)$ vectors be drawn. In the next section we shall explore the

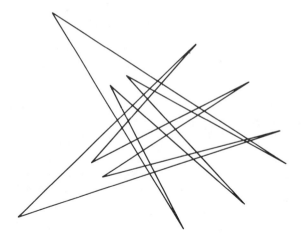

Figure 7.11   The intersection of two star-shaped polygons.

striking possibility that it may not be necessary to spend this much time if we
only want to know whether P and Q intersect at all.

7.2.3 Intersection of line segments

One of the major patterns in the study of computational geometry is that a
large collection of seemingly unrelated problems can be solved by the same
method if only their common algorithmic features can be isolated. The present
section shows how a diverse set of applications can be unified and reduced to
determining whether or not a set of N line segments in the plane are pairwise
disjoint.

PROBLEM I.2.1 (LINE-SEGMENT INTERSECTION TEST). Given N line
segments in the plane, determine whether any two intersect.

An algorithm for this problem would be quite instrumental for the applications of wire layout and component placement outlined in Section 7.1.3.
Before tackling Problem I.2.1 we describe additional significant applications
of it.

7.2.3.1 Applications

PROBLEM I.2.2 (POLYGON INTERSECTION TEST). Given two simple
polygons, P and Q, with M and N vertices, respectively, do they intersect?

As noted in Section 7.3 for convex polygons, if P and Q intersect, then
either P contains Q, Q contains P, or some edge of P intersects an edge of Q
(Figure 7.12).

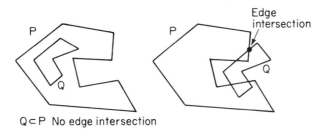

Q ⊂ P No edge intersection

Figure 7.12   Either $P \subset Q$, $Q \subset P$, or there is an edge intersection.

Since both P and Q are simple, any edge intersections that occur must be between edges of different polygons. Let $T(N)$ be the time required to solve Problem I.2.1. We can then detect any edge intersection between P and Q in $T(M + N)$ operations. If no intersection is found, we still must test whether $P \subset Q$ or $Q \subset P$.

If P is internal to Q, then *every* vertex of P is internal to Q, so we may apply the single-shot point-inclusion test of Theorem 2.1 in $O(N)$ time, using *any* vertex of P. If this vertex is found to lie outside Q, we can learn by the same method in $O(M)$ time whether $Q \subset P$. Therefore we have

Theorem 7.5. *Intersection of simple polygons is linear-time transformable to line-segment intersection testing*:

POLYGON INTERSECTION TEST $\propto_N$ LINE-SEGMENT
INTERSECTION TEST

We have already seen that the complexity of algorithms that deal with polygons can depend on whether the polygons are known to be simple or not. For example, the convex hull of a simple polygon can be found in $\theta(N)$ time (Theorem 4.12) but $\Omega(N \log N)$ is a lower bound for nonsimple polygons. It would be useful, therefore, to have an algorithmic test for simplicity.

PROBLEM I.2.3 (SIMPLICITY TEST). Given a polygon, is it simple?

Simple and nonsimple polygons are shown in Figure 7.13.
    Since a polygon is simple if and only if no two of its edges intersect, we have immediately that

SIMPLICITY TEST $\propto_N$ LINE-SEGMENT INTERSECTION TEST.

7.2.3.2 Segment intersection algorithms

Suppose we are given N intervals on the real line and wish to know whether any two overlap. This can be answered in $O(N^2)$ time by inspecting all pairs of intervals, but a better algorithm based on sorting comes to mind almost immediately. If we sort the $2N$ endpoints of the intervals and designate them as either right or left, then the intervals themselves are disjoint if and only if the

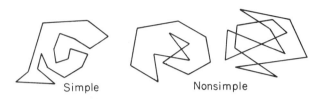

Figure 7.13   Simple and nonsimple polygons.

endpoints occur in alternating order: $L\,R\,L\,R\ldots R\,L\,R$ (Figure 7.14). This test can be performed in $O(N\log N)$ time.

The two questions we will want to deal with are whether this algorithm can be improved and whether it generalizes to two dimensions.

To show a lower bound, we will exhibit a correspondence between the segment overlap problem and a familiar and basic question in set theory, ELEMENT UNIQUENESS (given N real numbers, are they all distinct?—Section 5.2). We have shown earlier that this problem is easily solved in time $O(N\log N)$ by means of sorting, and although there is no simple way to show that sorting is required, by the results of Dobkin and Lipton (1976) and of Ben-Or (1983) we know that time $\Omega(N\log N)$ is necessary in the algebraic decision-tree computation model (Corollary 5.1). We now show that

$$\text{ELEMENT UNIQUENESS} \propto_N \text{INTERVAL OVERLAP.}$$

Given a collection of N real numbers x_i, these can be converted in linear time to N intervals $[x_i, x_i]$. These intervals overlap if and only if the original points were not distinct and this proves

Theorem 7.6. $\theta(N\log N)$ *comparisons are necessary and sufficient to determine whether N intervals are disjoint, if only algebraic functions of the input can be computed.*

How severe is the restriction to algebraic functions? For one thing, it forbids the use of the "floor" function, which, as we saw in Section 6.4, is a very powerful operation. (Recall that the use of this function allowed Gonzalez (1975) to solve the MAXGAP problem in time $\theta(N)$.) No techniques are

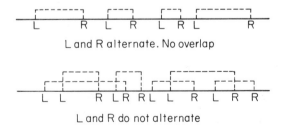

Figure 7.14   Detecting interval overlap.

known that would enable us to prove Theorem 7.6 if the "floor" function were allowed, but it may be conjectured that the lower bound would be unaffected by its introduction. Theorem 7.6 applies *a fortiori* in all dimensions. In particular, to test the disjointness of N segments in the plane requires $\Omega(N \log N)$ operations.

Let us explore what really happens when we sort to detect overlap. The motivation for doing this is that there is no natural total order on line segments in the plane, so a generalization based solely on sorting will have to fail. If we are able to understand the essential features of the algorithm, though, we may be able to extend it to the plane.

Overlap occurs if and only if two segments contain some common point. Each point on the real line has associated with it a set consisting of the intervals that cover it. This defines a function $C: \mathbb{R} \to \{0, 1\}^N$ from the reals to subsets of $\{1, \ldots, N\}$. The value of function C can change only at the $2N$ endpoints of the intervals. If the cardinality of $C(x)$ ever exceeds one, an overlap has occurred. To detect this, we first sort the endpoints and set up a rudimentary data structure that consists of just a single integer, the number of intervals covering the current abscissa. Scanning the endpoints from left to right, we INSERT an interval into the data structure when its left endpoint is encountered and DELETE it when its right endpoint is passed. If an attempt is ever made to INSERT when the data structure already bears the value one, an overlap has been found; otherwise, no overlap exists. Since the processing of each endpoint in this way takes only constant time, after sorting the checking process requires no more than linear time.

In two dimensions we are obliged to define a new order relation and make use of a more sophisticated data structure.[4] Consider two nonintersecting line segments s_1 and s_2, in the plane. We will say that s_1 and s_2 are *comparable at abscissa* x if there exists a vertical line by x that intersects both of them. We define the relation *above at* x in this way: s_1 is above s_2 at x, written $s_1 >_x s_2$, if s_1 and s_2 are comparable at x and the intersection of s_1 with the vertical line lies above the intersection of s_2 with that line.[5] In Figure 7.15 we have the following relationships among the line segments $s_1, s_2, s_3,$ and s_4:

$$s_2 >_u s_4, \qquad s_1 >_v s_2, \qquad s_2 >_v s_4, \qquad \text{and } s_1 >_v s_4.$$

Segment s_3 is not comparable with any other segment.

Note that the relation $>_x$ is a total order, which changes as the vertical line is swept from left to right. Segments enter and leave the ordering, but it always remains total. The ordering can change in only three ways:

---

[4] For purposes of discussion, we will assume that no segment is vertical and that no three segments meet in a point. If either of these conditions is not met, the algorithms we develop will be longer in detail but not in asymptotic running time.

[5] This order relation and the algorithm derived from it were developed by Dan Hoey. A proof of the validity of the algorithm was supplied by M. Shamos. These results were reported jointly in [Shamos–Hoey (1976)].

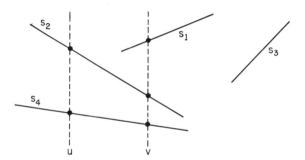

Figure 7.15   An order relation between line segments.

1. *The left endpoint of segment s is encountered.* In this case s must be added
   to the ordering.
2. *The right endpoint of s is encountered.* In this case s must be removed from
   the ordering because it is no longer comparable with any others.
3. *An intersection point of two segments s_1 and s_2 is reached.* Here s_1 and s_2
   exchange places in the ordering.

Note that a necessary condition for two segments s_1 and s_2 to intersect is
that there is some x for which s_1 and s_2 are consecutive in the ordering $>_x$.
This immediately suggests that the sequence of the intersections of the segment set with the vertical line by x (i.e., the relation $>_x$) contains *all* the
relevant information leading to the segment intersection, and that the natural
device for this class of problems is of the "plane-sweep" type (see Section
1.2.2, for a general formulation of this algorithmic technique). We recall that a
plane-sweep technique makes use of two basic data structures: the *sweep-line
status* and the *event-point schedule*.

As noted above, the sweep-line status is a description of the relation $>_x$,
that is, it is a sequence of items (segments). Referring to the previously
described mechanism, whereby this relation $>_x$ changes at a finite set of
abscissae in the plane sweep, it is clear that the data structure $\mathscr{L}$ implementing
the sweep-line status must support the following operations:

a. INSERT$(s, \mathscr{L})$. Insert segment s into the total order maintained by $\mathscr{L}$.
b. DELETE$(s, \mathscr{L})$. Delete segment s from $\mathscr{L}$.
c. ABOVE$(s, \mathscr{L})$. Return the name of the segment immediately above s in the
   ordering.
d. BELOW$(s, \mathscr{L})$. Return the name of the segment immediately below s in
   the ordering.

The postulated data structure is known as a *dictionary* (see Section 1.2.3), and
all of the above operations can be performed in time at most logarithmic in its
size. In reality by deploying a threaded dictionary (a straightforward use of
pointers, if the address of s is known) ABOVE(s) and BELOW(s) can be
performed in constant time.

As regards the event-point schedule, we note that, in order to report all

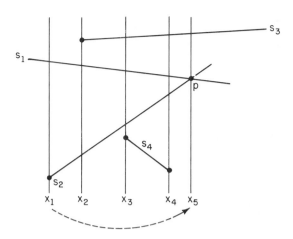

Figure 7.16   Intersection point p is detected for $x = x_1$ when segments s_1 and s_2 are found adjacent for the first time. However event abscissae x_2, x_3, and x_4 must be processed before point p at x_5 is handled.

intersections, we must be able to maintain the relation $>_x$ as the vertical line sweeps the plane. As noted earlier (see 1, 2, and 3 above), the relation $>_x$ is modified only at distinguished abscissae, which are either left or right endpoints of segments, or segment intersections. While the left and right endpoints of the segments are all given *a priori*, an intersection point—found by the plane sweep—dynamically produces an event, which, in general, must be recorded and handled by the algorithm at its correct time in the future. Notice, indeed, that several event abscissae may have to be processed between the time an intersection point is detected and the time the corresponding event is processed. In Figure 7.16 we have an illustration of this situation: here, intersection point p is detected at x_1 (when segments s_1 and s_2 become adjacent); however, abscissae x_2, x_3, and x_4 must be processed before its corresponding event x_5 is handled. We recognize therefore that, in order to report all the intersections, the data structure $\mathscr{E}$ implementing the event-point schedule must support the following operations ($\mathscr{E}$ stores the total order of the event points):

a. MIN($\mathscr{E}$). Determine the smallest element in $\mathscr{E}$ and delete it.
b. INSERT($x, \mathscr{E}$). Insert abscissa x into the total order maintained by $\mathscr{E}$.

In addition to this essential operation, we also require that $\mathscr{E}$ supports the operation

c. MEMBER($x, \mathscr{E}$). Determine if abscissa x is a member of $\mathscr{E}$.

The postulated data structure is a *priority queue* (see Section 1.2.3) and as is well-known, it supports all three above operations in time logarithmic in its size.

We are now ready to outline the algorithm: as the vertical line sweeps the plane, at each event point data structure $\mathscr{L}$ is updated and all pairs of segments that become adjacent in this update are checked for intersection. If an intersection is detected for the first time, it is both reported (printed out) and its abscissa is inserted into the event point schedule $\mathscr{E}$. Less informally we have (here $\mathscr{A}$ is a queue, internal to the procedure) [Bentley–Ottmann (1979)]:

procedure LINE SEGMENT INTERSECTION
1. **begin** sort the $2N$ endpoints lexicographically by x and y and place them into priority queue $\mathscr{E}$:
2. $\mathscr{A} := \varnothing$;
3.   **while** $(\mathscr{E} \neq \varnothing)$ **do**
4.   **begin** $p := \mathrm{MIN}(\mathscr{E})$;
5.   **if** (p is left endpoint) **then**
6.   **begin** $s :=$ segment of which p is endpoint;
7.   INSERT$(s, \mathscr{L})$;
8. $s_1 :=$ ABOVE$(s, \mathscr{L})$;
9. $s_2 :=$ BELOW$(s, \mathscr{L})$;
10.   **if** (s_2 intersects s) **then** $\mathscr{A} \Leftarrow (s_1, s)$;
11.   **if** (s_2 intersects s) **then** $\mathscr{A} \Leftarrow (s, s_2)$
          end
12.   **else if** (p is a right endpoint) **then**
13.   **begin** $s_1 :=$ ABOVE$(s, \mathscr{L})$;
14. $s_2 :=$ BELOW$(s, \mathscr{L})$;
15.   **if** (s_1 intersects s_2) **then** $\mathscr{A} \Leftarrow (s_1, s_2)$;
16.   DELETE$(s, \mathscr{L})$
            end
17.   **else** (*p is an intersection*)
18.   **begin** $(s_1, s_2) :=$ segments of which p is intersection (*with $s_1 =$ ABOVE(s_2) to the left of p*)
19. $s_3 :=$ ABOVE$(s_1, \mathscr{L})$;
20. $s_4 :=$ BELOW$(s_2, \mathscr{L})$;
21.   **if** (s_3 intersects s_2) **then** $\mathscr{A} \Leftarrow (s_3, s_2)$;
22.   **if** (s_1 intersects s_4) **then** $\mathscr{A} \Leftarrow (s_1, s_4)$;
23.   interchange s_1 and s_2 in $\mathscr{L}$
          end;
          (*the detected intersections must now be processed*)
24.   **while** $(\mathscr{A} \neq \varnothing)$ **do**
25.   **begin** $(s, s') \Leftarrow \mathscr{A}$;
26. $x :=$ common abscissa of s and s';
27.   **if** (MEMBER$(x, \mathscr{E}) =$ FALSE) **then**
28.   **begin** output (s, s');
29.   INSERT$(x, \mathscr{E})$
            end
        end
      end
    end.

That no intersection is missed by this algorithm follows from the observation that only adjacent segments may intersect and that all adjacencies are correctly examined at least once. Moreover, each intersection is reported exactly once, for, once its abscissa is inserted into $\mathscr{E}$, the test in line 27 prevents undesired repetitions.[6]

From the performance viewpoint, we note: line 1 (initial sorting) is completed in time $O(N \log N)$. Program blocks 6–11, 13–16, and 18–23 run each in time $O(\log N)$, since each operation on $\mathscr{L}$ can be performed within this time bound, in the worst-case, and a test for intersection uses constant time. For each event, i.e., each execution of the main *while*-loop at Line 3, these three program segments are mutually exclusive. If K denotes the number of intersections encountered by the algorithm, the main *while*-loop is executed exactly $(2N + K)$ times. The only item which still deserves careful attention concerns data structure $\mathscr{E}$, i.e., how many times is the test embodied in Line 27 going to be performed? Superficially, we note that a *single* intersection may be rediscovered a large number of times; however, if, from an accounting viewpoint, we charge *each* intersection pair to the execution of the main *while*-loop that produces it, we realize that the total number of intersection pairs produced is $O(N + K)$. Thus, Line 27 is executed $O(N + K)$ times, and each execution runs in time $O(\log (N + K)) = O(\log N)$, since $K \leq \binom{N}{2} = O(N^2)$. We conclude therefore that the time globally used by the main *while*-loop is $O((N + K) \log N)$, which clearly dominates the initial sorting step. We have therefore

Theorem 7.7 [Bentley–Ottmann (1979)]. *The K intersections of a set of N line segments can be reported in time $O((N + K) \log N)$.*

We remark—although this presentation reverses our usual pattern—that the above algorithm solves the following problem.

PROBLEM I.1.2 (LINE SEGMENT INTERSECTION). Given N line segments, determine all their intersections.

More recent research has improved upon the result of Theorem 7.7. Indeed, Chazelle (1983a) has shown that Problem I.1.2 can be solved in time $O(K + N \log^2 N / \log \log N)$; Mairson and Stolfi (1983) have shown that, given two sets A and B, each consisting of nonintersecting segments, so that $|A| + |B| = N$, time $O(K + N \log N)$ suffices to report all K intersections of segments of A with segments of B. These results will be further illustrated in Section 7.4, Notes and Comments, at the end of this chapter. Note that time $\Omega(K + N \log N)$ is a lower bound for this problem, since the term $\Omega(N \log N)$ is due to Theorem 7.6 and $O(K)$ is trivially due to the size of the output.

---

[6] To handle the hypothetical situation of intesections sharing the same abscissa, it is sufficient to design $\mathscr{E}$ to maintain the lexicographic order of the pairs (x, y) of the intersections.

If we abandon the idea of finding all intersections and content ourselves with the corresponding detection problem ("Find an intersection if one or more exist"), things are greatly simplified. Indeed, we may proceed as in the one-dimensional algorithm. Instead of holding just a single object, however, the data structure must be able to maintain the total order of the segments intersected by the sweep-line and coincides with the dictionary $\mathscr{L}$ used earlier. The main difference between the construction and the detection algorithms lies in the data structure for the event-point schedule, which in the latter case is a simple array, storing just the originally given $2N$ endpoints. We have the following algorithm.

> **procedure** LINE-SEGMENT INTERSECTION TEST
> **begin** sort the $2N$ endpoints lexicographically by x and y and place them in the
> real array POINT$[1:2N]$;
>   **for** $i := 1$ **until** $2N$ **do**
>   **begin** $p :=$ POINT$[i]$;
> $s :=$ segment of which p is endpoint;
>   **if** (p is a left endpoint) **then**
>   **begin** INSERT$(s, \mathscr{L})$;
> $s_1 :=$ ABOVE$(s, \mathscr{L})$;
> $s_2 :=$ BELOW$(S, \mathscr{L})$;
>   **if** (s_1 intersects s) **then** output (s_1, s)
>   **if** (s_2 intersects s) **then** output (s_2, s)
>   **end**;
>   **else** ($*p$ is the right endpoint of $s*$)
>   **begin** $s_1 :=$ ABOVE$(s, \mathscr{L})$;
> $s_2 :=$ BELOW$(s, \mathscr{L})$;
>   **if** (s_1 intersects s_2) **then** output (s_1, s_2);
>   DELETE$(s, \mathscr{L})$
>   **end**
>   **end**
> **end**.

The following theorem establishes the correctness of the algorithm.

Theorem 7.8. *Procedure* LINE-SEGMENT INTERSECTION TEST *finds an intersection if one exists.*

PROOF. Since the algorithm only reports an intersection if it finds one, it will never falsely claim that two segments cross, and we may turn our attention to the possibility that an intersection exists but remains undetected. We will show that the algorithm correctly finds the leftmost intersection point q_L. Indeed, if q_L coincides with a left extreme p of a segment, q_L is found when processing p. Therefore, assume that q_L is not the left extreme of a segment. To the left of q_L data structure $\mathscr{L}$ contains a correct representation of the relation $>_x$. Since there is a small nonempty interval of abscissae to the left of q_L where two segments intersecting at q_L become adjacent, then q_L is discovered as claimed.   □

The above algorithm, though simple, has some curious properties. Even though the leftmost intersection is always found, it is not necessarily the *first* intersection to be found. (The reader may test his understanding of the algorithm by characterizing exactly which intersection is found first.) Also, since the algorithm only performs $O(N)$ intersection tests, it may fail to find some intersections. The main *if*-block—the intersection test—is obviously carried out in time $O(\log N)$, in the worst case. Thus we conclude with the following theorem (note again, that optimality follows from Theorem 7.6).

Theorem 7.9. *Whether any two of N line segments in the plane intersect can be determined in* $\theta(N \log N)$ *time, and this is optimal.*

An immediate consequence of this result is the following.

Corollary 7.1. *The following problems can be solved in time* $O(N \log N)$, *in the worst case.*

PROBLEM I.2.2 (POLYGON INTERSECTION TEST). Do two given polygons intersect?

PROBLEM I.2.3 (SIMPLICITY TEST). Is a given polygon simple?

PROBLEM I.2.4 (EMBEDDING TEST). Does a straight-line embedding of a planar graph contain any crossing edges?

Further results have been obtained using variations of Algorithm LINE SEGMENT INTERSECTION and can be found in [Shamos–Hoey (1976)]. We just quote:

Corollary 7.2. *Whether any two of N circles intersect can be determined in* $O(N \log N)$ *time.*

7.2.4 Intersection of half-planes

The construction of the common intersection of N half-planes consists of finding the region of the solutions of a set of N linear inequalities (*constraints*) of the form

$$a_i x + b_i y + c_i \leq 0 \qquad i = 1, 2, \ldots, N. \tag{7.2}$$

A solution of (7.2) is conventionally called a *feasible* solution, and their locus is called *feasible region*. An illustration of the problem is given in Figure 7.17.

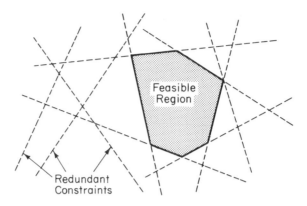

Figure 7.17   The feasible region of a set of linear constraints.

There is a simple quadratic algorithm for forming the intersection of N half-planes. Let us assume that we already have the intersection of the first i half-planes. This is a convex polygonal region of at most i sides, though it is not necessarily bounded. Intersecting this region with the next half-plane is a matter of slicing the region with a line and retaining either the right or left piece. This can be done in $O(i)$ time in the obvious way. The total work required is $O(N^2)$, but the algorithm has the advantage of being on-line.

Let us see if any improvement is possible with a divide-and-conquer approach for the problem formally defined as follows.

PROBLEM I.1.3 (INTERSECTION OF HALF-PLANES). Given N half-planes $H_1, H_2, \ldots, H_N$ form their intersection

$$H_1 \cap H_2 \cap \cdots \cap H_N.$$

Because the intersection operator is associative, the terms may be parenthesized any way we wish:

$$(H_1 \cap \cdots \cap H_{N/2}) \cap (H_{N/2+1} \cap \cdots \cap H_N). \tag{7.3}$$

The term in parentheses on the left is an intersection of $N/2$ half-planes and hence is a convex polygonal region of at most $N/2$ sides. The same is true of the term on the right. Since two convex polygonal regions each having k sides can be intersected in $O(k)$ time by Theorem 7.3, the middle intersection operation in (7.3) can be performed in $O(N)$ time. This suggests the following recursive algorithm:

procedure INTERSECTION OF HALF-PLANES

Input: N half-planes defined by directed line segments
Output: Their intersection, a convex polygonal region.

1. Partition the half-planes into two sets of approximately equal sizes.
2. Recursively form the intersection of the half-planes in each subproblem.

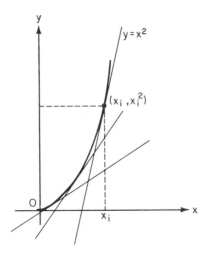

Figure 7.18   Illustration of how SORTING can be transformed to HALF-PLANES.

3. Merge the subproblems solutions by intersecting the two resulting convex polygonal regions.

If $T(N)$ denotes the time used to form the intersection of N half-planes by this algorithm, we have

$$T(N) = 2T(N/2) + O(N) = O(N \log N).   \qquad (7.4)$$

We summarize as follows.

Theorem 7.10. *The intersection of N half-planes can be found in $\theta(N \log N)$ time, and this is optimal.*

PROOF. The upper bound follows from equation (7.4). To prove the lower bound we show that

$$\text{SORTING} \propto_N \text{HALF-PLANES.}$$

Given N real numbers $x_1, \ldots, x_N$, let H_i be the half-plane containing the origin that is defined by the line tangent in the point (x_i, x_i^2) to the parabola $y = x^2$, i.e., the line $y = 2x_i x - x_i^2$ (see Figure 7.18). The intersection of these half-planes is a convex polygonal region whose successive edges are ordered by slope. Once this region is formed, we may read off the x_i's in sorted order.   □

The general result of Theorem 7.10 has a significant application expressed by the following.

Corollary 7.3. *The common intersection of N convex k-gons can be found in $O(Nk \log N)$ time.*

PROOF. It is straightforward to achieve $O(Nk \log Nk)$ time by intersecting the Nk left half-planes of the polygons. To reduce this time we will treat the polygons as N units rather than as a collection of Nk edges. Let $T(N, k)$ be the time sufficient to solve the problem. The intersection of $N/2$ convex k-gons is a convex polygon of at most $Nk/2$ sides. Two of these can be intersected in time ckN, for some constant c. So by recursively splitting the problem as in INTERSECTION OF HALF-PLANES, we have

$$T(N, k) = 2T(N/2, k) + ckN = O(Nk \log N). \qquad \square$$

7.2.5 Two-variable linear programming

The preceding formulation (7.2) of the problem of intersecting N half-planes is (deceptively) similar to the standard formulation of the linear programming problem [Gass (1969)] in two variables. Indeed, the latter is formulated as follows.

PROBLEM I.1.4 (2-VARIABLE LINEAR PROGRAMMING). Minimize $ax + by$, subject to

$$a_i x + b_i y + c_i \le 0, \qquad i = 1, \dots, N. \tag{7.5}$$

Again, the feasible region of the linear programming problem is the set of points (x, y) that satisfy the constraints in (7.5), and is obviously the intersection of N half-planes. The objective function defines a family of parallel lines $ax + by + \lambda = 0$, where λ is a real parameter. The lines of this family that support the feasible region pass through the vertices that minimize and maximize the objective function. (See Figure 7.19.)

Since we already known how to find lines of support of a convex polygon in $O(\log N)$ time (Section 3.3.6) we have the transformation

2-VARIABLE LINEAR PROGRAMMING $\propto_N$ HALF-PLANES

and, using the result of the preceding section (Theorem 7.10) we obtain

Theorem 7.11. *A linear program in two variables and N constraints can be solved in $O(N \log N)$ time. Once this has been done, a new objective function can be maximized or minimized in $O(\log N)$ time.*

At first we compare this performance with that of the Simplex algorithm [Gass (1969)]. Assuming we are given an initial feasible solution, Simplex operates by moving from vertex to vertex on the feasible region, spending $O(N)$ time for each move. It is easy to see that, in the worst case, Simplex will have to visit every vertex, for a total of $O(N^2)$ time. (In this respect it is very similar to Jarvis's algorithm in Section 3.3.2.) Furthermore, in order to maximize a new objective function, Simplex must inspect every constraint, so it will use $O(N)$ time. In other words, Simplex is not optimal.

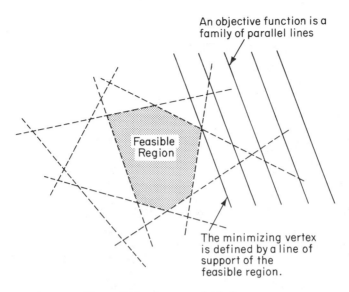

Figure 7.19   A two-variable linear program.

We must point out that explicit construction of the feasible polytope is not a viable approach to linear programming in higher dimensions because the number of vertices can grow exponentially with dimension (see Section 3.1).

One of the striking features of the Simplex algorithm is that, while its worst case is known to be exponential in dimension (and quadratic in two variables), its expected behavior is almost always excellent in practice. A similar behavior is exhibited by a method based on the half-plane intersection for a wide class of inputs [Bentley–Shamos (1978)]. The same principle is invoked that was presented in Section 4.1.1, namely, if the expected sizes of the subproblem solutions are small, the merge step of the divide and conquer algorithm can be performed in sublinear time. This will be the case if many of the half-planes are *redundant*, i.e., do not form edges of the intersection polygon. We now show that most of the half-planes in a random problem can be expected to be redundant.

It is intuitively straightforward and fairly easy to exhibit a reasonable probability distribution for random points in the plane; it is less obvious how to model a random selection of half-planes. However, by resorting to the fundamental geometric transform, introduced in Section 1.3.3, known as *polarity* (of which we shall make an important use in Section 7.3), we make the following observations. Polarity is an involutory point ↔ straight-line transform. With each line we associate a half-plane (the one which contains the origin). Suppose now we draw a set S of N points at random in some domain (say, the unit disk) and we dualize them to half-planes. The points on the hull of S dualize to a set of lines, each of which contains an edge of the common intersection of the half-planes. Thus, recalling the results quoted in Section

4.1.1, the expected number of *nonredundant* half-planes in a set of N half-planes is in this case $O(N^p), p < 1$, and similar results hold for other reasonable random models. It follows that a linear average-case algorithm for intersecting N half-planes results. This leads immediately to an $O(N)$ expected-time algorithm for linear programming in two variables. We see that the expected number of redundant half-planes—those that do not define faces of the feasible region—is very large, which may, in part, account for the excellent observed behavior of the Simplex Method.

However, we should not content ourselves with an $O(N)$ *expected* time algorithm that constructs the feasible region, since—as we noted in Section 7.1.4—such construction *is not necessary* to solve the linear programming problem! For many years this observation did not bear any fruit, until recently an extremely clever technique, which capitalizes on it, has been independently discovered by Megiddo (1983) and by Dyer (1984). This technique (which is readily adaptable to three dimensions and has been generalized by Megiddo (1983) to an arbitrary number of dimensions) not only discards redundant constraints (i.e., those that are also irrelevant to the half-plane intersection task) but also those constraints that are guaranteed not to contain a vertex extremizing the objective function (referred to as the *optimum vertex*). The technique is based on applying a linear transformation to the points of the plane so that the objective function becomes equal to one of the two coordinates, say, the ordinate of the plane. At this point the problem reduces to finding the extreme value of a piece-wise linear convex function of the abscissa. The key feature is that, since all we want is the identification of the extremizing value x_0, we need not explicitly construct this convex function, which remains implicitly defined by a set of linear constraints.

More specifically, the original problem[7]

$$\begin{cases} \text{minimize} & ax + by \\ \text{subject to} & a_i x + b_i y + c_i \leq 0 \qquad i = 1, 2, \ldots, N \end{cases} \qquad (7.6)$$

can be transformed by setting $Y = ax + by$ and $X = x$, as follows (here, since both a and b are not simultaneously 0, we assume without loss of generality $b \neq 0$):

$$\begin{cases} \text{minimize} & Y \\ \text{subject to} & \alpha_i X + \beta_i Y + c_i \leq 0 \qquad i = 1, 2, \ldots, N \end{cases} \qquad (7.7)$$

where $\alpha_i = (a_i - (a/b)b_i)$, and $\beta_i = b_i/b$. In this new form we have to compute the smallest Y of the vertices of the convex polygon P (feasible region) determined by the constraints (see Figure 7.20). To avoid the construction of the entire boundary of P, we proceed as follows. Depending upon whether β_i is zero, negative, or positive we partition the index set $\{1, \ldots, N\}$ into sets I_0, I_-, I_+, respectively. All constraints whose index is in I_0 are vertical lines (i.e.,

---

[7] In the present formulation we assume, without loss of generality, that the objective function is to be *minimized*.

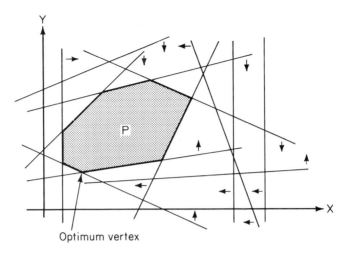

Figure 7.20   After the coordinate transformation, we must find the smallest ordinate of the feasible region.

parallel to the Y-axis) and determine the feasible interval for X as follows (see Figure 7.21):

$$u_1 \le X \le u_2$$
$$u_1 = \max\{-c_i/\alpha_i \colon i \in I_0\}$$
$$u_2 = \min\{-c_i/\alpha_i \colon i \in I_0\}.$$

On the other hand letting $-(\alpha_i/\beta_i) \triangleq \delta_i$ and $-(c_i/\beta_i) \triangleq \gamma_i$ all constraints in I_+ are of the form

$$Y \le \delta_i X + \gamma_i \qquad i \in I_+$$

so that they collectively define a piecewise linear upward-convex function $F_+(x)$ of the form

$$F_+(X) \triangleq \min_{i \in I_+}(\delta_i X + \gamma_i).$$

Similarly, the constraints in I_- collectively define a piecewise linear downward-convex function $F_-(x)$ of the form

$$F_-(X) = \max_{i \in I_-}(\delta_i X + \gamma_i).$$

Then we obtain the transformed constraint $F_-(X) \le Y \le F_+(X)$, and, since we have a minimizing linear program, $F_-(X)$ is our objective function. Our problem so becomes

$$\begin{cases} \text{minimize} & F_-(X) \\ \text{subject to} & F_-(X) \le F_+(X). \\ & u_1 \le X \le u_2. \end{cases}$$

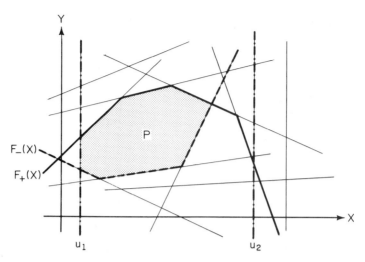

Figure 7.21   Illustration of $F_-(X)$, $F_+(X)$, u_1, and u_2 in the reformulation of the linear program.

The new situation is illustrated in Figure 7.21, where the relationship between u_1, u_2, $F_-(X)$ and $F_+(X)$ and the boundary of P is shown.

The primitive used by the technique is the evaluation, at a selected value X' of X, of $F_-(X')$, $F_+(X')$, and of their slopes on either side of X'. (We denote by $f_-^{(L)}(X')$ and $f_-^{(R)}(X')$ the slopes of $F_-(X')$ to the left and right of X', respectively; $f_+^{(L)}$ and $f_+^{(R)}$ are analogously defined.) We now show that this primitive, called *evaluation*, can be executed in time $O(N)$. Referring, for brevity, to $F_-(X)$ we have

$$F_-(X') = \max_{i \in I_-}(\delta_i X' + \gamma_i)$$

which can be evaluated in time proportional to $|I_-| = O(N)$. If there is only one value i_0 of i that achieves $F_-(X')$, then $f_-^{(L)}(X') = f_-^{(R)}(X') = \delta_{i_0}$; otherwise (if there are two such values i_1 and i_2) we have that $f_-^{(L)}(X') = \min(\delta_{i_1}, \delta_{i_2})$ and $f_-^{(R)}(X') = \max(\delta_{i_1}, \delta_{i_2})$ since F_- is downward-convex.

We now claim that given a value X' of X in the range $[u_1, u_2]$, we are able to reach one of the following conclusions in time $O(N)$:

(i) X' is infeasible, and there is no solution to the problem;
(ii) X' is infeasible and we know on which side of X' (right or left) any feasible value of X may lie;
(iii) X' is feasible and we know on which side of X' the minimum of $F_-(X)$ lies;
(iv) X' achieves the minimum of $F_-(X)$.

Indeed, if the function $H_-(X) = F_-(X) - F_+(X)$ is positive at X', then X' is infeasible. By considering the slopes of $F_-(X)$ and $F_+(X)$ at X' we have (see

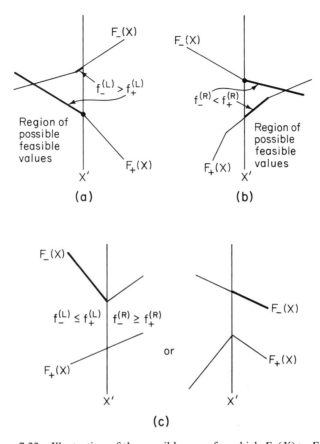

Figure 7.22   Illustration of the possible cases for which $F_-(X) > F_+(X)$.

Figure 7.22): if $f_-^{(L)}(X') > f_+^{(L)}(X')$, then $H(X)$ is increasing at X' and a feasible X can only be to the left of X' (Figure 7.22(a), Case (ii)); if $f_-^{(R)}(X') < f_+^{(R)}(X')$ then, by an analogous argument, a feasible X can only be to the right of X' (Figure 7.22(b), Case (ii)); finally if $f_-^{(L)}(X') \le f_+^{(L)}(X')$ and $f_-^{(R)}(X') \ge f_+^{(R)}(X')$ then $H(X)$ achieves its minimum at X' and the problem is not feasible (Figure 7.22(c), Case (i)).

Suppose now that $H(X') \le 0$, i.e., X' is feasible. Again, we leave it now as an exercise to decide on the basis of the four slopes $f_-^{(L)}(X'), f_-^{(R)}(X'), f_+^{(L)}(X')$, and $f_+^{(R)}(X')$ between cases (iii) and (iv).

A strategy now begins to emerge. We should try to choose abscissa X' where the evaluation takes place so that, if the algorithm does not immediately terminate, at least *a fixed fraction* α of the currently active constraints can be eliminated from contention, or "pruned" (i.e., each pruned constraint is assured not to contain the extremal vertex). If this objective is achieved, then in $\log_{1/(1-\alpha)} N$ stages the size of the set of active constraints becomes sufficiently small to allow a direct solution of the problem. With this hypothesis, at

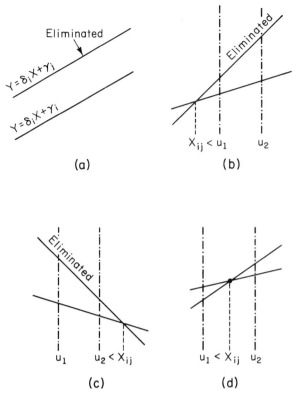

Figure 7.23   For each pair of constraints we check if any of the four illustrated cases occurs, in which case one constraint is eliminated.

the i-th stage the size of the active constraint set is at most $(1 - \alpha)^{i-1} N$ and the required processing is completed in time at most $K(1 - \alpha)^{i-1} N$ for some constant K. Thus the overall running time $T(N)$ is upper bounded by

$$T(N) \le \sum_{i=1}^{\log_{1/(1-\alpha)} N} K(1 - \alpha)^{i-1} N < \frac{KN}{\alpha},$$

i.e., it is linear in N and thus optimal! We shall now show that the value $\alpha = \frac{1}{4}$ can be achieved.

At the generic stage, let I_- and I_+ be the index sets as defined earlier, with $|I_+| + |I_-| = M$ (i.e., the stage has M active constraints, partitioned into two sets S_+ and S_-). We partition each of S_+ and S_- into pairs of constraints, with possibly at most one member per set remaining single. Let $i, j \in I^+$ and refer to Figure 7.23. If $\delta_i = \delta_j$, the corresponding straight lines are parallel and one can be immediately eliminated (in this case the one with larger value of γ) (Figure 7.23(a)). Otherwise, letting X_{ij} denote the abscissa of their intersection, we have the following three cases: If $X_{ij} < u_1$ we eliminate the constraint with the larger value of δ (Figure 7.23(b)); if $X_{ij} > u_2$ we eliminate the constraint with the smaller value of δ (Figure 7.23(c)); if $u_1 \leq X_{ij} \leq u_2$, we retain X_{ij} with no elimination. We then process in an analogous manner the constraint set S_-. For all pairs, neither member of which has been eliminated, we compute the abscissa of their intersection.

Thus, if k constraints have been immediately eliminated, we have obtained a set of $\lfloor (M - k)/2 \rfloor$ intersection abscissae, and this task is completed in $O(M)$ time. Next we apply to the set of abscissae a linear time median finding algorithm [Blum–Floyd–Pratt–Rivest–Tarjan (1973); Schönhage–Paterson–Pippenger (1976)] and obtain their median $\bar{X}$. The value $\bar{X}$ is the abscissa where the "evaluation" takes place; this operation, obviously, runs in time $O(M)$. If $\bar{X}$ is not the extremizing abscissa, then half of the computed X_{ij}'s lie in the region which is known not to contain the optimum. For each X_{ij} in this region, we can eliminate one member of the pair in a straight-forward manner. (For example, if X_{ij} is to the left of $\bar{X}$ and the optimum is to the right of it, then of the two lines intersecting at X_{ij} we eliminate the one with the smaller slope, etc.) This concludes the stage, with the result that at least $k + \lceil \lfloor (M - k)/2 \rfloor / 2 \rceil \geq \lfloor M/4 \rfloor$ constraints have been eliminated. This substantiates the earlier claim that $\alpha = \frac{1}{4}$, and we have the following surprising result.

Theorem 7.12. *A linear program in two variables and N constraints can be solved in optimal $\theta(N)$ time.*

As mentioned earlier this technique can be modified for the corresponding three-dimensional problem yielding again an optimal $\theta(N)$-time algorithm. The reader is referred to the previously cited papers by Megiddo and by Dyer for the details of this remarkable method.

Remark 1. The general approach outlined above can be applied, as shown by Megiddo, to slightly different situations; one such example is the calculation of the minimum enclosing circle of a N-point set $S = \{p_1, p_2, \ldots, p_N\}$, with $p_i = (x_i, y_i)$. As we discussed in Section 6.4, the determination of the minimum enclosing circle of S is expressed as the computation of

$$\min_{x,y} \max_{i=1}^{N} (x_i - x)^2 + (y_i - y)^2. \tag{7.8}$$

Introducing a variable z, expression (7.8) can be reformulated as the following mathematical programming problem:

$$\begin{cases} \text{minimize} & z \\ \text{subject to} & z \geq (x_i - x)^2 + (y_i - y)^2 \quad i = 1, 2, \ldots, N. \end{cases} \tag{7.9}$$

(Note that the z-coordinate of the solution gives the square of the radius of the circle.) The problem (7.9) is not a linear programming problem, since the constraints are quadratic. However the generic constraint can be rewritten as

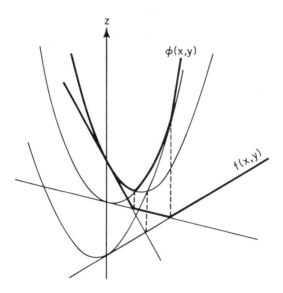

Figure 7.24   Illustration of a two-dimensional cut of the surfaces $z = f(x, y)$ and $z = \varphi(x, y) = f(x, y) + (x^2 + y^2)$.

$$z \geq -2x_i x - 2y_i y + c_i + (x^2 + y^2), \tag{7.10}$$

where $c_i = x_i^2 + y_i^2$. We observe that the right side of (7.10) consists of a linear expression $-2x_i x - 2y_i y + c_i$, dependent upon i, and of a common quadratic term $(x^2 + y^2)$. The set of linear constraints

$$z \geq -2x_i x - 2y_i y + c_i \qquad i = 1, 2, \ldots, N \tag{7.11}$$

defines a polyhedron, whose boundary is described by $z = f(x, y)$, with f a downward-convex function. Since also $(x^2 + y^2)$ is a downward-convex function, so is $x^2 + y^2 + f(x, y) \triangleq \varphi(x, y)$ (being the sum of two downward-convex functions). It is a simple exercise to verify that the surface $z = \varphi(x, y)$ resembles a polyhedral surface, in the sense that it has vertices resulting from the intersection of three[8] arcs of parabola lying in vertical planes (edges) and that the faces are portions of surfaces of paraboloids. For a two-dimensional illustration, see Figure 7.24. It is also a simple analysis to verify that the previously described method of constraint elimination is entirely applicable to this situation. The only significant difference is that the vertex w achieving the minimum value of z (among the vertices) may not be desired minimum (while it is always the minimum in the linear programming case). Thus the search is completed by inspecting the "edges" of $z = \varphi(x, y)$ incident on w, and if necessary, by locating the minimum on one of these edges.

The smallest enclosing circle can therefore be obtained by solving a variant of a three-dimensional linear programming problem. Combining this result

---

[8] Here we exclude, for simplicity, the presence of four cocircular points in S.

with the analogous of Theorem 7.12 for three dimensions we have the following improvement over Theorem 6.14

Corollary 7.4. *The smallest enclosing circle for an N-point set can be computed in optimum time $\theta(N)$.*

Remark 2. A particularly interesting application of linear programming is the problem of linear separability (considered here in E^d).

Definition 7.2. Two sets of points S_1 and S_2 in E^d are said to be *linearly separable* if there exists a linear $(d-1)$-dimensional variety π such that S_1 and S_2 lie on opposite sides of it. (In two dimensions π is a line, while in three dimensions it is an ordinary plane.)

We now show that linear separability is a linear programming problem. Specifically, given two sets of points in d-space, $S_1 = \{(x_1^{(i)}, \ldots, x_d^{(i)}): i = 1, \ldots, |S_1|\}$ and $S_2 = \{(x_1^{(j)}, \ldots, x_d^{(j)}): j = |S_1| + 1, \ldots, |S_1| + |S_2|\}$, with $|S_1| + |S_2| = N$ we seek a plane $p_1 x_1 + \cdots + p_d x_d + p_{d+1} = 0$ satisfying the conditions

$$\begin{cases} p_1 a_1^{(i)} + \cdots + p_d a_d^{(i)} + p_{d+1} \leq 0, & \text{for } 1 \leq i \leq |S_1| \\ p_1 a_1^{(i)} + \cdots + p_d a_d^{(i)} + p_{d+1} \geq 0, & \text{for } |S_1| + 1 \leq i \leq N. \end{cases}$$

This is clearly a linear program, which, by the preceding results can be solved in time $O(N)$ in two and three dimensions.

7.2.6 Kernel of a plane polygon

In Section 2.2.1 we showed that finding a point in the kernel of a star-shaped polygon is an essential step in the preprocessing needed to answer the corresponding inclusion question. At that time, we postponed the development of a kernel algorithm until the necessary tools were available. The problem is stated as follows.

PROBLEM I.1.5 (KERNEL). Given an N-vertex (not necessarily simple) polygon in the plane, construct its kernel.

We first note that the kernel is the intersection of N half-planes. Indeed, each edge of a polygon P determines a half-plane in which the kernel must lie. (Figure 7.25). These half-planes are referred to as the interior half-planes (or left half-planes, with reference to a counterclockwise traversal of the boundary). It has been shown [Yaglom–Boltyanskii (1961), p. 103] that the kernel of a polygon is the intersection of its left half-planes. Thus immediately we have

$$\text{KERNEL} \propto_N \text{HALF-PLANES},$$

and using the results of Section 7.2.4 (Theorem 7.10) we obtain

Corollary 7.5. *The kernel of an N-gon can be found in $O(N \log N)$ time.*

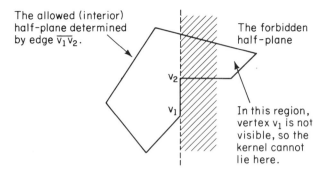

Figure 7.25   Each edge of P determines an allowed half-plane.

Note though, that the lower bound of $\Omega(N \log N)$ proved in Theorem 7.10 does not apply to the kernel problem because the edges of a simple polygon cannot be in arbitrary positions, and the transformability from sorting fails. In other words, there is no reason to believe that any more than linear time is required to find the kernel; moreover, we shall now show that such an algorithm exists [Lee–Preparata (1978)].

The input polygon P is represented by a sequence of vertices $v_0, v_1, \ldots, v_{N-1}$, with $N \geq 4$, and $e_i = \overline{v_{i-1} v_i}$ for $i = 1, 2, \ldots, N - 1$ is the edge of the polygon connecting vertices v_{i-1} and $v_i (e_0 = \overline{v_{N-1} v_0})$. For ease of reference we shall describe P by a circular list of vertices and intervening edges as $v_0 e_1 v_1 e_2 \cdots e_{N-1} v_{N-1} e_0 v_0$. We also assume that the boundary of P is directed counterclockwise, that is, the interior of P lies to the left of each edge. A vertex v_i is called *reflex* if the internal angle at v_i is $> 180°$, and *convex* otherwise; without loss of generality, we assume there is no internal angle equal to $180°$. If the kernel $K(P)$ is nonempty, the output will also be in the form of a sequence of vertices and edges of $K(P)$.

The algorithm scans the vertices of P in order and constructs a sequence of convex polygons $K_1, K_2, \ldots, K_{N-1}$. Each of these polygons may or may not be bounded. We shall later show (Lemma 7.1) that K_i is the common intersection of the half-planes lying to the left of the directed edges $e_0, e_1, \ldots, e_i$. This result has the obvious consequences that $K_{N-1} = K(P)$ and that $K_1 \supseteq K_2 \supseteq \cdots \supseteq K_i$; the latter implies that there is some $r_0 > 1$ such that K_i is unbounded or bounded depending upon whether $i < r_0$ or $i \geq r_0$, respectively.

Notationally, if points w_i and w_{i+1} belong to the line containing the edge e_{s_i} of P, then $w_i e_{s_i} w_{i+1}$ denotes the segment of this line between w_i and w_{i+1} and directed from w_i to w_{i+1}. When a polygon K_i is unbounded, two of its edges are half-lines; so, Λew denotes a half-line terminating at point w and directed like edge e, while $we\Lambda$ denotes the complementary half-line.

During the processing, the boundary of K is maintained as a doubly-linked list of vertices and intervening edges. This list will be either linear or circular, depending upon whether K_i is unbounded or bounded, respectively. In the first case, the first and last item of the list will be called the *list-head* and *list-tail*, respectively.

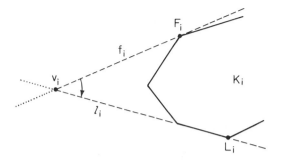

Figure 7.26 Illustration of the definition of F_i and L_i.

Among the vertices of K_i we distinguish two vertices F_i and L_i, defined as follows. Consider the two lines of support of K_i through vertex v_i of P (Figure 7.26). Let f_i and l_i be the two half-lines of these lines that contain the points of support, named so that the clockwise angle from f_i to l_i in the plane wedge containing K_i is no greater than π (Figure 7.26). Vertex F_i is the point common to f_i and K_i which is farthest from v_i, L_i is similarly defined. These two vertices play a crucial role in the construction of K_{i+1} from K_i.

If P has no reflex vertex, then P is convex and trivially $K(P) = P$. Thus let v_0 be a reflex vertex of P. We can now describe the kernel algorithm.

Initial Step. (*K_1 is set equal to the intersection of the half-planes lying to the left of edges e_0 and e_1, see Figure 7.27*)

$$K_1 := \Lambda e_1 v_0 e_0 \Lambda;$$

$$F_1 := \text{point at infinity of } \Lambda e_1 v_0;$$

$$L_1 := \text{point at infinity of } v_0 e_0 \Lambda.$$

General Step. (From K_i to K_{i+1}.) We assume that the vertices of K_i are

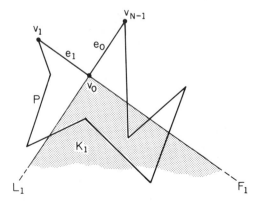

Figure 7.27 Illustration of polygon K_1.

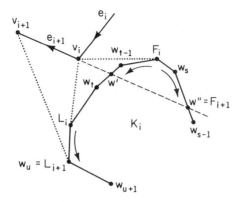

Figure 7.28   Action of the general step when v_i is reflex and F_i is on or to the right of $\Lambda e_{i+1}v_{i+1}$.

numbered consecutively as w_1, w_2, ..., counterclockwise. We distinguish several cases.

(1) *Vertex v_i is reflex* (see Figures 7.28 and 7.29)

    (1.1) *F_i lies on or to the right of $\Lambda e_{i+1}v_{i+1}$* (Figure 7.28). We scan the boundary of K_i *counterclockwise from F_i* until either we reach a unique edge $\overline{w_{t-1}w_t}$ of K_i intersecting $\Lambda e_{i+1}v_{i+1}$ or we reach L_i without finding such an edge. In the latter case, we terminate the algorithm ($K(P) = \varnothing$). In the former case, we take the following actions:

      (i) we find the intersection w' of $\overline{w_{t-1}w_t}$ and $\Lambda e_{i+1}v_{i+1}$;

      (ii) we scan the boundary of K_i *clockwise from F_i*, until either we reach an edge $\overline{w_{s-1}w_s}$ intersecting $\Lambda e_{i+1}v_{i+1}$ at a point w'' (this is guaranteed if K_i is bounded) or (only when K_i is unbounded) we reach the list-head without finding such an edge. In the first case, letting $K_i = \alpha w_s \cdots w_{t-1}\beta$ (where α and β are sequences of alternating edges and vertices), we set $K_{i+1} := \alpha w''e_{i+1}w'\beta$; in the second case ($K_i$ is unbounded) we must test whether K_{i+1} is bounded or unbounded. If the slope of $\Lambda e_{i+1}v_{i+1}$ is comprised between the slopes of the initial and final half-lines of K_i, then $K_{i+1} := \Lambda e_{i+1}w'\beta$ is also unbounded. Otherwise we begin scanning the boundary of K_i *clockwise from the list-tail* until an edge $\overline{w_{r-1}w_r}$ is found which intersects $\Lambda e_{i+1}v_{i+1}$ at a point w''; letting $K_i = \gamma w_{t-1}\delta w_r \eta$ we set $K_{i+1} := \delta w''e_{i+1}w'$ and the list *becomes circular*.

    The selection of F_{i+1} is done as follows: if $\Lambda e_{i+1}v_{i+1}$ has just one intersection with K_i, then $F_{i+1} := $ (point at infinity of $\Lambda e_{i+1}v_{i+1}$), otherwise $F_{i+1} := w''$. To determine L_{i+1}, we scan K_i counterclock-

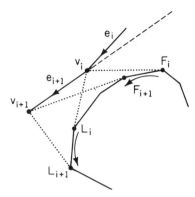

Figure 7.29   Action of the general step when v_i is reflex and F_i lies to the left of $\Lambda e_{i+1} v_{i+1}$.

wise from L_i until either a vertex w_u of K_i is found such that w_{u+1} lies to the left of $v_{i+1}(\overrightarrow{v_{i+1} w_u})\Lambda$, or the list of K_i is exhausted without finding such vertex. In the first case $L_{i+1} := w_u$; in the other case (which may happen only when K_i is unbounded) $L_{i+1} := L_i$.

(1.2) F_i *lies to the left of* $\Lambda e_{i+1} v_{i+1}$ (Figure 7.29). In this case $K_{i+1} := K_i$, but F_i and L_i must be updated. To determine F_{i+1}, we scan K_i *counterclockwise from* F_i until we find a vertex w_t of K_i such that w_{t+1} lies to the right of $v_{i+1}(\overrightarrow{v_{i+1} w_t})\Lambda$; we then set $F_{i+1} := w_t$. The determination of L_{i+1} is the same as in case (1.1).

(2)   *Vertex v_i is convex* (see Figures 7.30 and 7.31).

(2.1) L_i *lies on or to the right of* $v_i e_{i+1} \Lambda$. (Figure 7.30). We scan the boundary of K_i *clockwise from* L_i until either we reach a unique edge $\overline{w_{t-1} w_t}$ intersecting $v_i e_{i+1} \Lambda$ or we reach F_i without finding such an edge. In the latter case, we terminate the algorithm ($K(P) = \varnothing$). In

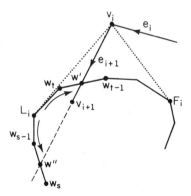

Figure 7.30   Action of the general step when v_i is convex and L_i lies on or to the right of $v_i e_{i+1} \Lambda$.

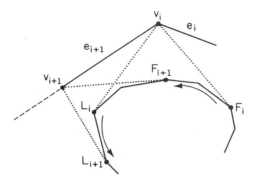

Figure 7.31   Action of the general step when v_i is convex and L_i lies to the left of $v_i e_{i+1} \Lambda$.

the former case, we take the following actions:

(i) we find the intersection w' of $\overline{w_{t-1} w_t}$ and $v_i e_{i+1} \Lambda$;
(ii) we scan the boundary of K_i *counterclockwise from* L_i until either we reach an edge $\overline{w_{s-1} w_s}$ intersecting $v_i e_{i+1} \Lambda$ at point w'' (guaranteed if K_i is bounded) or (only when K_i is unbounded) we reach the list-tail without finding such an edge. Letting $K_i = \alpha w_t \cdots w_{s-1} \beta$, in the first case we let $K_{i+1} := \alpha w' e_{i+1} w'' \beta$; in the the second case (K_i is unbounded) we must test whether K_{i+1} is bounded or unbounded. If the slope of $v_i e_{i+1} \Lambda$ is comprised between the slopes of the initial and final half-lines of K_i, then $K_{i+1} := \alpha w' e_{i+1} \Lambda$ is also unbounded. Otherwise we begin scanning the boundary of K_i *counterclockwise from the list-head* until an edge $\overline{w_{r-1} w_r}$ is found which intersects $v_i e_{i+1} \Lambda$ at a point w''; letting $K_i = \gamma w_{r-1} \delta w_t \eta$ we set $K_{i+1} := \delta w' e_{i+1} w''$ and the list *becomes circular*.

The selections of F_{i+1} and L_{i+1} depend upon the position of v_{i+1} on the half-line $v_i e_{i+1} \Lambda$ and upon whether $v_i e_{i+1} \Lambda$ has one or two intersections with K_i. We distinguish these two cases:

(2.1.1) $v_i e_{i+1} \Lambda$ intersects K_i in w' and w''. If $v_{i+1} \in [v_i e_{i+1} w']$ then F_{i+1} is selected as in case (1.2). Otherwise F_{i+1} is set to w'. If $v_{i+1} \in [v_i e_{i+1} w'']$ then L_{i+1} is set to w''. Otherwise L_{i+1} is selected as in case (1.1) except that we scan K_{i+1} *counterclockwise from* w''.
(2.1.2) $v_i e_{i+1} \Lambda$ intersects K_i in just w'. If $v_{i+1} \in [v_i e_{i+1} w']$, F_{i+1} is selected as in case (1.2); otherwise $F_{i+1} := w'$. L_{i+1} is set to the point at infinity of $v_i e_{i+1} \Lambda$.

(2.2) L_i *lies to the left of* $v_i e_{i+1} \Lambda$ (Figure 7.31). In this case $K_{i+1} := K_i$. F_{i+1} is determined as in (1.2). If K_i is bounded then L_{i+1} is determined as in case (1.1), otherwise $L_{i+1} := L_i$.

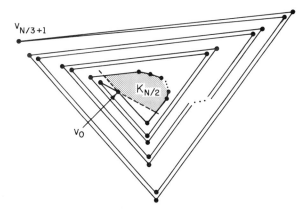

Figure 7.32 An instance where the unmodified algorithm terminates after $\theta(N^2)$ time. The N-sided polygon $K_{N/2}$ is shown shaded. Then the algorithm wraps around $K_{N/2}$ a total of $N/12$ times before reaching $V_{N/3+1}$, at which point it is discovered that $K(P) = \varnothing$.

The correctness of the algorithm is asserted by the following lemma, where we let H_j denote the half-plane lying to the left of line $\Lambda e_j \Lambda$.

Lemma 7.1. *The polygon K_{i+1} is the intersection of $H_0, H_1, \ldots, H_{i+1}$ for $i = 0, 1, \ldots, N-2$.*

PROOF. By induction. Note that K_1 is by definition the intersection of H_0 and H_1 (initial step of the algorithm). Assume inductively that $K_i = H_0 \cap H_1 \cap \cdots \cap H_i$. Then in all cases contemplated in the general step we constructively intersect K_i and H_{i+1}, thereby establishing the claim. □

While Lemma 7.1 guarantees that the preceding algorithm correctly constructs $K(P)$, a minor but important modification of the general step is needed to achieve efficiency. In fact, there are cases (see Figure 7.32) of polygons with empty kernel, for which $O(N^2)$ time could be used before termination. This can be avoided by an additional test to detect the situation illustrated in Figure 7.32. The details of this test are omitted here for brevity, and the reader is referred to [Lee–Preparata (1978)] for a complete account. it suffices to say that, with the modification in question, the algorithm is guaranteed to stop after its march on the boundary of P wraps around any point of the partial kernel K_i for an angle of 3π.

We now consider the performance of the algorithm. It is convenient to analyze separately the two basic types of actions performed by the kernel algorithm. The first concerns updating the kernel, by intersecting K_i with $\Lambda e_{i+1} \Lambda$ to obtain K_{i+1}; the second concerns updating F_i and L_i and consists of counterclockwise or *forward* scans of K_i to obtain the new vertices of support (notice however that in some cases, as (1.1) and (2.1), the update of K_i implicitly yields updates for one or the other of the support vertices).

We begin by considering intersection updates. In case (1.1), (when the algorithm does not terminate) we scan K_i starting from F_i both clockwise and counterclockwise (this scan also finds F_{i+1}). Let v_i be the total number of edges visited before finding the two intersections w' and w''. This process actually removes $v_i - 2$ edges from K_i (those comprised between w_s and w_{t-1} in Figure 7.28), and since each of the removed edges is collinear with a distinct edge of P, we have $\Sigma(v_i - 2) \leq N$. Thus, Σv_i, the total number of vertices visited by the algorithm in handling case (1.1), is bounded above by $3N$, i.e., it is $O(N)$. The same argument, with slight modifications, can be made for case (2.1).

Next, we consider those updates of the support vertices "F" and "L" which are not implicitly accomplished in the intersection process. These updates occur for "L" in all cases (1.1), (1.2), (2.1). (2.2) and for "F" in cases (1.2) and (2.2). Note that in all of these cases the vertices of support *advance* on the boundary of K_i. Let us consider, for example, the update of L in case (1.1): the other cases can be treated analogously. Consider the set of edges of K_{i+1} that the algorithm visits before determining L_{i+1}; the visit to the edge immediately following L_{i+1} is referred to as an *overshoot*. We immediately see that in handling case (1.1) the number of overshoots is globally $O(N)$, since there is at most one overshoot per vertex of P. Next, we claim that, ignoring overshoots, any edge is visited at most twice. In fact, assume that, when processing v_i, an edge is being visited for the third time. Due to the forward scan feature, this implies that the boundary of P wraps around K_i at least twice, i.e., there is some point $q \in K_i$ around which the boundary march wraps for angle $\geq 4\pi$, contrary to an earlier statement.

Thus the work performed in handling case (1.1)—as well as cases (1.2), (2.1) and (2.2)—is $O(N)$. Finally, the updates of "F" and "L" are all accomplished implicitly in finding w' and w''. Therefore, we conclude that the entire algorithm runs in time proportional to the number of vertices of P and we have proved

Theorem 7.13. *The kernel of an N-vertex polygon can be computed in optimal $\theta(N)$ time.*

7.3 Three-Dimensional Applications

7.3.1 Intersection of convex polyhedra

The problem is stated as follows.

PROBLEM I.1.6 (INTERSECTION OF CONVEX POLYHEDRA). Given two convex polyhedra P with L vertices and Q with M vertices, form their intersection.

By an argument identical to that of Theorem 7.2, the intersection $P \cap Q$ is itself a convex polyhedron. The crudest approach to the computation of $P \cap Q$, consists of testing each facet of P against each facet of Q to see if they intersect. If there are intersections, then it can be simply constructed; if no intersection is found, then $P \cap Q$ is empty, or one polyhedron is internal to the other. Since the surface of either polyhedron has the topology of a planar graph, by Euler's theorem on planar graphs it is easy to verify that such approach may take as much as $O((L + M)^2)$ time.

Once this approach is dismissed as unattractive, the next attempt may be the possible adaptation to three dimensions of the two methods for intersecting two convex polygons, illustrated in Section 7.2.1. The method of Shamos–Hoey (1976) (partition the plane and compute the intersection of P and Q in each region of the partition) could be generalized by slicing the space by a family of parallel planes, each passing by a vertex of $P \cup Q$. However, in each of the resulting slices of space (or "slabs," as they are frequently called), the portion of either polyhedron is not a simple geometric object as in the planar case. Indeed it is a "pancake" bounded on each side by two plane polygons, each of which may have $O(|\text{facets}|)$ edges. Thus, this approach would also lead to an $O(L + M)^2)$ algorithm. As for the method of [O'Rourke–Chien–Olson–Naddor (1982)], no generalization seems possible.

An entirely different approach—which, of course, can be successfully specialized to two dimensions—was proposed by Muller–Preparata (1978) and is discussed below. An alternative very attractive approach [Hertel–Mehlhorn–Mäntylä–Nievergelt (1985)], based on a space sweep, has been very recently proposed and will be sketched in the Notes and Comments section.

The central idea of the Muller–Preparata method is that if a point p in the intersection is known, the intersection can be obtained by known techniques via dualization. Indeed, if a point p in $P \cap Q$ exists, by a simple translation we may bring the origin to coincide with p. Thus, we may assume that both P and Q contain the origin of space *in their interior*. Letting x_1, x_2, and x_3 be the coordinates of the space, with each facet f of either P or Q we associate a half-space

$$n_1(f)x_1 + n_2(f)x_2 + n_3(f)x_3 \leq d(f). \tag{7.12}$$

Since any point of $P \cap Q$ satisfies all inequalities (7.12) for each facet f of either polyhedron, and the origin $(0, 0, 0)$ is a point in $P \cap Q$, it follows that we may normalize all inequalities so that $d(f) = 1$. (Note that we have assumed the origin in the interior of both P and Q.)

If we now interpret the triplet $(n_1(f), n_2(f), n_3(f))$ as a point, we have effectively established an involutory transformation between facet planes (i.e., a plane containing a facet) and points. This transformation is the conventional dualization (polarity) with respect to the unit sphere with center at the origin, discussed in Section 1.3.3. In this transformation, points at distance l from the origin are mapped to planes at distance $1/l$ from the origin and *vice versa*.

Notationally, the symbol $\delta(\ )$ is used to denote the transformation, i.e., $\delta(p)$ is the plane dual of a point p and $\delta(\pi)$ is the point dual of a plane π.

The dualization of a convex polyhedron is meaningful only if P contains the origin. Indeed in this case we have the following important property.

Lemma 7.2. *Let P be a convex polyhedron containing the origin and let S be the set of points that are the duals of the facet planes of P. Then each point p in the interior of P dualizes to a plane $\delta(p)$ that has no intersection with conv(S).*

PROOF. Let $a_i x_i + b_i x_2 + c_i x_3 \leq 1$ be the inequality defining the half-space bounded by facet f_i of P, for $i = 1, \ldots, |S|$. (Recall this half-space contains the origin.) Next consider a point $\bar{p} = (\bar{x}_1, \bar{x}_2, \bar{x}_3)$ in the interior of P. Then we have

$$a_i \bar{x}_1 + b_i \bar{x}_2 + c_i \bar{x}_3 < 1 \qquad i = 1, 2, \ldots, |S|.$$

We can now interpret these expressions as follows. The dual plane of $\bar{p}$, $\delta(\bar{p})$, has equation $\bar{x}_1 x_1 + \bar{x}_2 x_2 + \bar{x}_3 x_3 = 1$; if we consider point $\delta(\pi_i)$, the dual of the facet plane π_i containing f_i, the above expressions mean that all points $\delta(\pi_i)$ for $i = 1, 2, \ldots, |S|$, lie on the same side of $\delta(\bar{p})$ and none lies on it, i.e., $\delta(\bar{p})$ has no intersection with conv(S). $\square$

This result leads to the following conclusion.

Theorem 7.14. *If P is a convex polyhedron containing the origin, then so is its dual $P^{(\delta)}$.*

PROOF. Let π_i be the plane containing facet f_i. Define the point set $S \triangleq \{\delta(\pi_i): i = 1, \ldots, |S|\}$ and consider its convex hull CH(S). We claim that each point of S is a vertex of CH(S). Indeed, if $\delta(\pi_i)$ is *in the interior* of conv(S), then π_i has no intersection with P by Lemma 7.2, contrary to the assumption that π_i contains facet f_i. Thus CH(S) is the dual $P^{(\delta)}$ of P. $\square$

The previous proof also shows that, if P contains the origin, any plane external to P maps to a point internal to $P^{(\delta)}$ and conversely. Although the arguments can be developed for all polyhedra satisfying the hypothesis of Theorem 7.14, in the following discussion we assume that the origin is strictly *in the interior* of P (and, hence, of $P^{(\delta)}$). The reader may verify that this implies that both P and $P^{(\delta)}$ are bounded.

Let $V_P^{(\delta)}$ and $V_Q^{(\delta)}$ be the vertex sets of $P^{(\delta)}$ and $Q^{(\delta)}$, respectively. If we recall that each member of $V_P^{(\delta)} \cup V_Q^{(\delta)}$ is the dual of a facet plane of either polyhedron, and observe that $P \cap Q$ is the locus of the points that simultaneously lie in all the half-spaces determined by these facets and containing the origin, we conclude that $P \cap Q$ is the dual of the convex hull of $V_P^{(\delta)} \cup V_Q^{(\delta)}$. Hence, to compute $P \cap Q$ we may use the algorithm of Preparata–Hong (see Section 3.4.3) to find conv$(V_P^{(\delta)} \cup V_Q^{(\delta)})$ in time $O(N \log N)$ (where $N \triangleq L + M$), and

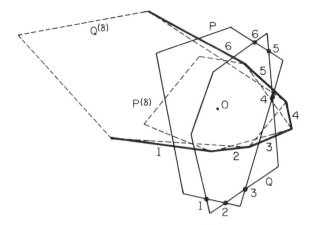

Figure 7.33   Illustration of the intersection of two convex polygons via the dualization approach. The two given polygons are shown in thin solid lines and their duals
(also convex polygons) in broken lines. The support lines of the convex hull are shown
in heavy lines, each of which dualizes to a vertex of the intersection (dual elements are
identically labelled).

upon taking the dual of the result we obtain the desired polyhedron. A two-
dimensional illustration of the approach is given in Figure 7.33.

A remark is here in order. The preceding techniques involve processing the
set of triplets $\{(n_1(f), n_2(f), n_3(f)): f$ is a facet either of P or of $Q\}$. We
intuitively interpret these triplets as points, but we could equally well have
interpreted them as planes. The only reason why the former interpretation is
preferred is that the technique receives in this case the full support of intuition,
which is otherwise less conspicuous in the latter. Therefore, polarity dualization does not transform the nature of the geometric objects (as the inversion
transform does—see Section 6.3.1.1) but it does aid our intuition. Thus,
$P \cap Q$ can be constructed by routine algorithms, once a point in it is known.
In general, when $P \cap Q$ is nonempty, how can we find one such point? We
shall proceed as follows.

Given a convex polyhedron P, we shall consider the set of its planes of
support parallel to the x_3-axis, briefly referred to as *vertical*. The intersection
of P with its vertical planes of support is, in general, an annular region $R(P)$ of
the surface of P which, in the absence of degeneracies, reduces to a cycle of
edges. The projection of $R(P)$ on the (x_1, x_2) plane is a convex polygon P^*
(Figure 7.34), which is the convex hull of the projections of the points of P on
this plane.

The region $R(P)$ is easily computed. For any face f_i of P the normal to f_i is
the vector $(n_1(f_i), n_2(f_i), n_3(f_i))$. It is perpendicular to f_i and points toward the
interior of P. Given any edge e of P, let f_i and f_j be its adjacent faces. Then
$e \in R(P)$ if and only if

$$n_3(f_i) \cdot n_3(f_j) \le 0. \tag{7.13}$$

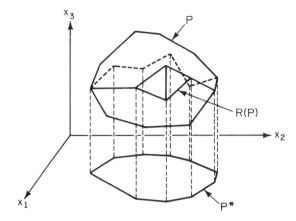

Figure 7.34   A convex polyhedron P, the annular region $R(P)$ and the projection polygon P^*.

Therefore, we begin by scanning the edge set of P until we find an edge e which belongs to $R(P)$, by verifying condition (7.13). At this point, we select one of the two vertices of e, call it v. Among the edges incident on v there are either one or two new edges, different from e, which belong to $R(P)$ and can be easily found. (We are assuming of course that the surface of P—a planar graph—be represented by the DCEL data structure described in Section 1.2.3.2.) Thus we can advance in the construction of $R(P)$, which will be completed upon re-encountering the initial edge e. Once $R(P)$ has been computed, P^* is trivially obtained. Thus we have the first two steps of the algorithm:

Step 1. Construct P^* and Q^*. (This step runs in time $O(N)$.)
Step 2. Find the intersection of the polygons P^* and Q^* (using any of the linear-time algorithms described in Section 7.1). If $P^* \cap Q^* = \varnothing$ halt, for $P \cap Q$ is also empty. (This step runs in time $O(N)$.)

If $P^* \cap Q^* \neq \varnothing$, let $\mathscr{D}$ be the closed (convex) domain bounded by $P^* \cap Q^*$. For each point $p = (\bar{x}_1, \bar{x}_2)$ in $\mathscr{D}$, consider the vertical line through p. The intersection of this with P is a segment $e_P(p)$, and similarly $e_Q(p)$ is defined ($e_P(p)$ and $e_Q(p)$ could each degenerate to a single point). Our objective is to decide if there is a point p in $\mathscr{D}$ for which $e_P(p)$ and $e_Q(p)$ overlap (see Figure 7.35). Assuming, without loss of generality, that there is a point $u \in \mathscr{D}$ for which the bottom of $e_P(u)$ lies above the top of $e_Q(u)$, respectively, we define the *near-sides* of P and Q as the loci of bottom$[e_P(u)]$ and top$[e_Q(u)]$, and a function $d(u)$—called the *vertical distance*—as follows:

$$d(u) = \text{bottom}[e_P(u)] - \text{top}[e_Q(u)]. \tag{7.14}$$

Thus we must decide whether there is a point $u \in \mathscr{D}$ such that $d(u) \leq 0$. This question can certainly be answered if we determine a point $\bar{u}$ which achieves

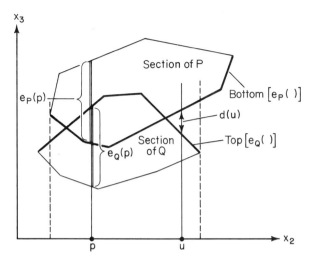

Figure 7.35 Cross-section of P and Q in the plane $x_1 = \bar{x}_1$. Illustration of the definitions of $e_P(\ )$, $e_Q(\ )$, and $d(\ )$.

the minimum value of d, i.e.,

$$d(\bar{u}) = \min_{u \in \mathscr{D}} d(u).$$

Obviously, in practice, the search for u will be actually completed only if $d(\bar{u}) > 0$, in which case the two polyhedra do not intersect. Otherwise, the search may stop as soon as a point $v \in \mathscr{D}$ is found for which $d(v) \leq 0$.

The approach we shall now describe for the solution of this problem is due to Dyer (1980), and it represents a considerable technical simplification of the original method [Muller–Preparata (1978)] (although it does not exhibit a significant performance improvement).

First of all, we note that the function $d(u)$ is convex in $\mathscr{D}$. Indeed, let S_P be the portion of the surface of P, which is the locus of bottom$[e_P(u)]$ for $u \in \mathscr{D}$, while S_Q is the locus of top$[e_Q(u)]$. In other words, S_P and S_Q are the *opposing sides* of the two polyhedra. If we now project S_P on the plane $x_3 = 0$ we obtain a planar graph G_P with convex regions; similarly we define G_Q.

Imagine superimposing G_P and G_Q. The resulting graph G^* is the intersection of two planar maps, and its vertices are both the original vertices of G_P and G_Q (conveniently called *true vertices*) and the vertices resulting from the intersections of edges of G_P with edges of G_Q (conveniently called *pseudovertices*). Thus the domain $\mathscr{D}$ is subdivided into convex regions by G^*. Notice that inside any region of G_P the function bottom$[e_P(u)]$ is linear in the (x_1, x_2) coordinates at u; similarly, for the function top$[e_Q(u)]$ inside any region of G_Q. Thus in any region induced by G^* in $\mathscr{D}$, the function $d(u)$ is linear in $x_1(u)$ and $x_2(u)$. Moreover, bottom$[e_P(u)]$ is convex-downward and top$[e_Q(u)]$ is convex-upward; it follows that $d(u)$ is a convex-downward function. We

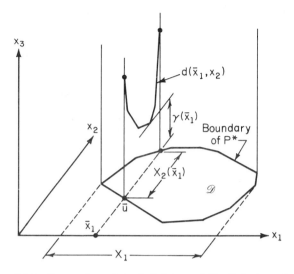

Figure 7.36   Illustration of the definitions of X_1, $X_2(\bar{x}_1)$, and $\gamma(x_1)$.

conclude that the minimum of d occurs at a vertex of G^*. Note that the number of vertices of G^* could be $O(N^2)$: in fact, it is not hard to construct two planar graphs, each with v vertices, so that, when superimposed, $(v-1)^2$ intersections of edges are obtained. Referring to Figure 7.36, let X_1 be the interval

$$\left[\min_{u \in \mathcal{D}} x_1(u), \max_{u \in \mathcal{D}} x_1(u)\right].$$

Also, let $X_2(\bar{x}_1)$ be the interval

$$\left[\min_{x_1(u)=\bar{x}_1} x_2(u), \max_{x_1(u)=\bar{x}_1} x_2(u)\right], \quad \text{for } \bar{x}_1 \in X_1$$

(the segment obtained by intersecting $\mathcal{D}$ with the line $x_1 = \bar{x}_1$). Given that $d(x_1, x_2) = d(u)$ is convex in $\mathcal{D}$, by a well-known result in convex analysis [Rockafellar (1970)], the function

$$\gamma(x_1) \triangleq \min_{x_2 \in X_2(x_1)} d(x_1, x_2), \qquad x_1 \in X_1$$

is convex in X_1. Since

$$\min_{x_1 \in X_1} \gamma(x_1) = \min_{u \in \mathcal{D}} d(u),$$

our problem is now reduced to the evaluation of the minimum of $\gamma(x_1)$ in the interval X_1.

First of all, we note that the determination of $\gamma(\bar{x}_1)$, for a given $\bar{x}_1$ is accomplished in time $O(N)$. Considering one polyhedron at a time, say P, we determine in a straightforward manner in time $O(L)$ the value of bottom$[e_P(\bar{u})]$ for a point $\bar{u}$ in the plane $x_3 = 0$ at the intersection of the line x_1

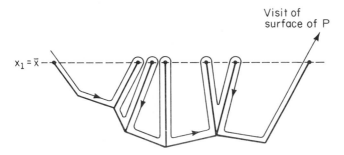

Figure 7.37   The evaluation of bottom $[e_P(\ )]$ on the polygon obtained by intersecting P with the plane $x_1 = \bar{x}_1$ is done by a routine visit on the DCEL.

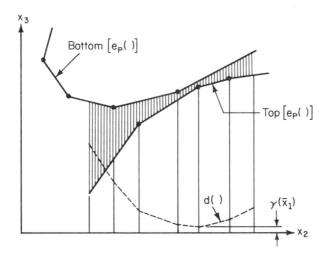

Figure 7.38   Illustration of $d(v)$ and $\gamma(\bar{x}_1)$. Shown is intersection of P and Q with the plane $x_1 = \bar{x}_1$.

$= \bar{x}_1$ with the boundary of P^*. Then by a routine visit of the surface of P on the DCEL we can determine, still in time $O(L)$, the values of bottom$[e_P(\ )]$ corresponding to each edge of P intersecting the plane $x_1 = \bar{x}_1$ (see Figure 7.37). By the same procedure we can process Q in time $O(M)$. The functions bottom$[e_P(v)]$ and top$[e_Q(v)]$ for $x_1(v) = \bar{x}_1$ are illustrated in Figure 7.38. The function $d(v)$, for $v \in \mathcal{D}$ and $x_1(v) = \bar{x}_1$, is then computed in time $O(L + M)$ $= O(N)$ and its minimum, $\gamma(\bar{x}_1)$, also found in this process.

The next question to be addressed is the evaluation of the extremum of a unimodal function[9] (note that a convex function is unimodal). It is a simple

---

[9] A function $f(x)$ is said to be *downward-unimodal* in an interval $[x_1, x_2]$ if, for some $x_0 \in [x_1, x_2]$, $f(x)$ is nonincreasing in $[x_1, x_0]$ and nondecreasing in $[x_0, x_2]$. An *upward-unimodal* function is analogously defined.

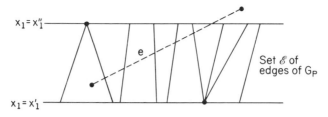

Figure 7.39 Projection on the plane $x_3 = 0$ of the near-side of P relative to the strip $[x'_1, x''_1]$. This strip contains no vertex of G_P in its interior.

and well-known fact that the extremum of a unimodal function defined at N discrete points of the real line can be obtained by a modified binary search with $O(\log N)$ evaluations of the function and $O(\log N)$ comparisons.

We can now state the third step of the search, where we assume, without loss of generality, that $L \leq M$;

Step 3. Sort in increasing order the set $\{x_1 : x_1 = x_1(v),\ v$ is a vertex on the near-side of $P\}$, and let V be the sorted sequence. Since $\gamma(x_1)$ is unimodal on the sequence V, determine a unique interval $[x'_1, x''_1]$ of consecutive terms of V, which is guaranteed to contain the minimum of $\gamma(x_1)$ for $x_1 \in X_1$. (Obviously the search stops if a value $\bar{x}_1$ is found for which $\gamma(\bar{x}_1) \leq 0$.)

To analyze the running time of Step 3, notice first that $O(L \log L)$ time is used by the initial sorting of the abscissae. Next, for a given x_1, the evaluation of $\gamma(x_1)$ can be accomplished in time $O(N)$, by the technique described earlier. Finally, due to the unimodality of γ in X_1, by modified binary search we obtain $[x'_1, x''_1]$ with $O(\log L)$ evaluations of γ. Thus, the global running time is $O(L \log L) + O(N \log L) = O(N \log L)$.

We now observe that $[x'_1, x''_1]$ does not contain in its interior the abscissa of any vertex in the near-side of P, by its own construction. Thus, G_P does not contain any vertex in the strip $x'_1 \leq x_1 \leq x''_1$, that is, the set $\mathscr{E}$ of edges of G_P intersecting this strip can be ordered (see Figure 7.39). Consider now any edge e of G_Q, which has a nonempty intersection with the strip $x'_1 \leq x_1 \leq x''_1$. The function $d(x_1, x_2)$ along this edge is downward convex, and its minimum occurs at a vertex of G^* (this vertex is either an extreme of e or the intersection of e with an edge of $\mathscr{E}$). Since the latter edges are ordered, this minimum can be determined very efficiently. Indeed suppose we organize the members of $\mathscr{E}$ in a binary search tree, where the primitive operation is the determination of the intersection v of two edges and the evaluation of the function $d(v)$ (this primitive runs in $O(1)$ time). Then, by the preceding discussion of unimodal functions, $O(\log L)$ time suffices to process an edge of G_Q. Thus we have the final step of the search:

Step 4. For each edge e of G_Q intersecting the strip $x'_1 \leq x_1 \leq x''_1$ determine the minimum of $d(x_1, x_2)$ along it. The minimum of these minima is the absolute minimum of d in $\mathcal{D}$. (Again, the search stops if an edge is found whose corresponding minimum of d is nonpositive.)

Since the number of edges processed by Step 4 is no larger than the number of edges of Q (which is $O(M)$), and the processing of a single edge is completed in time $O(\log L)$, Step 4 is completed in time $O(M \log L)$. Noting that, since both $L < N$ and $M < N$, in $O(N \log N)$ time either we have obtained a point u for which $\delta(u) \leq 0$, and therefore a point in $P \cap Q$, or concluded that $P \cap Q$ is empty. Combining this result with the earlier result on the determination of $P \cap Q$ once one of its points is known, we have

Theorem 7.15. *Given two convex polyhedra, P with L vertices and Q with M vertices, and letting $N = L + M$, in time $O(N \log N)$ we can compute their intersection $P \cap Q$.*

Unfortunately, the only lower bound we can establish for this problem is the trivial $\Omega(N)$. Thus the result of Theorem 7.15 is not known to be optimal. At this stage it is reasonable to conjecture that a linear-time method may exist for this problem.

7.3.2 Intersection of half-spaces

The most natural approach to solving a three-dimensional problem is an attempt to generalize the best-known method for the corresponding two-dimensional problem. In our case, the two-dimensional analogue is the intersection of half-planes, which was solved in Section 7.2.4 by a divide-and-conquer approach, whose "merge" step was the intersection of two convex polygons. Correspondingly, in three dimensions, the merge step becomes the intersection of two convex polyhedra, which we studied in the preceding section. However, whereas two convex polygons with respectively L and M vertices (where $L + M \triangleq N$) can be intersected in time $\theta(N)$, two convex polyhedra with the same parameters are intersected in time $O(N \log N)$. Thus the generalization of the two-dimensional instance would yield an $O(N \log^2 N)$ algorithm. Contrasting this result with the $\Omega(N \log N)$ lower bound, established in Section 7.2.4, may suggest that some alternative approach should be explored before declaring that there is a gap between the bounds we are unable to fill.

Indeed, we shall now show that there exists an optimal method [Preparata–Muller (1979); Brown (1978)], which is not of the divide-and-conquer type and makes use both of the polyhedron intersection algorithm and of the separating plane algorithm (see Section 7.2.5). As in the described intersection of polyhedra, the method rests heavily on the use of dualization with respect to the unit sphere. We must point out that both to simplify the drawings and in some case just to support intuition (otherwise we should try to represent a four-dimensional object!), all of our examples will refer to two-dimensional instances. It must be stressed, however, that this does not affect the validity of the arguments.

The general problem is formulated as follows.

PROBLEM I.1.7 (INTERSECTION OF HALF-SPACES). Find the set of solutions (x, y, z) (*feasible points*) of the set of N linear inequalities

$$a_i x + b_i y + c_i z + d_i \leq 0 \qquad i = 1, 2, \ldots, N \qquad (7.15)$$

with a_i, b_i, c_i, and d_i real and not all simultaneously 0.

The desired solution will be the description of the convex polyhedral surface bounding the domain of the feasible points. Qualitatively, the solution may have one of these forms of increasing dimensionality:

1. Empty (the set (7.15) is inconsistent).
2. A point.
3. A line segment.
4. A convex plane polygon.
5. A convex polyhedron.

Cases 2, 3, and 4 are degenerate; the interesting cases are 1 and 5.

A simple case occurs when, for every $i = 1, \ldots, N$, we have $d_i \leq 0$. Indeed in this instance the origin $(0, 0, 0)$ satisfies each constraint, i.e., it is contained in the common intersection of the half-spaces. Then, by dualizing the plane π_i of equation $a_i x + b_i y + c_i z + d_i = 0$ to the point $(a_i/d_i, b_i/d_i, c_i/d_i)$ and computing, in time $O(N \log N)$, the convex hull of the resulting point set, we obtain a convex polyhedron, whose dual is the sought intersection.

The situation is considerably more complex when the d_i are arbitrary. For simplicity, we assume in this presentation that no $d_i = 0$ so that the index set $\{1, \ldots, N\}$ can be partitioned into two subsets, I_+ and I_-, where $i \in I_+$ if $d_i > 0$ and $i \in I_-$ if $d_i < 0$. Also, without loss of generality, we assume $d_i = \pm 1$.[10]

To understand the mechanism, it is convenient to view the problem in homogeneous coordinates, by applying the transformation

$$x = \frac{x_1}{x_4}, \qquad y = \frac{x_2}{x_4}, \qquad z = \frac{x_3}{x_4}. \qquad (7.16)$$

As we saw in Section 1.3.2, where homogeneous coordinates were introduced, a point (x, y, z) in E^3 may be interpreted as a line of points (cx, cy, cz, c), in E^4, with $c \in \mathbb{R}$. If we denote the coordinates of E^4 as x_1, x_2, x_3, and x_4, in this interpretation the ordinary space E^3 is the hyperplane $x_4 = 1$ of E^4 (See Figure 7.40 for an analogue in one less dimension). Also if we project from the origin the plane $x_4 = 1$ onto the unit hypersphere S^4 centered at the origin,

---

[10] The general problem admitting the degeneracy $d_i = 0$ is treated in [Preparata–Muller (1979)].

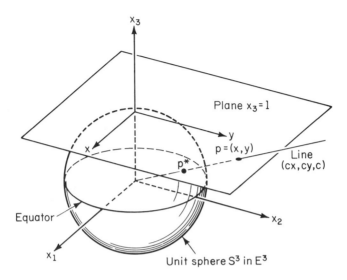

Figure 7.40 Illustration of the correspondence between ordinary coordinates in E^2 (the plane $x_3 = 1$) and homogeneous coordinates in E^3. Each point of E^2 corresponds to a line by the origin in E^3.

each point p of E^3 corresponds to a unique point p^* of S^4 (with $x_4 > 0$) obtained by intersecting S^4 with the segment from the origin O to p. The points of S^4 with $x_4 > 0$ and $x_4 < 0$ form, respectively, the *positive and negative open hemispheres*; the points for which $x_4 = 0$, the *equator*, correspond to the plane at infinity of E^3.

With coordinate transformation (7.16) each half-space $a_i x + b_i y + c_i z + d_i \leq 0$ in E^3 becomes a half-space

$$a_i x_1 + b_i x_2 + c_i x_3 + d_i x_4 \leq 0$$

of E^4, whose bounding hyperplane passes by the origin and where the vector $\mathbf{v}_i = (a_i, b_i, c_i, d_i)$ is normal to the hyperplane and *pointing away* from the half-space. The hyperplane intersects S^4 in a "great circle" C_i. Moreover, the constraint $a_i x + b_i y + c_i z + d_i \leq 0$ specifies a unique hemisphere bounded by this great circle C_i. This hemisphere contains the origin of E^3 if and only if $d_i = -1$. Thus, given a set of N linear constraints in E^3, their common intersection corresponds to a (possibly empty) connected domain $\mathscr{D}$ on the surface of S^4 (see Figure 7.41). The domain $\mathscr{D}$ is partitioned into two (possibly empty) connected domains $\mathscr{D}_+$ and $\mathscr{D}_-$, respectively in the positive and negative hemispheres (see Figure 7.42). Suppose now we project $\mathscr{D}$ from the origin: then $\mathscr{D}_+$ projects to a convex domain $\mathscr{D}_+^*$ in the hyperplane $x_4 = 1$ (our ordinary space E^3) while $\mathscr{D}_-$ projects to a convex domain $\mathscr{D}_-^*$ in the hyperplane $x_1 = -1$. The points of $\mathscr{D}_+^*$ are the common intersection (or *common interior*) of the half-spaces specified by the constraints. On the other hand,

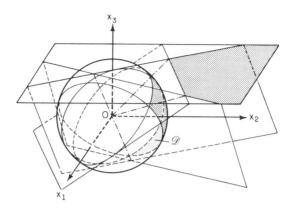

Figure 7.41   The intersection of the hemispheres of S^3 corresponding to the constraints (half-planes in E^2) is a connected domain $\mathscr{D}$ on the surface of S^3.

the points of $\mathscr{D}_-^*$ have a curious interpretation. Since $x_4 = -1$, $\mathscr{D}_-^*$ is the set of points $(x_1, x_2, x_3, -1)$ for which

$$a_i x_1 + b_i x_2 + c_i x_3 - d_i \leq 0, \qquad i = 1, 2, \ldots, N.$$

If we now map each such point to a point in $x_4 = 1$ by a symmetry with respect to the origin of E^4 (this symmetry replaces x_i with $-x_i$, for $i = 1, 2, 3$), we obtain the set of points for which $a_i x_1 + b_i x_2 + c_i x_3 + d_i \geq 0$, $i = 1, \ldots, N$, or

$$a_i x + b_i y + c_i z + d_i \geq 0, \qquad i = 1, 2, \ldots, N. \tag{7.17}$$

If we compare these inequalities with the constraints (7.15), we recognize that they define the *common exterior* of the given half-spaces. Therefore, in $x_4 = 1$ the combined set of points may consist of two separate unbounded convex

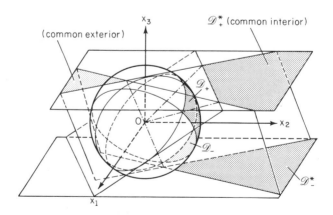

Figure 7.42   Relationships between the $\mathscr{D}_+$, $\mathscr{D}_-$ domains and the common interior and exterior of the given constraints.

sets, one being the common interior and the other the common exterior of the given half-spaces, as illustrated two-dimensionally in Figure 7.42. We call this the *hyperbolic* case. If all points of the solution project into the positive (or the negative) open hemisphere, the corresponding set in E^3 is bounded and we call this the *elliptic* case. The *parabolic* case occurs when there is a single, but unbounded set in E^3, i.e., the case when equatorial as well as positive (or negative) points occur in S^4. In all three cases the sets of points form what we shall call a *generalized* convex polyhedron. The elliptic case alone corresponds to a conventional bounded polyhedron.

Our objective is the simultaneous computation of both the common interior and the common exterior. Notice that the domain $\mathcal{D}$ is always contained in some hemisphere of S^4, bounded by the hyperplane π whose equation is $p_1 x_1 + p_2 x_2 + p_3 x_3 + p_4 x_4 = 0$. The ray normal to π, $(\rho p_1, \rho p_2, \rho p_3, \rho p_4)$ for $\rho \geq 0$, pierces S^4 in a point u. If we apply a rotation R of E^4 which brings the origin of E^3 to u, then the *entire* $\mathcal{D}$ will be in the positive hemisphere. We are then reduced to the elliptic case, which is solved by the method outlined earlier. Once we have obtained the intersection polyhedron, by applying the inverse rotation R^{-1}, we transform it back into the sought common interior and common exterior. Thus, the goal is to obtain the vector $\mathbf{p} = (p_1, p_2, p_3, p_4)$.

Since $\mathcal{D}$ is contained in the half-space $p_1 x_1 + p_2 x_2 + p_3 x_3 + p_4 x_4 \geq 0$, letting $\mathbf{v}_i = (a_i, b_i, c_i, d_i)$, we must have

$$\mathbf{p} \cdot \mathbf{v}_i^T \leq 0, \qquad i = 1, 2, \ldots, N,$$

(where "$\cdot$" denotes, as usual, "inner product"). This can be rewritten as

$$a_i p_1 + b_i p_2 + c_i p_3 + d_i p_4 \leq 0, \qquad i = 1, 2, \ldots, N$$

that is,

$$\begin{cases} a_i p_1 + b_i p_2 + c_i p_3 + p_4 \leq 0, & \text{for } d_i = +1 \\ -a_i p_1 - b_i p_2 - c_i p_3 + p_4 \geq 0, & \text{for } d_i = -1. \end{cases}$$

These conditions are now interpreted. Let $p_1 x + p_2 y + p_3 z + p_4 = 0$ denote a plane in E^3. Depending upon whether $d_i = +1$ or $d_i = -1$, either point (a_i, b_i, c_i) or point $(-a_i, -b_i, -c_i)$ is the dual of the plane $a_i x + b_i y + c_i z + d_i = 0$ in the usual polarity. Thus the plane $a_i x + b_i y + c_i z + d_i = 0$ *separates* the dual images of the planes whose indices are in I_+ and I_- respectively.

This immediately suggests a method to find $\mathbf{p}$:

1. Form the dual sets S^+ of $\{\pi_i : i \in I_+\}$ and S^- of $\{\pi_i : i \in I_-\}$. Construct $\text{CH}(S^+)$ and $\text{CH}(S^-)$. [In time $O(N \log N)$].
2. If $\text{CH}(S^+) \cap \text{CH}(S^-) \neq \emptyset$ (with interior), then the set of inequalities is inconsistent. Otherwise construct a separating plane $p_1 x + p_2 y + p_3 z + p_4 = 0$. [In time $O(N)$ using the method mentioned in Remark 2, Section 7.2.5.]

At this point the appropriate rotation of E^4 can be constructed and applied—in time $O(N)$—, after which the techniques described earlier complete the task in additional time $O(N \log N)$. Thus we have

Theorem 7.16. *The common interior and the common exterior of a set of N half-spaces in E^3 can be constructed in time $\theta(N \log N)$ (i.e., in optimal time).*

The optimality follows from the remark that the $\Omega(N \log N)$ lower bound obtained for the intersection of N half-planes applies trivially to the problem in question.

7.4 Notes and Comments

The techniques and the results illustrated in the preceding sections are not a full catalog of the abundant literature on geometric intersection problems. In this section, we attempt to remedy at least in part to this deficiency by a brief sketch of a number of additional significant results.

An approach proposed by Chazelle and Dobkin (1980) to intersection problems is analogous to repetitive-mode query processing. Specifically, rather than considering two geometric objects and computing their intersection, they propose to consider a large set of such objects and preprocess them so that their pairwise intersection can be tested very rapidly (Problem Type I.3). The objects they study are points, lines, planes, polygons, and polyhedra; given any two objects in these categories (e.g., a (polygon, polygon) pair, or a (plane, polyhedron) pair, etc.) they give "poly-log" algorithms for testing whether their intersection is empty. As an example, the case (point, polyhedron) is readily solved as a planar point location problem (Section 2.2), by projecting stereographically the polyhedron and the point to a suitable plane; on the other hand, the solution to this problem provides, by straightforward dualization (Section 1.3.3), the solution of the problem (plane, polyhedron).

Another class of intersection problems, of which an important case—intersection of segments—has been presented in Section 7.2.3, is the report of all K intersecting pairs in a set of N objects. The goal is the development of algorithms whose running time is $O(f(N) + K)$, i.e., it consists of a fixed overhead $O(f(N))$ and is otherwise proportional to the size of the reported set. The algorithm in Section 7.2.3 fails to achieve this goal since it runs in time $O((K + N) \log N)$. It is plausible that more efficient techniques can be obtained when the objects obey specific constraints. With reference to line segments, a first step in this direction was taken by Nievergelt and Preparata (1982), who considered the intersection of two planar maps (i.e., the segments are the edges of two planar graphs embedded in the plane); they showed that, if the faces of each graph are convex, time $O(N \log N + K)$ is attainable. In reality, the constraint of convexity was unnecessary, since a more general result was later proved by Mairson and Stolfi (1983): two collections A and B, each consisting of nonintersecting segments, and such that $|A| + |B| = N$, can be intersected in time $O(N \log N + K)$, where K is the number of intersections. An additional instance of this problem occurs when A and B are each a class of parallel segments, for which an even simpler technique achieving the same time bound will be presented in Chapter 8. All of these techniques are of the plane-sweep type; in the general case, the dynamic reinsertion of an intersection in the event point schedule automatically determines an $O(K \log K)$ cost. In a radical departure from sweep-line methods, Chazelle (1983a) has recently proposed a segment intersection algorithm based on a hierarchical approach and running in time $O(K + N \log^2 N/\log \log N)$.

If rather than reporting the intersections of N line segments the objective is simply to obtain their number (segment intersection counting) the problem is considerably different. There exists an obvious transformation, since any intersection reporting algorithm can be used to solve the corresponding counting problem; however, efficiency can be poor since the number K of intersections can be $\theta(N^2)$. The best result independent of K known to date is an algorithm due to Chazelle (1983a) running in time $O(N^{1.695} \cdot \log N)$ and using linear space.

In Section 7.2.3.2 we have seen that to detect if N line segments in the plane are disjoint requires $\Omega(N \log N)$ operations. What happens if the N segments obey the constraint of being the edges of a polygon? In this case, the detection problem becomes the test for simplicity of a polygon, and it is interesting to explore if the added condition leads to algorithmic simplifications. So far, SIMPLICITY TEST has eluded a complexity characterization. Recently some progress has been reported on a related problem, POLYGON INTERSECTED EDGES, i.e.: given an N-gon P, report all its edges that are intersected (by some other edge). For this problem, by using the general theory of Ben-Or (Section 1.4), Jaromczik has shown that $\Omega(N \log N)$ operations are necessary [Jaromczik (1984)]. Unfortunately, this bound does not affect SIMPLICITY TEST, since the transformation of the latter to POLYGON INTERSECTED EDGES goes in the wrong direction. Thus, the complexity of SIMPLICITY TEST remains an outstanding open problem in computational geometry.

The optimal-time algorithm that constructs the intersection of two convex polygons, reported in Section 7.2.1, is a plane sweep. Its brute generalization to three dimensions fails for the reasons presented at the beginning of Section 7.3.1 However, the appropriate generalization has been recently found by Hertel et al. [Hertel–Mehlhorn–Mäntylä–Nievergelt (1985)]; their method is not only an important example of space-sweep algorithm, but it also represents an alternative to the polygon intersection technique of Section 7.3.1, and a good competitor for elegance (if not for performance). To give an idea of the technique, the two-dimensional analog is a plane sweep that computes the points that are intersections of the boundaries of two polygons P_1 and P_2: then a march, alternating between the boundaries of P_1 and P_2 at the intersection points, constructs the boundary of $P_1 \cap P_2$. The two edges of P_i ($i = 1, 2$) intersected by the sweep-line in E^2 become a cycle of polygons (called a *crown*) intersected by the sweep-plane in E^3 and the role of intersection points in E^2 is taken by polygonal cycles in E^3. They show that a space sweep efficiently updates the crowns and detects at least one vertex of each polygonal cycle, from which the entire boundary can be readily constructed. The algorithm is shown to run in time $O(N \log N)$. The corresponding detection problem is certainly no harder than the construction problem, and is likely to be simpler. Indeed, Dobkin and Kirkpatrick (1984) have devised a $\theta(N)$ algorithm for this problem, and also credit Dyer for an independent discovery of an analogous result.

The construction of the kernel of a simple polygon (Section 7.2.6), is a representative of an interesting large class, called *visibility problems* (indeed, the kernel is the locus of points from which all vertices are visible). Such problems occur in a variety of applications in graphics, scene analysis, robotics, etc. The basic concept is that two points p_1 and p_2 are (mutually) visible if $\overline{p_1 p_2}$ does not intersect any "forbidden" curve. Visibility is normally defined with respect to a point p. In the plane, if the forbidden curves are segments, the locus of points visible from p, called the *visibility polygon* of p, is a (possibly unbounded) star-shaped polygon. The visibility polygon is always bounded if p is internal to a simple polygon P and the boundary of P is the forbidden curve. If the set of segments in the plane consists of a single simple polygon, at least two linear time algorithms are known [ElGindy–Avis (1981); Lee (1983b)]. The technique used is similar to Graham's scan in that an initial point visible from p is first determined and then, by using the property of simplicity, the vertices of P are scanned and the portions of P that are not visible from p are eliminated. In the case where the segment set consists of m disjoint convex polygons with N edges in total and K denotes the output size, this problem can be solved in $O(m \log N + K)$ [Lee–Chen (1985)]. Edelsbrunner *et al.* [Edelsbrunner–Overmars–Wood (1983)] also examine this problem and discuss the maintenance of the visibility polygons when insertions or deletions are allowed, as well as the issue of coherence when the viewing direction or the points are allowed to move. By set-theoretic operations on visibility polygons one may define special types of visibility, and the reader is referred to Toussaint (1980a) for a concise account of this topic.

7.5 Exercises

1. Let P_1 and P_2 be two convex polygons, whose numbers of vertices sum to N. Design a plane-sweep algorithm that computes the intersection of P_1 and P_2 in time $O(N)$ as follows:
   (a) Determine the set $\mathscr{I}$ of all intersections of the boundaries of P_1 and P_2;
   (b) By using $\mathscr{I}$, construct the boundary of $P_1 \cap P_2$.

2. Given a collection S of N segments in the plane, devise an algorithm that decides if there is a line l intersecting all members of S (stabbing line) and, if so, construct it.

3*. Given a collection S of N segments in the plane, and a query segment s^*, proprocess S to obtain an efficient algorithm to test if s^* intersects any member of S.

4. Let P be a convex polyhedron. A query consists of testing if a segment s in E^3 has a nonempty intersection with P. Assuming to operate in the repetitive mode, preprocess P and design an efficient query algorithm.

5. Given two convex polyhedra P and Q, whose numbers of vertices sum to N, can you think of an algorithm that *tests* if P and Q intersect in time $\theta(N)$?

6. Modify the algorithm for intersecting two convex polyhedra (given in Section 7.3.1) to test if two polyhedra P and Q are linearly separated and, if so, construct a separating plane π. The algorithm should run in time linear in the total number of vertices of P and Q.

CHAPTER 8
The Geometry of Rectangles

The methods developed earlier to intersect planar convex polygons are certainly applicable to a collection of rectangles. Similarly, a collection of segments, each of which is either parallel or orthogonal to any other member of the collection, can be handled by the general segment intersection technique described in the preceding chapter. However, the very special nature of segments having just two orthogonal directions, or of rectangles whose sides are such segments, suggests that perhaps more efficient *ad hoc* techniques can be developed to treat such highly structured cases. And in the process of studying this class of problems one may acquire more insight into the geometric properties possessed by the objects mentioned above—rectangles and orthogonal segments—and may succeed in characterizing the wider class to which the techniques are still applicable.

All this would be an interesting, albeit a little sterile, exercise, were it not that rectangles and orthogonal segments are the fundamental ingredients of a number of applications, some of which have acquired extreme importance in the wake of current technological breakthroughs, notably Very-Large-Scale-Integration (VLSI) of digital circuits. Therefore, before entering the detailed discussion of the algorithmic material, it is appropriate to dwell a moment on some significant applications.

8.1 Some Applications of the Geometry of Rectangles

8.1.1 Aids for VLSI design

Masks used in the fabrication of integrated circuits are frequently expressed as

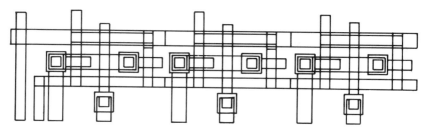

Figure 8.1   Typical geometry of an integrated circuit.

a collection of rectangles with sides parallel to two orthogonal directions. Each such rectangle is either a portion of a wire, or an ion implantation region, or a "via" contact, etc., as illustrated in Figure 8.1 [Lauther (1978); Baird (1978); Mead–Conway (1979)].

These rectangles are laid out in the plane according to certain design rules, which specify minimum spacings between types of rectangles, minimum overlaps between other types, and so on. The conditions on clearances and overlaps arise from the conflicting requirements to minimize the total area used (an obvious economic constraint) and to ensure the necessary electrical relationship (insulation or contact) in spite of possible production fluctuations. (Clearly, the more parsimonious is the clearance, the more likely is the occurrence of an undesired contact.)

Therefore, an important task in the design of an IC-mask (integrated circuit mask) is the verification that the design rules have been complied with, briefly referred to as "design-rule checking." For example, in Figure 8.2 one wishes to ensure a minimum clearance λ between rectangle R and the neighboring rectangles. To verify this situation one may enlarge R by means of a crown of width λ, and verify whether the enlarged rectangle R' intersects any of the surrounding rectangles. Thus the verification of spacing is transformed to a rectangle intersection problem. Variations of this approach can be used to verify other design rules, such as overlaps, containments, and the like.

Whereas the above steps concern, in a sense, the "syntactic" correctness of

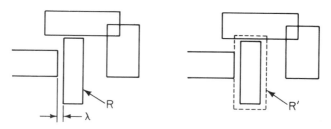

Figure 8.2   A minimum clearance of value λ around rectangle R can be verified by enlarging R by λ (to form λ') and checking for overlaps.

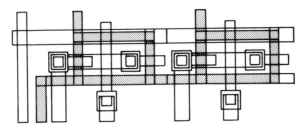

Figure 8.3   The identification of the connected component of a set of rectangles can be used in a circuit extraction task.

the mask, another important task is the verification that the mask is the functionally correct realization of the specified circuits. The first step in this task is the interpretation of the geometry of the mask as an electric circuit, that is, the identification of subsets of rectangles whose members are all electrically connected. This activity is referred to as *circuit extraction*. In somewhat more abstract terms, one may imagine a graph, whose vertices correspond to the rectangles, so that there is an edge between two vertices if the corresponding rectangles have a nonempty intersection. In this formulation, circuit extraction is solved by finding the *connected components* of this graph. However, the determination of the connected components achieves somewhat less than required by circuit extraction, since what is really needed is an outline of the boundary (the so-called *contour*) of each connected component (see Figure 8.3).

Thus, we see that the design of integrated circuits is a source of a wealth of geometric problems involving rectangles, of which we have only illustrated some significant representatives.

8.1.2 Concurrency controls in databases

An interesting geometric model has been recently developed for the concurrent access to a database by several users [Yannakakis–Papadimitriou–Kung (1979)]. Typically, a user *transaction* is a sequence of steps, each of which accesses an item of the database (a *variable*) and updates it. The update of a variable, however, has a certain duration. In this time interval it is necessary to prevent any other user from accessing the same variable. (In more concrete terms, for example, if the variable in question is the seat availability of a commercial flight, it is obvious that, in order to maintain the correct count, all transactions involving this variable must occur sequentially.) A very common method for resolving these conflicts is *locking* [Eswaran–Gray–Lorie–Traiger (1976)], whereby a user "locks" a variable at the beginning of an update and "unlocks" it at its completion. A locked variable is not accessible by any other user.

This situation can be modeled as follows. The history of each user is a sequence of steps of the types "lock x," "update x," and "unlock x," where x is a database variable. If there are d users, we identify each of them with an axis of the d-dimensional cartesian space E^d, and for each axis we establish a one-to-one correspondence between transaction steps and points with integer coordinates. In this manner an update of a variable is identified with an interval (locking interval) on this axis, bounded by the points corresponding to the lock and unlock steps, respectively. Also, if $k \leq d$ users have a transaction involving the same variable, the cartesian product of the corresponding locking intervals is a k-dimensional rectangle. The situation is illustrated for two users U_1 and U_2 in Figure 8.4(a). The state of the database is represented by a poing in the plane of coordinates U_1 and U_2, indicating the progress made in the history of each user; for example, point $(8, 3)$ in Figure 8.4(a) indicates that step 8 of U_1 and step 3 of U_2 are currently being executed. It is clear that in a locked-transaction system the state point cannot be internal to any rectangle associated with a variable.

As we indicated earlier, each user may access several variables. Each access to a variable x involves the subsequence of steps "lock x," "update x," and "unlock x"; the sequence of steps of a given user (user's activity) is displayed by a set of points on the user axis, as shown in Figure 8.4(b) for two users.

A *schedule* illustrates the actual evolution of the database activity and is an arbitrary merge of the activities of each user. Since each activity is a sequence, a feasible schedule is represented by a monotone nondecreasing staircase curve in the plane (U_1, U_2) of Figure 8.4(b). An initial portion of such curve describes a *partial* schedule.

A *serial* schedule is one where users are processed sequentially, that is, a user gains access to the database and maintains it until the completion of his activity. Clearly, for d users there are $d!$ serial schedules (and, for two users, there are just two schedules, shown in Figure 8.4(b)). A database system is considered *safe* if the effect of any legal schedule (i.e., a schedule avoiding the rectangles) is the same as that of some serial schedule. This condition can be expressed in purely geometrical terms [Yannakakis–Papadimitriou–Kung (1979)], by requiring that any safe schedule be *homotopic* (continuously tramsformable avoiding the rectangles) to a serial schedule.

On the other hand, the system may reach a condition of impasse, called *deadlock*, in which it would remain indefinitely without an external intervention. For example, this is the situations represented by the partial schedule c shown in Figure 8.4(b), described by the sequence $S_{21}S_{11}S_{22}S_{12}S_{23}$. Step S_{12} (lock x_1) cannot be executed because x_1 is already locked (by S_{21}); so S_{12} must wait and attention is given to user U_2, who demands execution of S_{23}: lock x_2. Again, x_2 is already in the locked state (by S_{11}), so that no further processing may take place, since both users are trying to execute a nonexecutable step.

The conditions of *safeness* and *deadlockfreedom* are effectively visualized in

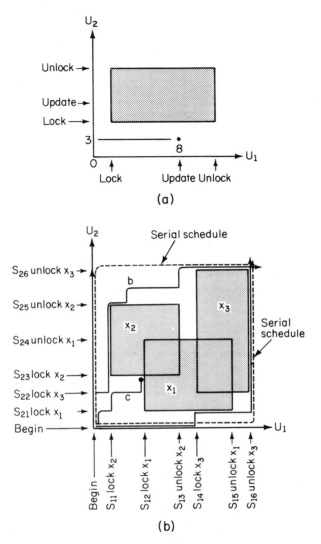

Figure 8.4 Modeling a locked transaction database system. (a) Two users, one variable. (b) Two users, three variables.

the transaction diagram. First of all, no curve representing a schedule can penetrate into the interior of the union of the transaction rectangles. This union will appear as a collection of connected components (see Figure 8.5). But this condition alone does not guarantee that the system be free of deadlocks. Indeed, we should characterize a region, containing the transaction rectangles, such that any schedule (i.e., any monotone nondecreasing curve) external to it is guaranteed to be deadlock free. Such region is sketched

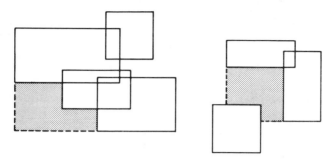

Figure 8.5 Illustration of the SW-closure of a union of rectangles. Any schedule avoiding this region is deadlockfree.

in Figure 8.5, where the shaded areas have been added ot the union of the transaction rectangles. It is intuitively clear, at this point, that such extension of the union of rectangles satisfies the given requirement. This region has been called the SW-*closure* (south-west closure) of the union of rectangles [Yannakakis–Papadimitriou–Kung (1979); Lipski–Papadimitriou (1981); [Soisalon-Soininen–Wood (1982)], and will be technically defined in Section 8.6 of this chapter.

8.2 Domain of Validity of the Results

Although a collection of rectangles was the original setting that motivated most of the research in the area, and practically all results are cast in terms of orthogonal directions (typically, the coordinate axes), it is appropriate at this point to try to identify the widest range of validity of the theory.

We start with two collections of segments S_1 and S_2, each one consisting of parallel segments and with the property that the directions of S_1 and S_2 are mutually orthogonal. The straight lines containing the segments determine a grid of the plane. We now represent points in homogeneous coordinates (x_1, x_2, x_3) (see Sections 1.3.2 and 7.3.2) and apply a generic nonsingular projective transformation to the points of the planes, that is, a transformation described by a 3×3 nonsingular matrix $\mathscr{A}$. It is well-known (see, for example, [Ewald (1971)] or [Birkhoff–MacLane (1965)]) that this transformation maps points to points and lines to lines, and *preserves incidence*, that is, any such transformation does not alter the structure of the grid. If the matrix $\mathscr{A}$ is of the form

$$\mathscr{A} = \left[\begin{array}{cc|c} & & 0 \\ & \mathscr{B} & 0 \\ \hline 0 & 0 & a \end{array} \right]$$

with $\mathscr{B}$ a 2×2 nonsingular matrix and $a \neq 0$, then the line at infinity maps to itself, and the two originally orthogonal directions become now arbitrary

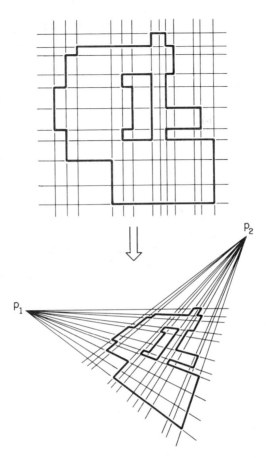

Figure 8.6   A nonsingular projective transformation applied to a plane figure formed with orthogonal segment.

directions. If the nonsingular matrix $\mathscr{A}$ does not have the block form given above, then one or both of the points at infinity of the two original directions are mapped to points at finite, whence even the parallelism of the lines of each collection of segments disappears. A typical situation is illustrated in Figure 8.6.

The preceding analysis, in a sense, characterizes the range of validity of the so-called "geometry of rectangles" and motivates the following definitions.

Definition 8.1. A set $\mathscr{R}$ of quadrilaterals is said to be *isothetic*[1] if the sides of each member of $\mathscr{R}$ belong to lines of two pencils of lines with centers at points

---

[1] Isothetic means "equally placed," from the Greek words *isos* (equal) and *tithenai* (to place). There has been a considerable terminological uncertainty on this notion, ranging from "recti-linear" to "orthogonal" to "aligned" to "iso-oriented." While the latter are no misnomers (except, possibly, for "rectilinear"), perhaps "isothetic" is the word closest to the notion.

p_1 and p_2, respectively (p_1 and p_2 possibly on the line at infinity of the plane) and all members of $\mathscr{R}$ lie entirely on the same side of the line by p_1 and p_2.

With this clarification, for ease of presentation we shall hereafter refer to the case in which p_1 and p_2 define orthogonal directions.

The algorithms to be illustrated in this chapter concern the determination of various properties of sets of rectangles (or of orthogonal segments). A fundamental classification can be made depending on the basis of the mode of operation, that is, depending upon whether we operate in a static or in a dynamic environment. In the first case the data of the problem are entirely available before computation begins; in the other case, we also allow dynamic updates of the data set, by means of insertions and deletions.

The two modes of operation—static and dynamic—involve, as is to be expected, the use of substantially different data structures. While operating in the static mode is to-date reasonably well understood, the more complex dynamic mode is still in an active stage of development. Due to this situation, it is appropriate to examine in detail just the static mode algorithms; reference to ongoing research on dynamic mode algorithms will be made at the end of the chapter.

The following presentation will be mostly concerned with the two-dimensional environment. Extension to the more general d-dimensional case will be made or mentioned whenever appropriate.

8.3 General Considerations on Static-Mode Algorithms

A collection of isothetic rectangles (or, equivalently, two orthogonal collections of parallel segments) is characterized by the unique property that the plane can be subdivided into strips (for example, vertical strips), in each of which the environment becomes unidimensional. (Analogously, E^d can be partitioned into slabs whose environment is $(d-1)$-dimensional.) Specifically, if one considers the abscissae of vertical sides of rectangles, in the plane strip between two consecutive such abscissae all vertical sections are identical, and consist of the ordinates of the horizontal sides crossing the strip. Thus the vertical sections of the plane are subdivided into equivalence classes (each class being the set of sections between two consecutive abscissae of vertical rectangle sides). In addition, we realize that the section pertaining to a given strip appears as a minor modification of the ones pertaining to either one of the two adjacent strips.

Several of the computationally interesting problems concerning a set of rectangles (such as area of union, perimeter of union, contour of union, report of intersections, etc.) have a very useful property. A generic vertical line determines two half-planes; the solution of the given problem is related in a generally simple manner to the solutions of the analogous subproblems in each of the two half-planes. This simple relation is sometimes just set-theoretic union (intersection report), or arithmetic sum (area, perimeter), or a simple concatenation of appropriate components (contour). In all of these cases the intersection of the bisecting line with the set of rectangles contains *all* the relevant information for the combination of the partial solutions. Moreover, the static mode of operation suggests that the solution obtained for, say, the half-plane to the left of a bisecting line is *final* (i.e., not modifiable by possible *future* updates), so that the global solution can be obtained *incrementally* by extending to the right the current solution. We recognize that this is the paradigmatic situation where a plane-sweep technique is called for, the sweep-line being parallel to one direction and the motion occurring along the perpendicular direction.

Turning now our attention to the generic vertical section of the rectangle set—which is just a sequence of ordinates—we note that, since the intersected segments are parallel, the ordinate sequence of the section *is always a subsequence* of the sorted sequence of ordinates of the horizontal segments. Through a preliminary normalization that sorts the set of ordinates and replaces each ordinate with its rank in the sorting, the sequence of ordinates can be likened to a sequence of consecutive integers. Thus, in order to maintain the sweep-line status, it seems unnecessary to resort to a general priority queue, when use can be made of more efficient structures tailored to the situation. Such structures are all determined (at least as a skeleton) by the set of ordinates of *all* horizontal segments, and each particular section can be obtained—in the "skeleton" metaphor—by "fleshing out" the appropriate portion of the skeleton. An example of such structure is the *segment tree*, which we already discussed and used earlier in this text (see Section 1.2.3.1); another will be the *interval tree* [McCreight (1981); Edelsbrunner (1980)], to be described in Section 8.8.1.

For the convenience of the reader we recall that each node v of the segment tree is characterized by an interval $[B[v], E[v]]$, and by some additional parameters, necessary to carry out specific computations. Of these parameters, $C[v]$—the node *count*—is common to all applications, since it denotes the multiplicity of the segments currently assigned to v; other parameters, on the other hand, are specific to the application under consideration.

The primitive operations on the segment tree are insertion and deletion. Specifically, if $[b, e]$ denotes an interval, these primitives, acting on a node v, are INSERT$(b, e; v)$ and DELETE $(b, e; v)$. The general form of INSERT, to be used throughout the next sections, is as follows:

> **procedure** INSERT$(b, e; v)$
> 1. **begin if** $(b \leq B[v])$ **and** $(E[v] \leq e)$ **then** $C[v] := C[v] + 1$
> 2.   **else begin if** $(b < \lfloor (B[v] + E[v])/2 \rfloor)$ **then** INSERT$(b, e; \text{LSON}[v])$;
> 3.   **if** $(\lfloor (B[v] + E[v])/2 \rfloor < e)$ **then** INSERT$(b, e; \text{RSON}[v])$
>   **end**;

4.   UPDATE(v) (*update specific parameters of v*)
      end.

The crucial feature, which will distinguish the different applications, is the function UPDATE(v) in line 4, whose instances will be discussed case by case. Analogous considerations apply to the procedure DELETE($b, e; v$).

   In conclusion, the plane-sweep approach and the use of interval-oriented data structures are the natural tools for tackling a variety of static computational problems in the geometry of rectangles. The following sections are devoted to a detailed presentation of specific techniques.

8.4 Measure and Perimeter of a Union of Rectangles

In a note [Klee (1977)] that raised considerable interest and can be traced as the origin of a whole research topic, V. Klee posed the following question: "Given N intervals $[a_1, b_1], \ldots, [a_N, b_N]$ in the real line, it is desired to find the measure of their union. How efficiently can that be done?"

   In this formulation, we have the simplest (unidimensional) instance of the measure-of-union problem. (The computation of the measure-of-union in arbitrary dimension will be referred to as Problem R.1.) Klee readily exhibited an $O(N \log N)$ algorithm to solve it, based on presorting the abscissae a_1, $b_1, \ldots, a_N, b_N$ into an array $X[1:2N]$. The additional property of $X[1:2N]$ is that if a_i is placed in $X[h]$, b_j in $X[k]$, and $a_i = b_j$ then $h < k$ (that is, a right endpoint is placed after a left endpoint of the same value). The computation is completed by a simple linear time scan of $X[1:2N]$ (the following algorithm is an adaptation of Klee's suitable for subsequent generalizations):

      procedure MEASURE OF UNION OF INTERVALS
1. **begin** $X[1:2N] :=$ sorted sequence of abscissae of intervals;
2. $X[0] := X[1]$;
3. $m := 0$; (*m is the measure*)
4. $C := 0$; (*C is the multiplicity of overlapping segments*)
5.   **for** $i := 1$ **until** $2N$ **do**
6.   **begin if** $(C \neq 0)$ **then** $m := m + X[i] - X[i-1]$;
7.   **if** $(X[i]$ is a left endpoint) **then** $C := C + 1$ **else** $C := C - 1$
            end
      end.

   In his original paper, Klee also noted that, although sorting is the key to the above result, it is *a priori* not required, and asked whether there exists a solution involving $o(N \log N)$ steps.

   The answer to this question was readily provided by Fredman and Weide (1978) for the linear decision-tree model (which encompasses the above al-

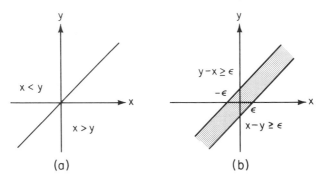

Figure 8.7   Illustration in E^2 of the distinction between the membership sets of ELEMENT UNIQUENESS (a) and ε-CLOSENESS (b).

gorithm); the general technique of Ben-Or (Section 1.4) extends the result to general algebraic decision trees. Characteristically, the type of argument fits a by now familiar pattern, encountered in connection with convex hulls (Section 3.2), maxima of vectors (Section 4.1.3), element uniqueness (Section 5.2), and maximum gap (Section 6.4). The ubiquitous underlying structure, the symmetric group of degree N, is again the key of the argument.

The computational prototype to be used here is the following.

PROBLEM R.2 (ε-CLOSENESS). Given $N + 1$ real numbers $x_1, x_2, \ldots, x_N$, and $\varepsilon > 0$ determine whether any two x_i and $x_j (i \neq j)$ are at distance less that ε from each other.

First, we show that the transformation

$$\varepsilon\text{-CLOSENESS} \propto_N \text{MEASURE OF UNION OF INTERVALS}$$

can be easily established. Indeed, construct interval $[x_i, x_i + \varepsilon]$ for $i = 1, 2, \ldots, N$. These intervals form the input for the algorithm MEASURE OF UNION, which returns as a result a real value m. Then clearly no two elements of $\{x_1, \ldots, x_N\}$ are at distance less than ε from each other if and only if $m = N\varepsilon$.

There is a close similarity between ε-CLOSENESS and ELEMENT UNIQUENESS, discussed in Section 5.2. However, ELEMENT UNIQUENESS is not a special case of ε-CLOSENESS, since the value of $\varepsilon = 0$, required to obtain the specialization, is not allowed. Rather, their relationship is shown in Figure 8.7, in a two-dimensional instance: in the context of ELEMENT UNIQUENESS the disjoint connected components of the membership set W (see Sections 1.4 and 5.2) are open sets and two of them may be separated just by their common frontier, whereas in the context of ε-CLOSENESS the components are closed sets and the separation is always positive. Except for this difference, the argument establishing that the membership set W of ε-CLOSENESS has $N!$ disjoint connected components is

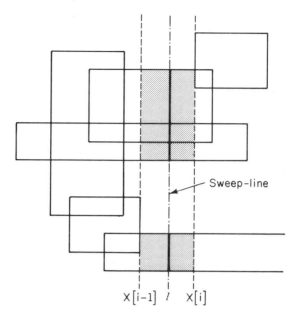

Figure 8.8   The plane-sweep approach to the measure-of-union problem in two dimensions.

analogous to the one developed in Section 5.2, so that we have the following corollary to Theorem 1.2.

Corollary 8.1. *In the algebraic decision-tree model any algorithm that determines whether any two members of a set of N numbers differ from each other by less than ε requires $\Omega(N \log N)$ tests.*

The preceding result establishes the optimality of Klee's result for the measure problem in one dimension, but leaves open the question of how well we can do for $d \geq 2$.

Bentley (1977)[2] attacked this problem and succeeded in developing an optimal measure-of-union algorithm for $d = 2$. The technique is a natural, albeit clever, modification of the unidimensional method. Specifically, in one dimension the length of the interval $[X[i-1], X[i]]$ is added to the measure (line 6 of Algorithm MEASURE OF UNION OF INTERVALS) depending upon whether there is at least one segment that spans it or not, or, equivalently, the parameter C is nonzero or zero, respectively. As a consequence, all we need to maintain is the value of C (Line 7). In two dimensions (see Figure 8.8) the plane strip between $X[i-1]$ and $X[i]$ contributes to the measure-of-union of rectangles the quantity $(X[i] - X[i-1]) \times m_i$, where m_i is the length of the intercept of an arbitrary vertical line in the strip with the union of

---

[2] Also reported in [van Leeuwen–Wood (1981)].

rectangles itself. Thus the quantity m_i (which was constant and equal to 1 in one dimension) is the key parameter of a two-dimensional technique. If m_i can be maintained and measured in time not exceeding $O(\log N)$, then the times for presorting and scanning would be equalized and an optimal $\theta(N \log N)$ algorithm would result.

Since the ordinates of the horizontal sides of rectangles are known *a priori*, the objective can be realized by means of the *segment tree* (see Section 8.3). Referring to the general format of the segment tree primitives INSERT and DELETE presented in Section 8.3, for the present application we define the following additional node parameter:

$m[v]$:= the contribution of the interval $[B[v], E[v]]$ to the quantity m_i.

The computation of $m[v]$ is specified by the following procedure (to be called in line 4 of procedure INSERT):

```
        procedure UPDATE(v)
begin if (C[v] ≠ 0) then m[v] := E[v] − B[v]
        else if (v is not a leaf ) then m[v] := m[LSON[v]] + m[RSON[v]]
            else m[v] := 0
end.
```

It is clear that $m_i = m$[root of segment tree]. The general parameter $C[v]$ and the specific parameter $m[v]$ can both be easily maintained—in constant time per node—when a segment (a vertical side of a rectangle) is inserted into or deleted from the segment tree. Thus, the maintenance of the segment tree and the determination of m_i can be jointly effected in time $O(\log N)$ per vertical rectangle side, realizing the previously stated objective. We can now formulate the algorithm, where b_i and e_i are respectively the minimum and maximum ordinates of the vertical side at abscissa $X[i]$. (Note that this algorithm is a direct generalization of the one-dimensional one given earlier.)

```
        procedure MEASURE OF UNION OF RECTANGLES
1. begin X[1:2N] := sorted sequence of abscissae of vertical sides;
2.        X[0] := X[1];
3.        m := 0;
4.        construct and initialize segment tree T for ordinates of rectangle
          sides;
5.        for i := 1 until 2N do
6.            begin m* := m[root(T)];
7.                    m := m + m* × (X[i] − X[i − 1]);
8.                    if (X[i] is a left-side abscissa) then
                            INSERT(b_i, e_i; root(T))
9.                    else DELETE(b_i, e_i; root(T))
            end
        end.
```

We summarize the discussion with the following theorem.

Theorem 8.1. *The measure of the union of N isothetic rectangles in the plane can be obtained in optimal time* $\theta(N \log N)$.

If we reconsider the last algorithm, we readily recognize it in the plane-sweep category, (see Section 8.3), where the array $X[1:2N]$ provides the *event-point schedule* and the segment tree gives the *sweep-line status*. This realization immediately offers a further generalization of the technique to more than two dimensions. Indeed the sweep approach transforms the original d-dimensional problem (on N hyperrectangles in E^d) into a "sequence" of N $(d-1)$-dimensional subproblems (each on at most N hyperrectangles in E^{d-1}). For $d = 3$, the subproblems become two-dimensional and, by Theorem 8.1, each can be solved in $O(N \log N)$ time; thus, the measure of union problem in three dimensions can be solved in time $O(N^2 \log N)$. This establishes the basis for an induction argument which leads to the following corollary.

Corollary 8.2. *The measure of the union of N isothetic hyperrectangles in* $d \geq 2$ *dimensions can be obtained in time* $O(N^{d-1} \log N)$.

What is grossly unsatisfactory about the outlined method for $d \geq 3$ is the fact that there is a "coherence" between two consecutive sections in the sweep that we are unable to exploit. (For $d = 2$, the segment tree takes full advantage of the section-to-section coherence by efficiently constructing the current section as a simple update of the preceding one.) This idea was further pursued in three dimensions by van Leeuwen and Wood (1981), who proposed to use as data structure the "quad-tree" [Finkel–Bentley (1974)]. The quad-tree could be viewed as a two-dimensional generalization of the segment tree, but, curiously enough, it was developed earlier than the more specialized structure. We shall now briefly review its organization.

The *quad-tree* is a way to organize a rectangular grid of $N \times M$ cells, determined by $(N + 1)$ horizontal lines and $(M + 1)$ vertical lines. (Correspondingly, the segment tree is a way to organize N contiguous intervals.) To simplify the discussion, we assume that $N = M = 2^k$. The quad-tree T is a quaternary tree where with each node we associate a $2^i \times 2^i$ portion (called a 2^i-square) of the grid ($i = 0, 1, \ldots, k$). The 2^i-square associated with a given node v of T (for $i > 0$) is subdivided into four 2^{i-1}-squares by vertical and horizontal bisections (refer to Figure 8.9); each of these four squares (referred to as NW, NE, SE, and SW squares) is associated with one of the four offsprings of v. The construction is initialized by associating the entire grid with the root of T, and terminates when the quadrisection process yields 2^0-squares (the leaves of T). Since T has N^2 leaves, its construction takes $O(N^2)$ time and uses $O(N^2)$ space.

We now examine how a rectangle R can be stored in a quad-tree. The basic grid is formed by the $2N$ abscissae and $2N$ ordinates of the given n rectangles, and the quad-tree T is constructed on this grid. Each node v of T has an

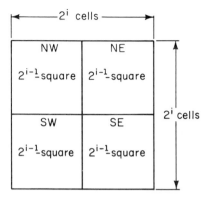

Figure 8.9   Illustration of the partition of the plane-grid accomplished by a node of the quad-tree corresponding to a 2^i-square.

occupancy integer parameter $C[v]$, which is initially set to 0 (we refer to this condition as the *skeletal* quad-tree). A rectangle R contributes 1 to $C[v]$ if and only if (i) R contains the square of v and (ii) no ancestor of v in T enjoys the same property. Clearly, this specifies a unique way to partition R into squares, each of which is stored in a node of v. While in the segment tree each segment is partitioned into at most $O(\log N)$ intervals, it is relatively simple to show[3] that a rectangle is partitioned into at most $O(N)$ squares. Thus an insertion into and deletion from a quad-tree, as well as the determination of the measure-of-union (area), can each be accomplished at a cost of $O(N)$ operations. We conclude therefore that the section-to-section update in a space-sweep application will cost $O(N)$ time, and, since there are $2N$ sections, a total $O(N^2)$ running time obtains for the measure-of-union problem in the three-dimensional space. This technique—rather than the two-dimensional one based on the segment-tree—can be used as the basis of induction, thereby proving the following result.

Theorem 8.2. *The measure of the union of N isothetic hyperrectangles in $d \geq 3$ dimensions can be obtained in time $O(N^{d-1})$.*

Although it seems rather difficult to improve upon this result, no conjecture about its optimality has been formulated.

Remark. The algorithm for computing the measure of a union F of rectangles can be appropriately modified to solve the following problem.

PROBLEM R.3 (PERIMETER OF UNION OF RECTANGLES). Given the union F of N isothetic rectangles, find the *perimeter* of F, i.e., the length of its boundary.

---

[3] The reader is referred to the paper by van Leeuwen–Wood (1981) for a proof.

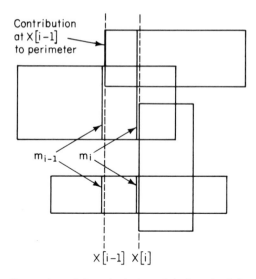

Figure 8.10   Illustration of the calculation of the length of the vertical sides.

The technique is based on the following observation. Suppose that at the i-th step a rectangle is either inserted into or deleted from the segment tree (conventionally, we say that a rectangle is inserted or deleted when the plane sweep reaches either the left or the right side of the rectangle, respectively). Refer to Figure 8.10 and recall that m_i gives the total length of the vertical section immediately to the left of $X[i]$ (see line 6 of the algorithm MEASURE OF UNION OF RECTANGLES). The total length of the vertical sides of the boundary of F at $X[i-1]$ is given by the difference between the lengths of a vertical section of F immediately to the left of $X[i-1]$ and of one immediately to the right of it. While the first is by definition m_{i-1}, the second is equal to m_i, since the section is invariant in the interval $[X[i-1], X[i]]$. Thus the sought total length is given by $|m_i - m_{i-1}|$.[4] In addition, we must consider the contribution of the horizontal boundary sides in a vertical strip. In $[X[i-1], X[i]]$ this is obviously given by $(X[i] - X[i-1]) \times \alpha_i$, where the integer α_i is the number of horizontal sides of the boundary of F in the vertical strip under consideration. The parameter α_i is very similar in nature to m_i. Indeed, let us define for each node v of the segment tree three new specific parameters, an even integer parameter $\alpha[v]$ and two binary parameters $\mathrm{LBD}[v]$ and $\mathrm{RBD}[v]$. Denoting by $\mathscr{I}$ the current vertical section of F ($\mathscr{I}$ is a disjoint collection of intervals), these parameters are defined as follows:

$$\alpha[v] := \text{twice the number of disjoint portions of } \mathscr{I} \cap [[B[v], E[v]];$$

---

[4] Note that it is possible to have $X[i-1] = X[i]$, i.e., two vertical sides occur at the same abscissa. In this case we adopt a conventional lexicographic ordering (for example, on the ordinates of their lower extremes), and imagine that the two sides are fictitiously separated by ε in the x-direction.

LBD[v] := 1 or 0, depending upon whether or not $B[v]$ is the lower
            extreme of an interval in $\mathscr{I} \cap [B[v], E[v]]$;
RBD[v] := 1 or 0, depending upon whether or not $E[v]$ is the upper
            extreme of an interval in $\mathscr{I} \cap [B[v], E[v]]$.

These three specific parameters are initialized to 0 and their maintenance is
carried out by the following specialization of the subroutine UPDATE, to be
called in line 4 of procedure INSERT (see Section 8.2):

procedure UPDATE(v)
1. **begin if** ($C[v] > 0$) **then**
2.   **begin** $\alpha[v] := 2$;
3.   LBD(v] := 1;
4.   RBD[v] := 1
           end
5.   **else begin** $\alpha[v] := \alpha[\text{LSON}[v]] + \alpha[\text{RSON}[v]] - 2 \cdot \text{RBD}[\text{LSON}[v]]$
                     $\cdot \text{LBD}[\text{RSON}[v]])$;
6.   LBD[v] := LBD[LSON[v]];
7.   RBD[v] := RBD[RSON[v]]
         end
  end.

The correctness of this UPDATE procedure is readily established. If $C[v] \geq 1$,
i.e., $[B[v], E[v]]$ is fully occupied, there is just one term in $\mathscr{I} \cap [B[v], E[v]]$,
whence $\alpha[v] = 2$, $\text{LBD}[v] = \text{RBD}[v] = 1$, as in lines 2–4. If $C[v] = 0$, then
$\mathscr{I} \cap [B[v], E[v]]$ contains as many terms as the sum of the terms in its two
offspring nodes, except when the section $\mathscr{I}$ contains an interval bridging the
point $E[\text{LSON}[v]] = B[\text{RSON}[v]]$; this last case is characterized by
$\text{RBD}[\text{LSON}[v]] = \text{LBD}[\text{RSON}[v]] = 1$. This establishes the correctness of
lines 5–7. It is also easily realized that the additional parameters α, LBD, and
RBD are maintained at a constant cost per visited node.

It is now evident that $\alpha_i = \alpha[\text{root}(T)]$. The algorithm that computes the
perimeter of F is then obtained by a simple modification of the corresponding
measure-of-union algorithm. Notice, however, that while the measure-of-
union algorithm accumulates the area of F in the just completed vertical strip
(so that the segment-tree update follows the area accumulation), here the
situation is slightly different. Referring to Figure 8.11, the contribution of the
current step to the perimeter consists of two items: the horizontal edges in the
strip $[X[i-1], X[i]]$—contributing $\alpha_i \times (X[i] - X[i-1])$—and the vertical
edges at abscissa $X[i]$—contributing $|m_{i+1} - m_i|$. Thus the value of α_i must be
extracted from the segment tree *before* the update of the latter, while m_{i+1} is to
be extracted *after* the update. Thus we have

procedure PERIMETER OF UNION OF RECTANGLES
begin $X[1:2N] :=$ sorted sequence of abscissae of vertical sides;
    $X[0] := X[1]$;
    $m_0 := 0$;

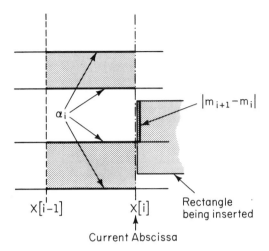

Figure 8.11   The contribution of the current step to the perimeter consists of horizontal edges and vertical edges.

$p := 0$;
construct and initialize segment tree T of ordinates of rectangle sides;
for $i := 1$ **until** $2N$ **do**
    begin $\alpha^* := \alpha[\text{root}(T)]$;
        if $(X[i]$ is a left side abscissa) **then** INSERT$(b_i, e_i; \text{root}(T))$
        else DELETE$(b_i, e_i; \text{root}(T))$;
        $m^* := m[\text{root}(T)]$;
        $p := p + \alpha^* \times (X[i] - X[i-1]) + |m^* - m_0|$;
        $m_0 := m^*$
    end
end.

In conclusion, we have

Theorem 8.3. *The perimeter of a union of N isothetic rectangles can be computed
in time $O(N \log N)$.*

8.5 The Contour of a Union of Rectangles

The same general approach (a plane-sweep technique supported by the segment tree) can be profitably used to solve another interesting problem: the
determination of the contour of the union F of N isothetic rectangles
$R_1, \ldots, R_N$. Again, $F = R_1 \cup \ldots \cup R_N$ may consist of one or more disjoint
connected components, and each component may or may not have internal
holes (note that a hole may contain in its interior some connected components
of F). The contour (boundary) of F consists of a collection of disjoint cycles

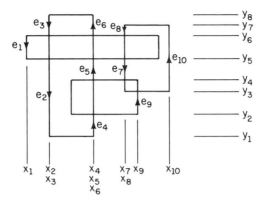

Figure 8.12   An instance of the problem.

composed of (alternating) vertical and horizontal edges. By convention, any edge is directed in such a way that we have the figure on the left while traversing the edge; this is equivalent to saying that a cycle is oriented clockwise if it is the boundary of a hole, and counterclockwise if it is an external boundary of a connected component. So we have

PROBLEM R.4 (CONTOUR OF UNION OF RECTANGLES). Given a collection of N isothetic rectangles find the contour of their union.

We note at first that the perimeter—for which a rather simple algorithm was described in the preceding section—is nothing but the length of the contour. However, we shall see that it is significantly simpler to obtain the perimeter than the contour itself, as a collection of disjoint cycles.

The algorithm is best introduced, informally, with the aid of an example. The technique has two main phases. In the first phase we find the set V of vertical edges of the contour (edges e_1 through e_{10} in Figure 8.12); in the second phase we link these vertical edges by means of horizontal edges to form the oriented cycles of the contour [Lipski–Preparata (1980)].

We denote by $(x; b, t)$ a vertical edge of abscissa x having b and t $(b < t)$ as its bottom and top ordinates, respectively; similarly, $(y; l, r)$ represents a horizontal edge of ordinate y having l and r $(l < r)$ as its left and right abscissae, respectively.

In order to obtain the set V we scan the abscissae corresponding to vertical sides of rectangles from left to right. At a generic abscissa c, the section $\mathscr{I}(c)$ of the vertical line $x = c$ with F is a disjoint union of intervals. This section remains constant between two consecutive vertical sides of the rectangles, and is updated in the scan each time one such side is reached. If s is a left vertical side of some rectangle R at abscissa c, the portion of the contour of F contributed by s is given by $s \cap \overline{\mathscr{I}(c_-)}$, where $\mathscr{I}(c_-)$ is the union of intervals immediately to the left of abscissa c, and $\overline{\mathscr{I}(c_-)}$ is its set-theoretic comple-

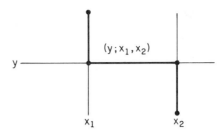

Figure 8.13   Each horizontal edge is adjacent to two vertical edges.

ment. Similarly, if s is the right vertical side of R at c, then the contribution of s to the contour is $s \cap \overline{\mathscr{I}(c_+)}$, with straightforward notation. Storing and updating the section $\mathscr{I}$ and an efficient determination of either $s \cap \mathscr{I}(c_-)$ or $s \cap \mathscr{I}(c_+)$ represent the most delicate part of the algorithm, which we shall address later.

Thus, for the time being, we assume that the set V has been obtained, and, in conformity with the adopted convention on the orientation of the contour, each of the edges in V is directed either upward or downward depending on whether it originates from a right or left side of a rectangle, respectively.

We now observe that, once the set V of vertical edges is available, the horizontal edges can be obtained in a straightforward manner. Indeed if there is a horizontal edge $(y; x_1, x_2)$ (see Figure 8.13), then there are two vertical edges at abscissae x_1 and x_2, respectively, each having an endpoint at ordinate y. Suppose then that for vertical edge $e_j \in V$, with $e_j = (x_j; b_j, t_j)$ we generate the pair of triples $\langle x_j, b_j; t_j \rangle$ and $\langle x_j, t_j; b_j \rangle$. Each of these triples is to be interpreted as a point (given by the two leftmost coordinates), and the two triples correspond to the two endpoints of e_j; the third term of each triple is simply a reference to the other endpoint of e_j. We now sort the set of triples *lexicographically in ascending order (first on the ordinate, next on the abscissa)*. In our example we obtain the following sequence.

$$\langle x_2, b_2; t_2 \rangle \langle x_4, b_4; t_4 \rangle \langle x_4, t_4; b_4 \rangle \langle x_9, b_9; t_9 \rangle \langle x_9, t_9; b_9 \rangle \langle x_{10}, b_{10}; t_{10} \rangle$$
$$\langle x_5, b_5; t_5 \rangle \langle x_7, b_7; t_7 \rangle \langle x_1, b_1; t_1 \rangle \langle x_2, t_2; b_2 \rangle \langle x_5, t_5; b_5 \rangle \langle x_7, b_7; t_7 \rangle$$
$$\langle x_1, t_1; b_1 \rangle \langle x_3, b_3; t_3 \rangle \langle x_6, b_6; t_6 \rangle \langle x_8, b_8; t_8 \rangle \langle x_8, t_8; b_8 \rangle \langle x_{10}, t_{10}; b_{10} \rangle$$
$$\langle x_3, t_3; b_3 \rangle \langle x_6, t_6; b_6 \rangle.$$

The reason for the chosen arrangement is the following. Let $a_1, a_2, \ldots, a_p$ (p even) be the resulting sequence. It is easy to see that the horizontal edges of the contour can now be obtained as the segments joining the vertices represented by a_{2k-1} and a_{2k}, for $k = 1, 2, \ldots, p/2$. More exactly, let $a_{2k-1} = \langle l, y; y_1 \rangle$, $a_{2k} = \langle r, y; y_2 \rangle$. The horizontal edge $(y; l, r)$ is assigned the direction from left to right if either the edge corresponding to the triple $\langle l, y; y_1 \rangle$ is directed downward and $y < y_1$, or if $\langle l, y; y_1 \rangle$ is directed upward and $y_1 < y$; otherwise we direct $(y; l, r)$ from right to left. In our example, the pair $(a_1, a_2) =$

$(\langle x_2, b_2; t_2 \rangle, \langle x_4, b_4; t_4 \rangle)$, with $b_2 = b_4$, gives rise to the horizontal edge $(b_2; x_2, x_4)$ which is oriented from left to right because the edge e_2, corresponding to $\langle x_2, b_2; t_2 \rangle$, is directed downward and $b_2 < t_2$; by contrast, the pair $(a_{17}, a_{18}) = (\langle x_8, t_8; b_8 \rangle, \langle x_{10}, t_{10}; b_{10} \rangle)$, with $t_8 = t_{10}$, gives rise to the edge $(t_8; x_8, x_{10})$, which is directed from right to left because e_8 is directed downward but $t_8 \not< b_8$. It is clear that by a single scan of the sequence $a_1, \ldots, a_p$ we may produce all the horizontal edges and doubly link each of them to its two adjacent vertical edges. The resulting lists give explicitly the cycles of the contour. Moreover, if we identify each cycle by means of its vertical edge of minimal abscissa, this edge determines with its direction whether the cycle is the boundary of a hole (when the edge is directed upward), or is a portion of the external boundary (when the edge is directed downward).

From a performance viewpoint, the reader should not worry about the recourse to a sorting operation of $2p$ items (the lexicographic sorting described above). Indeed, by a preliminary sorting of the abscissae and ordinates of the rectangle sides (an $O(N \log N)$ operation), we may normalize the coordinates and replace them by consecutive integers. At this point, the $2p$ triples to be lexicographically sorted consist of integers. By using standard bucket sorting techniques [Knuth (1973)], the desired sorting is obtained in time $O(p)$; additional time $O(p)$ is used to generate the horizontal edges and the contour cycles.

We now turn our attention to the efficient implementation of the first phase of the algorithm, that is, the generation of the set V of vertical contour edges. As it is to be expected, the proposed technique is a plane sweep supported by the segment tree T. Referring to the general scheme discussed in Section 8.3, the specific node parameter used in the application is STATUS$[v]$, which provides a rough classification of the measure of $\mathscr{I} \cap [B[v], E[v]]$. Specifically, STATUS may assume one of three values as follows:

$$
\text{STATUS}[v] = \begin{cases} \text{full,} & \text{if } C[v] > 0, \\ \text{partial,} & \text{if } C[v] = 0 \text{ but } C[u] > 0 \\ & \quad \text{for some descendant } u \text{ of } v, \\ \text{empty,} & \text{if } C[u] = 0 \text{ for each } u \text{ in the subtree rooted at } v. \end{cases}
$$

With this definition, the current section $\mathscr{I}$ is the union of all segments $[B[v], E[v]]$ over all nodes of the segment tree whose STATUS is full.

Given a segment $s = (x; b, e)$ (a vertical side of a rectangle), the set $s \cap \bar{\mathscr{I}}$ is illustrated in Figure 8.14; $s \cap \bar{\mathscr{I}}$ is the sequence of the *gaps* between consecutive members of $\mathscr{I}$ in the window $[b, e]$. The side s, when stored in the segment tree, is partitioned into $O(\log N)$ fragments in the well-known manner.

It is convenient to obtain $s \cap \bar{\mathscr{I}}$ as the union of the contributions of each of these fragments. (Recall that each such fragment corresponds to a node v of T for which $b \le B[v] < E[v] \le e$.) It is easy to realize that the contribution *contr*(v) to $s \cap \bar{\mathscr{I}}$ of a node v, corresponding to a fragment of s, is given by:

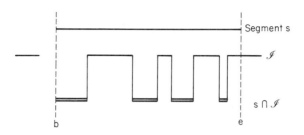

Figure 8.14   Illustration of the set of intervals $s \cap \bar{\mathscr{I}}$. (Note that s is represented as a horizontal segment.)

$$contr(v) = [B[v], E[v]] \cap \bar{\mathscr{I}} = \begin{cases} \emptyset & \text{if STATUS}[v] \text{ is full or} \\ & \text{STATUS}[u] \text{ is full for} \\ & \text{some ancestor } u \text{ of } v \\ & \text{in } T \\ [B[v], E[v]] & \text{if STATUS}[v] \text{ is empty} \\ contr(\text{LSON}[v]) \cup contr(\text{RSON}[v]) \\ & \text{if STATUS}[v] \text{ is partial.} \end{cases}$$

It follows that the subtree rooted at v must be searched only if STATUS$[v]$ = partial. We shall return later to the implementation of this search.

Assuming for the time being that the sequence ($contr(v)$: v corresponds to a fragment of s) has been obtained, in order to ensure that each contour edge is produced as a single item, contiguous intervals have to be merged. The situation is illustrated in Figure 8.15. Here, a segment $s = (x; b, e)$ is decomposed by T into five fragments, each corresponding to a node of T. Each such node generates, in general, a collection of intervals for $s \cap \bar{\mathscr{I}}$. Any two intervals, which are contiguous and have been generated by distinct segment-tree nodes, have to be merged together. To implement this task, the intervals of $contr(v)$ are assembled in a STACK corresponding to a bottom-to-top scan of the figure in the plane. At the top of STACK there is always the upper

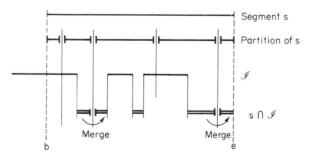

Figure 8.15   Merging of continguous intervals produced in the computation of $s \cap \bar{\mathscr{I}}$. (Again, s is displayed horizontally.)

extreme of the last inserted interval. If the lower extreme of the next interval to be added to STACK coincides with the top of STACK, then the latter is deleted from STACK prior to adding to STACK the upper extreme of the next segment, thereby realizing the desired merges.

We can now describe the search for the gaps of $\mathcal{I}$ in $s = (x; b, e)$. The original implementation of Lipski and Preparata has the standard structure, and makes use of a subroutine $\text{CONTR}(b, e; \text{root}(T))$, which reports the set $[b, e] \cap \bar{\mathcal{I}}$ and their common abscissa:

procedure CONTOUR OF UNION OF RECTANGLES
begin $X[1 : 2N] :=$ sorted sequence of abscissae of vertical sides; (*see
    comment below*)
    $\mathcal{A} := \varnothing; (*\mathcal{A}$ is the set of vertical edges*)
    construct and initialize segment tree T of ordinates of rectangle sides;
    for $i := 1$ **until** $2N$ **do**
      begin if $(X[i]$ is a left-side abscissa) **then**
          begin $\mathcal{A} := \text{CONTR}(b_i, e_i; \text{root}(T)) \cup \mathcal{A}$;
              $\text{INSERT}(b_i, e_i; \text{root}(T))$
        end
        else begin $\text{DELETE}(b_i, e_i; \text{root}(T))$;
              $\mathcal{A} := \text{CONTR}(b_i, e_i; \text{root}(T)) \cup \mathcal{A}$
        end
end.

It is appropriate to point out that the update of the set of vertical edges must precede the insertion of a left side, while it must follow the deletion of a right side. This has the consequence that if a right side and a left side share the same abscissa, processing of the left side must precede processing of the right side; therefore, the sorting step of the above algorithm must satisfy this condition.

Two items must still be discussed. The first is the update of the specific parameters (in this case, STATUS[v]) in INSERT and DELETE; the second is the subroutine CONTR.

As to the first item we have the following action, which is obviously executed in constant time per visited node:

procedure UPDATE(v)
begin if $(C[v] > 0)$ **then** STATUS[v] := full
    else if $(\text{LSON}[v] = \Lambda)$ **then** STATUS[v] := empty ($*v$ is a leaf*)
        else if $(\text{STATUS}[\text{LSON}[v]] = \text{STATUS}[\text{RSON}[v]] = \text{empty})$
            then STATUS[v] := empty
            else STATUS[v] := partial
end.

We now describe the subroutine CONTR:

    function CONTR($b, e; v$)
(*this function makes use of a STACK, which is external to it. STACK is

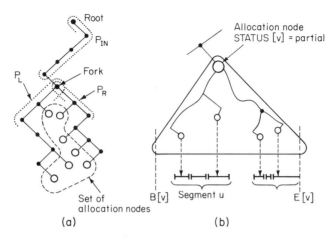

Figure 8.16   (a) Typical substructure of T visited by INSERT or DELETE. (b) For each partial-status allocation node reached by CONTR, the subroutine visits all the paths leading to end-nodes of segments in $\bar{\mathscr{I}} \cap [B[v], E[v]]$.

initialized as empty when the call $\text{CONTR}(b, e; \text{root}(T))$ is issued by the main procedure. This function pushes on STACK a sequence of segments representing $[b, e] \cap contr(v)$. The content of STACK is returned by the call $\text{CONTR}(b, e; \text{root}(T))*$)

    begin
1.   **if** $(\text{STATUS}[v] \neq \text{full})$ **then**
2.   **if** $(b \leq B[v])$ **and** $(E[v] \leq e)$ **and** $(\text{STATUS}[v] = \text{empty})$ **then**
        (*$[B[v], E[v]]$ is contributed*)
3.   **begin if** $(B[v] = \text{top}(\text{STACK}))$ **then** (*merge contiguous segments*)
4.   delete top(STACK)
5.   **else** $\text{STACK} \Leftarrow B[v]$ (*beginning of edge*);
6. $\text{STACK} \Leftarrow E[v]$ (*current termination of edge*)
     end
7.   **else begin if** $(b < (\lfloor B[v] + E[v] \rfloor / 2 \rfloor))$ **then**
8. $\text{CONTR}(b, e; \text{LSON}[v])$;
9.   **if** $(\lfloor (B[v] + E[v]) / 2 \rfloor < e)$ **then**
10. $\text{CONTR}(b, e; \text{RSON}[v])$
      end
   end.

The itinerary followed by procedure CONTR in the segment tree (i.e., the sequence of the visited nodes) substantially overlaps with that of the corresponding INSERT (or DELETE), with two significant differences, which we now discuss in the context of INSERT (with no loss of generality). It is well-known that the itinerary of INSERT has a typical structure, already discussed in Section 1.2.3.1 and repeated in Figure 8.16 for the reader's convenience. A (possibly empty) initial path P_{IN} leads to a node, referred to as the *fork*, from which two (possibly empty) paths P_L and P_R diverge; shown is also the set of the *allocation* nodes identifying the segmentation of the inserted interval. Subroutine CONTR visits P_{IN}, P_L, and P_R; however, if a node on any of these paths is a full-status node, then the traversal of the path is aborted (line 1 of CONTR). This occurs because the y-interval of segment s (or a portion thereof) is "covered" by $\mathscr{I}$ and its contribution to $s \cap \bar{\mathscr{I}}$ is empty. For any allocation node v with nonfull status reached by CONTR, if STATUS[v] = empty then the entire [$B[v]$, $E[v]$] is contributed; otherwise CONTR begins a search of the subtree rooted at v (lines 7–10) (this happens, of course, only if STATUS[v] = partial). This search is the most time consuming portion of the task, as we shall now analyze. Referring to Figure 8.16(b), for each segment u in $\bar{\mathscr{I}} \cap [B[v], E[v]]$ there are two unique nodes in the subtree rooted at v corresponding to the leftmost and rightmost fragments of u. A simple analysis shows that CONTR performs a preorder traversal of the paths leading from v to these nodes, performing a fixed amount of work at each node visited, and possibly at its siblings. Thus the total work is proportional to the total length of these paths. By a general property of binary trees [Lipski–Preparata (1980)] if there are v end-nodes in the subtree, then the desired total path length is bounded by $v \log(16N/v)$. Therefore, denoting by n_i the number of disjoint pieces in $s_i \cap \bar{\mathscr{I}}$, where s_i is the i-th rectangle side processed by the algorithm, we have that the total work of CONTR is bounded by

$$C \sum_{i=1}^{2N} n_i \log \frac{16N}{n_i} \le Cp \log \frac{32N^2}{p},$$

where p is the total number of contour edges and C is a constant. Combining this result with the straightforward analysis of INSERT and DELETE and with the result on the performance of the second phase of the technique (which links horizontal and vertical contour edges), we have

Theorem 8.4. *The p-edge contour of a union of N isothetic rectangles can be found in time $O(N \log N + p \log (N^2/p))$.*

To address the question of optimality, we note that $\Omega(N \log N + p)$ is the known lower bound for this problem. Indeed $O(p)$ is a trivial bound induced by the contour size, while an $\Omega(N \log N)$ lower bound is obtained by transforming SORTING to finding the contour of a (hole-free) union of rectangles as shown by the following.

Theorem 8.5. *The complexity of finding a contour of $F = R_1 \cup \cdots \cup R_N$, where F is without holes, is $\Omega(N \log N)$ under the decision-tree model.*

PROOF. We show that sorting of N numbers $x_1, \ldots, x_N$ is transformed in $O(N)$ time to our problem. Indeed, given x_i, let R_i be the Cartesian product of the x-interval $[0, x_i]$ and of the y-interval $[0, M - x_i]$, where

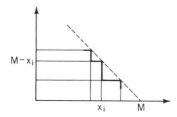

Figure 8.17   Transforming SORTING to finding the contour of a hole-free union of rectangles.

$$M = \max_{1 \le i \le N} x_i + 1$$

(see Fig. 8.17; without loss of generality we assume $x_1, \ldots, x_N > 0$). It is clear that $F = R_1 \cup \cdots \cup R_N$ is without holes and that from the contour of F we can obtain the sorted sequence of the x_i in $O(N)$ time.   □

The preceding algorithm falls short of the lower bound. Indeed, it is not difficult to detect some inefficiency in the search of the subtrees of allocation nodes with partial status, since the cost of traversing long paths has as payoff only their terminal nodes. This shortcoming was obviated by Güting (1984), who designed an ingenious data structure for reporting the gaps in each subtree in time proportional to their number, thereby achieving an asymptotically optimal algorithm. As may be expected, Güting's data structure is considerably complex in its maintenance. The reader is referred to the original paper for a treatment of the technique.

8.6 The Closure of a Union of Rectangles

We begin by giving a precise definition of the closure of a union of rectangles. We say that two points $p_1 = (x_1, y_1)$ and $p_2 = (x_2, y_2)$ in the plane are *incomparable* if they are not related in the "dominance" relation (see Section 4.1.3), i.e., if $(x_1 - x_2)(y_1 - y_2) < 0$. If we assume, without loss of generality, that $x_2 > x_1$, following a convenient terminology [Soisalon-Soininen and Wood (1982)] we call the SW- and NE-*conjugates* of p_1 and p_2 the two points $q_1 = (x_1, y_2)$ and $q_2 = (x_2, y_1)$, respectively (see Figure 8.18). Given a plane region R (not necessarily connected), two points are said to be connected in R if there exists a curve, totally contained in a component of R, connecting the two points. We then have

Definition 8.2. A region R of the plane is NE-*closed* if for any two incomparable points p_1 and p_2 connected in R, the NE-conjugate of p_1 and p_2 is also in R. Analogously, we define a SW-*closed* region of plane.

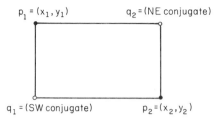

Figure 8.18   Two points p_1 and p_2, incomparable under dominance, and their conjugates q_1 and q_2.

Definition 8.3. The NE-*closure* of a region S of the plane, denoted NE(S), is the smallest NE-closed region R containing S. Analogously we define the SW-*closure*, SW(S), of S. The NESW-*closure* (or *closure*, for short) denoted NESW(S), is the smallest region R containing S, that is both SW-closed and NE-closed.

A theory of closure has been developed in the works of Yannakakis–Papadimitriou–Kung (1979), Lipski–Papadimitriou (1981), and Soisalon-Soininen and Wood (1982). For our purposes, it suffices to note some significant properties.

A curve Γ in the plane is said to be *x-monotonic* if for any two points (x_1, y_1) and (x_2, y_2) on Γ, $x_2 > x_1 \Rightarrow y_2 \geq y_1$. Also, given a forbidden domain D, two curves Γ_1 and Γ_2 in the plane are *homotopic* if they can be transformed continuously one to the other, without intersecting D.

If by $R(F)$ we denote the minimum enclosing isothetic rectangle of a given (not necessarily connected) plane domain F, then we have the obvious inclusions

$$F \subseteq \text{NESW}(F) \subseteq R(F).$$

If NESW(F) is connected, then it is easily shown that the NE- and SW-corners of $R(F)$ (see Figure 8.19(a)) belong to the closure of F. Also, for each connected component G of NESW(F) (see Figure 8.19(b)), the boundary of NESW(F) consists of two x-monotonic curves, respectively homotopic to the upper-left and to the lower-right boundary portions of $R(G)$. Each of these two curves is, in its family of homotopic curves, the extreme of x-monotonic curves not intersecting G. It is appropriate to refer to these two curves as the *upper* and *lower* contours of the closed region NESW(F). In general, for an arbitrary domain F, the closure NESW(F) of F consists of one or more components; if there are more than one component, any two of them must be separable by an x-monotonic curve (otherwise the closure property would be violated) (see Figure 8.19(c)).

Soisalon-Soininen and Wood also prove the following significant property, which holds for any figure F in the plane:

$$\text{NESW}(F) = \text{NE}(\text{SW}(F)) = \text{SW}(\text{NE}(F)). \tag{8.1}$$

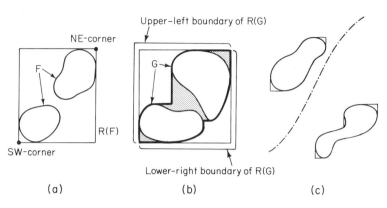

Figure 8.19   (a) A plane figure F and its minimum enclosing rectangle. (b) A connected component of the NESW-closure of a plane figure F. (c) Separability of two components of NESW (F).

In our applications, we are interested in the case in which F is a union of isothetic rectangles, so that we have the following.

PROBLEM R.5 (CLOSURE OF UNION OF RECTANGLES). Given a collection of N isothetic rectangles find the closure of their union.

Property (8.1) suggests two equivalent algorithms to compute the closure of a given figure F. For example the identity $\text{NESW}(F) = \text{SW}(\text{NE}(F))$ suggests to compute first the NE-closure of F, and subsequently to complete the task by computing the SW-closure of $\text{NE}(F)$. These two tasks are considerably simplified when F is the union of isothetic rectangles, since the upper and lower contours of any component fo $\text{NESW}(F)$ are each an x-monotonic staircase curve. Informally, the algorithm will act as follows. It consists of two plane sweeps, in opposite directions. The first, from left to right, constructs the NE-closure of the figure F (a union of isothetic rectangles), that is, it identifies the NE-closed components. The second plane sweep, from right to left, constructs the SW-closure of $\text{NE}(F)$, that is, starting from the connected components of $\text{NE}(F)$, it identifies the components of $\text{SW}(\text{NE}(F))$ (viz., of $\text{NESW}(F)$).

A component of $\text{NE}(F)$ is said to be *active* (in the plane sweep) if it is intersected by the sweep-line; clearly, a component of $\text{NE}(F)$ is active if and only if it contains at least one rectangle of F intersecting the sweep-line (referred to as an *active rectangle*).

Let us first examine the two basic situations to be handled by the plane sweeps (for concreteness, we make reference to the left-to-right sweep that constructs $\text{NE}(F)$). Obviously the event-points are the abscissae of the vertical sides of the rectangles. When the current event is a left-side, then we (i) initialize an active component, or (ii) extend an active component, or (iii) merge two or more active components; on the other hand, the completion

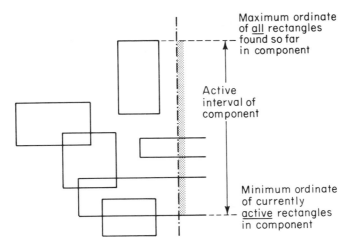

Figure 8.20   An active component and its active interval. (A shaded strip will be used here and hereafter to denote active intervals.)

(and deactivation) of an active component may only occur in coincidence with a right-side of a rectangle. An active component is characterized by an *active interval* (at the sweep-line) which is a plane strip comprised between the maximum ordinate of *all* rectangles found so far in the component and the minimum ordinate of the currently *active* rectangles in the components (a typical situation is shown in Figure 8.20). The active interval plays a crucial role in discriminating the three possible situations that may occur when the sweep line reaches a left side, as illustrated in Figure 8.21.

When the current event is a right side of a rectangle R, three situations may occur: (i) R ceases being an active rectangle without affecting the active interval (Figure 8.22(a)); or (ii) R ceases being an active rectangle and the

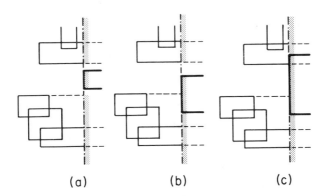

Figure 8.21   Situations that may arise when the plane sweep reaches a left side. (a) A new component is initiated; (b) an existing component is extended; (c) two (or more) components are bridged. (Shown are the active intervals *after* the insertion.)

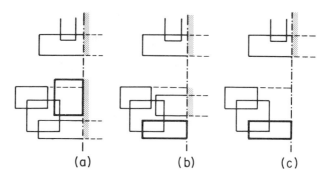

Figure 8.22 Situations that may arise when the plane sweep reaches a right side. (a) A rectangle is deactivated without affecting the active interval; (b) a rectangle is deactivated shrinking the active interval; (c) the component is terminated. (Again, shown are the active intervals *after* the deletion.)

active interval shrinks upward (Figure 8.22(b): this occurs only when R is the *bottommost* active rectangle of the component); or (iii) the component is terminated (Figure 8.22(c): this occurs only when R is the *only* active rectangle of the component).

To handle the previously described actions, the sweep-line status is implemented by a moderately complex data structure. There is a primary structure containing the extremes of the active intervals, organized as the leaves of a height-balanced tree T. Each leaf of T storing the lower endpoint of an active interval points to a secondary structure, itself a height-balanced tree, storing the active rectangles in the component. Intersections (of left sides of rectangles) and deletions (of right sides) are carried out as follows.

When a left side $[y_1, y_2]$ is to be processed, y_1 is first located in T and then the leaf sequence is traversed until y_2 is also located. This operation is carried out in time $O(\log N + h)$, where h is the number of merged active intervals (Figure 8.21(c)); y_2 is inserted also into a secondary structure and, if $h \geq 2$, the corresponding secondary structures are concatenated (as concatenable queues). The work pertaining to the concatenation of secondary structures is easily estimated for the overall execution of the algorithm by charging the $O(\log N)$ work of concatenating two such structures A_1 and A_2 to the rightmost active rectangle of A_1: since each rectangle can be charged at most once, the upper bound $O(N \log N)$ is established for this activity.

When a right side $[y_1, y_2]$ is to be processed, again y_1 is located in an active interval (in the primary structure T). Next, y_1 is also located in the appropriate secondary structure and deleted from it, and, if it coincides with the minimum element in the secondary structure, the lower extreme of the active interval is updated (refer to Figure 8.22(b)). Finally, if y_1 is the only member of the secondary structure, the corresponding active interval is deleted altogether from the primary structure. Clearly, processing of a right side can be carried out in time $O(\log N)$.

The right-to-left plane sweep that constructs $SW(NE(F))$ is analogous, so that we have

Theorem 8.6. *The NESW-closure of a union of N isothetic rectangles can be optimally constructed in time $\theta(N \log N)$ and space $\theta(N)$.*

PROOF (Sketch). Indeed, after a preliminary sorting of the abscissae of the rectangle sides, processing of the $2N$ sides also runs in time $O(N \log N)$ as shown above. The data structures are stored in space $O(N)$. These performance bounds are optimal, since the ELEMENT UNIQUENESS problem can be easily transformed to the closure of rectangles problem.   □

Remark. The above technique can be readily specialized to compute the connected components of a set of rectangles within the same space and time bounds.[5] Indeed, it is easily realized that all that is needed is the replacement of the (single) active interval of a component (as described above) with the collection of disjoint intervals represented by the intersections of the active rectangles of a component with the sweep-line.

8.7 The External Contour of a Union of Rectangles

In the preceding sections we have seen how the same basic technique—the plane sweep—can be adapted to solve a variety of different problems concerning a union of isothetic rectangles, i.e., area, perimeter, closure, and contour (Problems R.1, R.3, R.4, and R.5). As we noted earlier (refer to Section 8.5) the contour of a union of N rectangles may have $\theta(N^2)$ edges; this does not hold for a portion of the contour, identified by the following definition.

Definition 8.4. The *external contour* of a union F of isothetic rectangles is the boundary between F and the unbounded region of the plane.

We shall now show that the external contour has $O(N)$ edges. This result is readily established if one refers to two interesting supersets of the external contour, the first of which is defined as follows.

Definition 8.5. The *nontrivial contour* of a union F of isothetic rectangles is the set of contour cycles, such that each of them contains at least one vertex of the given rectangles.

---

[5] See also [Edelsbrunner–van Leeuwen–Ottmann–Wood (1981)].

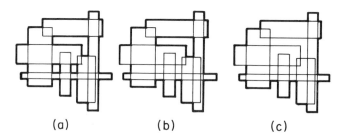

(a)   (b)   (c)

Figure 8.23   The contour (a), nontrivial contour (b), and external contour (c) of a union of rectangles.

The three notions of contour, nontrivial contour, and external contour are exemplified in Figure 8.23. It is straightforward to see that the external contour is a subset of the nontrivial contour.

It is convenient at this point to regard each edge as being lined by two arcs with opposite directions and lying on either side of the edge, as illustrated in Figure 8.24. The edges of the rectangles $R_1, \ldots, R_N$ partition the plane into regions, one of which is unbounded. With the introduction of lining arcs, the boundaries of all regions of the partition give rise to a collection of directed circuits of arcs; a circuit is called *external* or *internal* depending upon whether it is clockwise or counterclockwise. An arc of a circuit is said to be *terminal* if it contains an endpoint of the original segment (a rectangle side) to which it belongs. A circuit is said to be *nontrivial* or *trivial* depending upon whether or not it contains a terminal arc. The set of nontrivial circuits for the example of Figure 8.23 is illustrated in Figure 8.25. We can now introduce the following problems.

PROBLEM R.6 (NONTRIVIAL CONTOUR OF UNION OF RECTAN-GLES). Given a set of N isothetic rectangles, find the nontrivial contour of their union.

PROBLEM R.7 (EXTERNAL CONTOUR OF UNION OF RECTANGLES). Given a set of N isothetic rectangles, find the external contour of their union.

Figure 8.24   Conventions on directions of lining arcs of a given edge.

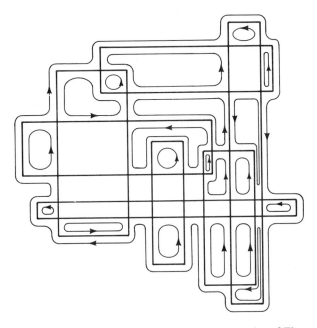

Figure 8.25 Nontrivial circuits for the union of rectangles of Figure 8.23.

It is straightforward to note the following nested inclusions.

External contour $\subseteq$ Nontrivial contour $\subseteq$ Nontrivial circuits.

An interesting property of these sets is that the number of arcs in the nontrivial circuits is $O(N)$, as established by the following theorem.

Theorem 8.7. *The total number of arcs of the nontrivial circuits of the union of N isothetic rectangles is $O(N)$.*

PROOF. We say that a vertex is of type i $(i = 1, 3)$ if the clockwise angle formed by the arcs meeting at the vertex (ordered according to the direction on the

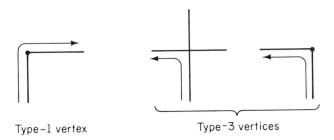

Figure 8.26 Types of vertices.

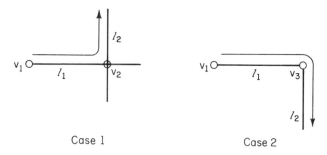

Case 1   Case 2

Figure 8.27   Cases arising in the advancing step of the algorithm.

circuit) is $i\pi/2$ (see Fig. 8.26). We also let v_i denote the number of vertices of type i on a given circuit. Then, for any circuit we have the following relation:

$$v_3 - v_1 = \begin{cases} 4, & \text{for an internal (counterclockwise) circuit,} \\ -4, & \text{for an external (clockwise) circuit.} \end{cases}$$

Consequently, the total number v of vertices (i.e., of arcs) is expressible as

$$v = v_1 + v_3 = 2v_1 \pm 4 \leq 2v_1 + 4.$$

By summing v over all nontrivial circuits we obtain (the summations are extended over all such circuits) for the total number of arcs t:

$$t \leq 2\Sigma v_1 + 4\Sigma 1$$

$$\leq 2n + 4 \cdot 4n \leq 18n. \qquad \square$$

This theorem indicates that if the (total) contour has $O(N^2)$ arcs, the increase is due to trivial circuits. Thus, if we want to obtain just the nontrivial, or the external, contours, we may wish to develop special techniques, which avoid the overhead represented by the trivial circuits. Before confronting this task, however, we note that $\Omega(N \log N)$ is a lower bound to the computational effort, since the argument used in the proof of Theorem 8.5 (in Section 8.5) applies to the present situation: indeed, in that proof, total, nontrivial, and external contours coincide.

A plane-sweep technique is not likely to cope with our problem, since in the sweep we can detect whether a circuit is trivial or nontrivial only at the completion of its construction (rather than at the beginning of it). Indeed, a recently proposed technique [Lipski–Preparata (1981)] performs a "march" along the circuits of the contour to be constructed, adding one arc at a time. Clearly, if each arc can be added in time $O(\log N)$, then, by Theorem 8.5 and 8.7, a time-optimal algorithm results. We shall now describe this technique.

The basic component of the march is an *advancing mechanism*, whereby, starting from the *current vertex* v_1 (see Figure 8.27), we march along the *current segment* l_1 in the assigned direction, and one of the following two cases

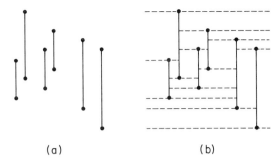

(a) (b)

Figure 8.28   The set V of vertical segments and the resulting horizontal adjacency map.

occurs:

1. There is a segment l_2, *closest* to v_1, which intersects l_1 and crosses the region to the left of l_1. In this case we make a left turn, i.e., the intersection v_2 becomes the current vertex and l_2 becomes the current segment;
2. We reach the endpoint v_3 of l_1, which is also the endpoint of a segment l_2 (obviously, l_2 does not cross the region to the left of l_1). In this case we make a right turn, i.e., v_3 becomes the current vertex and l_2 becomes the current segment.

The advancing step outlined above can be implemented by searching two suitable geometric structures, called the *horizontal and vertical adjacency maps*, which we now describe. We confine ourselves to the horizontal adjacency map (HAM), since the discussion is entirely applicable to the vertical adjacency map (VAM).

Consider the set V of the vertical segments that are the vertical sides of the given rectangles (Figure 8.28). Through each endpoint p of each member of V we trace a horizontal half-line to the right and one to the left; each of these half-lines either terminates on the vertical segment closest to p or, if no such intercept exists, the half-line continues to infinity. In this manner the plane is partitioned into regions, of which two are half-planes and all the others are rectangles, possibly unbounded in one or both horizontal directions. We shall refer to these regions as "generalized rectangles." Each generalized rectangle is an equivalence class of points of the plane with respect to their horizontal adjacency to vertical segments (whence the name "horizontal adjacency map"). A simple induction argument shows that the total number of regions in the HAM is at most $3|V| + 1$.

The horizontal adjacency map is just a special case of a subdivision of the plane induced by an embedded planar graph. Assuming that the current segment l_1 is horizontal (refer to Figure 8.27), to locate the current vertex v_1 in this map (i.e., to identify the generalized rectangle containing v_1) means to identify the vertical sides of this region. It is easy to convince ourselves that this is all that is needed to perform the advancing step. In fact suppose that $v_1 = (x_1, y_1)$ and that l_1 belongs to the line $y = y_1$ in the interval $[x_1, x_3]$. We locate in the HAM the point (x_1, y_1) and we obtain the abscissa x_2 of the closest vertical segment to the right of (x_1, y_1). If $x_2 \le x_3$, then we have a left turn (Case 1); if $x_2 > x_3$, then we have a right turn (Case 2). The other three possible cases for the current segment (to the left, above, and below the current vertex) are handled in exactly analogous ways. Thus the addition of one arc to a nontrivial circuit costs one interrogation of either adjacency map.

To perform an interrogation of an adjacency map (i.e., a planar point location) we may resort to one of the various $O(\log N)$-time search techniques described in Chapter 2, Section 2.2. Particularly suited to the situation is the so-called "(median-based) trapezoid method" (Section 2.2.2.4) for which, due to the nature of the planar map (defined by a set of parallel segments) the trapezoids specialize to rectangles. For $O(N)$ points (as is the case, since there are $4N$ points, the vertices of the given isothetic rectangles) this technique has a search time $O(\log N)$, on a data structure (a binary search tree) which can be constructed in time $O(N \log N)$. An undesirable aspect of this otherwise very simple technique is that it has a worst-case space requirement $O(N \log N)$, rather than $\theta(N)$. However, a probabilistic analysis [Bilardi–Preparata (1982)], tightly supported by extensive simulations, shows that for a wide range of hypotheses on the statistical distribution of the segments, the average space requirement is $\theta(N)$, with a very small, practically acceptable constant of proportionality (approximately 6).

The nontrivial contour (and, consequently, the external contour) can be obtained by first constructing, in a systematic way, the nontrivial circuits and then removing the unwanted terms. The generation of the nontrivial circuits can be effected by initializing the advancing step described earlier to a rectangle vertex and the arc issuing from it and external to the rectangle. Therefore each of the rectangle vertices becomes a "circuit seed." Initially, all vertices are placed in a pool (array) and are tagged. They are extracted one by one from the pool, to generate all the nontrivial circuits, and the process terminates when the pool becomes empty. Since there are $O(N)$ arcs in nontrivial circuits, and the construction of one such arc can be done in time $O(\log N)$, the set of nontrivial circuits can be constructed in time $O(N \log N)$, including the preprocessing required to construct the search trees of the adjacency maps.

To complete the construction of the nontrivial contour we must remove those circuits that do not separate F from its exterior. To discriminate between the nontrivial circuits that belong to the external contour and those that do not, a plane-sweep routine can be designed to run in time $O(N \log N)$ (see Exercise 1 at the end of this chapter).

Theorem 8.8. *The nontrivial contour and the external contour of a union of N isothetic rectangles can be constructed in optimal time $\theta(N \log N)$ with worst-case space $O(N \log N)$.*

8.8 Intersections of Rectangles and Related Problems

In the previous sections we have discussed some techniques for computing some global property of a set of isothetic rectangles, such as the measure, the perimeter, etc. In this section we shall instead consider the problem of computing two relations that are naturally defined on a set of rectangles. These two relations "intersection" and "inclusion" (or "enclosure," as it is sometimes called) are considered separately in the next subsections.

8.8.1 Intersections of rectangles

Two isothetic rectangles intersect if and only if they share at least one point. The simplest instance occurs in one dimension, where the "rectangles" are intervals on the line. Since a d-dimensional rectangle is the Cartesian product of d intervals, each on a different coordinate axis, two d-dimensional rectangles intersect if and only if their x_j-projections (two intervals) intersect, for $j = 1, 2, \ldots, d$. Thus, we realize that the one-dimensional case plays a fundamental role and we shall study it first.

Given two intervals $R' = [x'_1, x'_2]$ and $R'' = [x''_1, x''_2]$, the condition $R' \cap R'' \neq \varnothing$ is equivalent to one of the following mutually exclusive conditions:

$$x'_1 \leq x''_1 \leq x'_2; \tag{8.2}$$

$$x''_1 < x'_1 \leq x''_2. \tag{8.3}$$

The four possible configurations of interval extremes corresponding to the situation $R' \cap R'' \neq \varnothing$ are indeed encompassed by (8.2) and (8.3) above, as can be immediately verified. Thus to test whether R' and R'' intersect, we check if either the left extreme of R' falls in R'' or the left extreme of R'' falls in R'.

This characterization of the one-dimensional problem plays a crucial role in the solution of

PROBLEM R.8 (REPORT INTERSECTION OF RECTANGLES). Given a collection of N isothetic rectangles, report all intersecting pairs.

To solve this problem, it is natural to resort to a plane-sweep technique, where the event points are, as usual, the abscissae of the vertical sides of the rectangles. The sweep-line status is given by the intersections of the rectangles with the sweep-line; for ease of reference, the rectangles intersected by the sweep-line are again called *active rectangles* (see Section 8.6). In a left-to-right plane sweep, suppose the current event point is the left side of a (new) rectangle $R = [x_1, x_2] \times [y_1, y_2]$. Clearly the sweep-line abscissa x_1 (which is

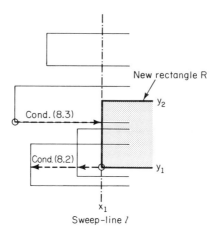

Figure 8.29 Situation arising when the left side of R is reached in the plane sweep. Only active rectangles are illustrated.

also the left extreme of the x-interval of R) belongs to the x-interval of each of the active rectangles; thus, all we need to do is to determine for which of the active rectangles either condition (8.2) or condition (8.3) holds for the y-interval as well (see Figure 8.29). This seems to involve a two-fold search, one for condition (8.2) and one for condition (8.3). The reciprocal nature of these two searches (one, the identification of all the intervals containing a point, the other, the identification of all the points contained in an interval) may suggest the use of two different data structures and the development of slightly complicated algorithms [Bentley–Wood (1980); Six–Wood (1980)].

A brilliant solution was achieved by the introduction [McCreight (1981); Edelsbrunner (1980)] of a new data structure, called the *interval tree* by Edelsbrunner. Although a sophisticated dynamic version of the interval tree is possible, we shall now describe a simple static version suited to our purposes.

If $[b^{(i)}, e^{(i)}]$ denotes the y-interval of rectangle $R^{(i)}$, we let $(y_1, y_2, \ldots, y_{2N})$ be the sorted sequence of the extremes of the N y-intervals. The (*static*) *interval tree* has a skeletal structure (called *primary* structure), statically defined for a given sequence of points (in our case, the sequence $(y_1, \ldots, y_{2N})$), while it may store an arbitrary subset (*active subset*) of the intervals whose extremes are in the set $\{y_1, y_2, \ldots, y_{2N}\}$. Thus, the *interval tree* T for the sequence $(y_1, y_2, \ldots, y_{2N})$ and interval set $I \subseteq \{[b^{(i)}, t^{(i)}]: i = 1, \ldots, N\}$ is defined as follows:

1. The root w of T has a discriminant $\delta(w) = (y_N + y_{N+1})/2$ and points to two (secondary) lists $\mathscr{L}(w)$ and $\mathscr{R}(w)$. $\mathscr{L}(w)$ and $\mathscr{R}(w)$ contain, respectively, the sorted lists of the left and right endpoints of the members of I containing $\delta(w)$. ($\mathscr{L}(w)$ and $\mathscr{R}(w)$ are sorted in ascending and descending order, respectively.)

2. The left subtree of w is the interval tree for the sequence $(y_1, \ldots, y_N)$ and

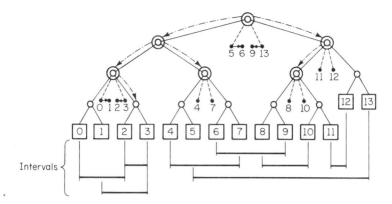

Figure 8.30   An example of the allocation of a set of intervals to the static interval tree. Regular arcs form the primary structure; broken-line arcs point to the secondary structures; dot-line arcs form the special superstructure $\mathcal{T}$.

the subset $I_L \subseteq I$ of the intervals whose right extreme is less than $\delta(w)$. Analogously, one defines the right subtree of w.

3. Each primary node of T is classified as either "active" or "inactive." A node is active if either its secondary lists are nonempty or it contains active nodes in both of its subtrees.

The static interval tree has some interesting properties, which follow directly from the definitions:

(i) The primary structure T is a balanced binary tree, the leaves of which are associated with the values $y_1, \ldots, y_{2N}$, and an inorder traversal of T yields the ordered sequence $(y_1, \ldots, y_{2N})$.

(ii) The secondary lists $\mathcal{L}(v)$ and $\mathcal{R}(v)$, for a nonleaf primary node v, must support insertions and deletions. They are conveniently realized as height-balanced or weight-balanced trees.

(iii) The active nodes can be uniquely connected as a binary tree $\mathcal{T}$, each arc of which corresponds to a (portion of a) path issuing from the root of T. Thus, each primary node v has two new pointers LPTR (left pointer) and RPTR (right pointer) that are used to realize the tree $\mathcal{T}$. If v is inactive, then LPTR$[v] = \Lambda$ and RPTR$[v] = \Lambda$; however, if v is active, then LPTR$[v] \neq \Lambda$ only if there are active nodes in the left subtree of v, and, similarly, RPTR$[v] \neq \Lambda$ only if there are active nodes in the right subtree of v. Note that more than half of the active nodes have nonempty secondary lists.

In Figure 8.30 we show an example of the allocation of intervals to a static interval tree.

We first consider the management of insertions and deletions in the static interval tree. Notice that both the primary structure and each of the secondary lists are binary trees at depth $O(\log N)$. Inserting interval $[b, e]$ consists of

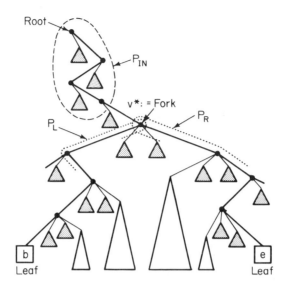

Figure 8.31   Interval tree search. The shaded triangles are secondary lists to be visited. Also to be visited (via LPTR and RPTR) are the indicated subtrees.

tracing a path from the root until the first node v^* of T is found such that $b \leq \delta(v^*) \leq e$ (fork node): at this point, the ordinate b is inserted into $\mathscr{L}(v^*)$ and the ordinate e is inserted into $\mathscr{R}(v^*)$. It is also a relatively simple exercise to see how the LPTRs and RPTRs can be maintained, at a constant cost per node, in this process. Analogously, with nonsignificant changes, we can implement deletions. Either operation runs in time $O(\log N)$.

Let us now analyze the interval tree search, that is, the detection of intersecting pairs of intervals when inserting an interval $[b, e]$ into T. Again, we trace a (possibly empty) path from the root to the first node v^* (fork) such that $b \leq \delta(v^*) \leq e$. Next we trace from v^* two diverging paths to two leaves in the subtree rooted at v^*, respectively associated with values b and e (see Figure 8.31). Let P_{IN} be the sequence of nodes from the root to the node preceding v^*; also, let P_{L} be the node sequence from v^* to leaf b, and P_{R} be the node sequence from v^* to e. We now consider the two main cases:

1. $v \in P_{\text{IN}}$. In this case, $[b, e]$ lies either to the left or to the right of $\delta(v)$. Let us assume, without loss of generality, that $e \leq \delta(v)$. Thus, before proceeding, we must check if $[b, e]$ intersects any of the active intervals that contain $\delta(v)$. For any such active interval $[b^{(i)}, e^{(i)}]$, this is done by detecting either if $b \leq b^{(i)} \leq e$ or if $b^{(i)} \leq b \leq e^{(i)}$. But since $b < e \leq \delta(v) < e^{(i)}$, the characteristic condition for $[b^{(i)}, e^{(i)}] \cap [b, e] \neq \varnothing$ is $b^{(i)} \leq e$, which is tested by scanning $\mathscr{L}(v)$ in ascending order and reporting all intervals for which the condition holds. This is done, efficiently, in time *proportional* to the number of reported intersections (with no search overhead).

2. $v \in P_{\text{L}}$ (and, analogously with the appropriate changes, if $v \in P_{\text{R}}$). If $\delta(v) \leq b$, then, arguing as for $v \in P_{\text{IN}}$, we must scan in descending order the

right list $\mathscr{R}(v)$ of v, (again, with no overhead). If $b < \delta(v)$, then $[b, e]$ is known to intersect not only *all* the active intervals assigned to v, but also *all* the intervals assigned to nodes in the right subtree of v. The first set is obtained by just reporting all the intervals whose right extreme is in $\mathscr{R}(v)$. The second set involves a visit of the active nodes of the right subtree. To efficiently accomplish this task, we shall use the arcs realized by LPTR and RPTR. By a previous argument (less than half of the active nodes in this subtree have empty secondary lists), the number of nodes visited is of the same order as the number of reported intervals.

The performance of the method outlined above is readily analyzed. First of all, the static interval tree for a collection of N intervals uses space $O(N)$, since there are $4N - 1$ primary nodes and at most $2N$ items to be stored in the secondary lists. The skeletal primary structure is constructed in time $O(N \log N)$, since a preliminary sorting of the abscissae is needed. Each interval is inserted or deleted in time $O(\log N)$, as noted earlier. The search of the tree uses time $O(\log N)$ to trace the search paths (see again Figure 8.31), while a constant amount of time is spent for each reported interval; the only overhead, to be charged to the corresponding primary node, is proportional to the number of secondary lists to be visited in this process. Notice also that if the tree search is performed only in conjunction with the insertion of a rectangle (i.e., when inserting its left side), each intersecting pair is reported exactly once. In conclusion, we summarize the previous discussion with the following theorem.

Theorem 8.9. *The s intersecting pairs in a collection of N isothetic rectangles can be reported in optimal time $\theta(N \log N + s)$, with an $O(N \log N)$ preprocessing time and using optimal space $\theta(N)$.*

PROOF. The time and space upper bounds are established by the preceding discussion. Thus we shall restrict ourselves to the consideration of optimality. The space optimality is trivial. As to the time optimality, note that ELEMENT UNIQUENESS, for which an $\Omega(N \log N)$-time lower bound holds in the decision-tree model (see Chapter 5), can be trivially transformed to our problem. Indeed, given N real numbers $\{z_1, \ldots, z_N\}$ and chosen an interval $[b, e]$, from z_i we construct the rectangle $[b, e] \times [z_i, z_i]$ and apply the rectangle intersection algorithm. If no intersection is reported, then all numbers are distinct. $\square$

8.8.2 The rectangle intersection problem revisited

As noted at the beginning of Section 8.8.1, two intervals $R' = [x'_1, x'_2]$ and $R'' = [x''_1, x''_2]$ intersect if either of the following conditions holds:

(i) $x'_1 \le x''_1 \le x'_2;$ (8.2)

(ii) $x''_1 < x'_1 \le x''_2.$ (8.3)

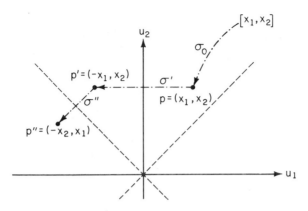

Figure 8.32   Mapping of an interval to two points p' and p'' in the plane.

It is easy to realize that the disjunction of the above two conditions is equivalent to the conjunction of the two conditions

$$x''_1 \le x'_2, \qquad x'_1 \le x''_2$$

which are trivially transformed into[5]

$$-x'_2 \le -x''_1, \qquad x'_1 \le x''_2. \tag{8.4}$$

The latter is the expression of a "dominance" relation $\prec$ in the plane (see Section 4.1.3) between point $(-x''_1, x''_2)$ and point $(-x'_2, x'_1)$. With this interpretation, the problem of determining all intersecting pairs in a collection of N intervals, is transformed as follows. Given a set $\mathscr{R} = \{R^{(j)} = [x_1^{(j)}, x_2^{(j)}]:$ $j = 1, \ldots, N\}$ of N intervals, we first define a function $\sigma_0 \colon \mathscr{R} \to E^2$ which maps the interval $[x_1, x_2]$ to the point (x_1, x_2) in the plane (refer to Figure 8.32; note that point (x_1, x_2) lies above the bisector $x_1 = x_2$ of the first quadrant). Next we define the map $\sigma' \colon E^2 \to E^2$, which is a symmetry of the plane with respect to the x_2-axis (i.e., point (x_1, x_2) is mapped to point $(-x_1, x_2)$). Finally we define a map $\sigma'' \colon E^2 \to E^2$, which is a symmetry of the plane with respect to the bisector $x_2 = -x_1$ of the second quadrant (i.e., point $(-x_1, x_2)$ is mapped to $(-x_2, x_1)$). For ease of reference, we call p' and p'' the composite functions $\sigma' \sigma_0$ and $\sigma'' \sigma' \sigma_0$, respectively, and establish the following equivalence, which is a straightforward consequence of the definitions of maps p' and p'' and of (8.4) above:

$$R^{(i)} \cap R^{(j)} \ne \varnothing \Leftrightarrow p'(R^{(i)}) \succ p''(R^{(j)}) \Leftrightarrow p'(R^{(j)}) \succ p''(R^{(i)}).$$

---

[5] Indeed, let $P_1 := (x'_1 \le x''_1)$, $P_2 := (x''_1 \le x'_2)$, $P_3 := (x''_1 < x'_1)$ and $P_4 := (x'_1 \le x''_2)$. Symbolically we have

$$P_1 P_2 \vee P_3 P_4 = T \Leftrightarrow (P_1 \vee P_3)(P_2 \vee P_3)(P_1 \vee P_4)(P_2 \vee P_4) = T$$

which, due to $\bar{P}_1 = P_3$, $\bar{P}_1 \vee P_4 = T$, $\bar{P}_3 \vee P_2 = T$, is transformed into

$$P_2 P_4 (P_2 \vee P_4) = P_2 P_4 = T.$$

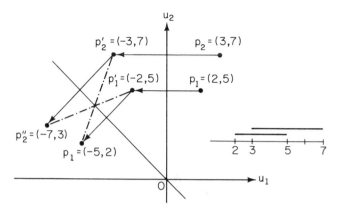

Figure 8.33   Dominance pairs, each equivalent to interval intersection.

The situation is illustrated with an example in Figure 8.33. Note that each pair of intersecting intervals gives rise to *two* pairs in the dominance relation.

Turning now our attention to Problem R.8, (report of all intersecting pairs of a set $\mathscr{R}$ of rectangles—a two-dimensional problem), given a rectangle $R = [x_1, x_2] \times [y_1, y_2]$, we first map it to the two points $P'(R) = (-x_1, x_2, -y_1, y_2)$ and $P''(R) = (-x_2, x_1, -y_2, y_1)$ in E^4, with coordinates u_1, u_2, u_3, u_4. Next we form the two sets of points in E^4:

$$S' = \{P'(R): R \in \mathscr{R}\}$$

$$S'' = \{P''(R): R \in \mathscr{R}\}.$$

The preceding discussion establishes that two rectangles R_1 and R_2 in $\mathscr{R}$ intersect if and only if $P'(R_1) > P''(R_2)$ (or, equivalently, if $P'(R_2) > P''(R_1)$). Thus the "all-intersecting pairs" Problem R.8 is a special instance of the following general dominance problem.[6]

PROBLEM R.9 (DOMINANCE MERGE). Given two sets of points S_1 and S_2 in E^d, find all pairs $p_1 \in S_1$ and $p_2 \in S_2$ such that $p_1 > p_2$.

The special nature of the problem is due to the fact that, in our situation, S' and S'' are separated by the two hyperplanes $u_1 + u_2 = 0$, $u_3 + u_4 = 0$ in the four-dimensional space. By applying an algorithm to be described in the next section, we could solve our four-dimensional dominance merge problem, and thus solve the "all-intersecting pairs" problem, in time $O(N \log^2 N + s)$, rather than in optimal time $\theta(N \log N + s)$, as provided by the technique of Section 8.8.1. The gap between the two results is more likely to be due to our inability to exploit the special nature of the problem instance, rather than to an inadequacy of the general algorithm for "dominance merge."

---

[6] Note that there is some redundancy in the solution not only because each pair will be reported twice, but because all trivial pairs (R, R) will also be reported.

8.8.3 Enclosure of rectangles

The rectangle enclosure problem (or containment or inclusion problem, as it is frequently referred to) is, in a sense, a special case of the intersection problem, and is defined as follows.

PROBLEM R.10 (RECTANGLE ENCLOSURE). Given a set S of N isothetic rectangles in the plane, report all ordered pairs of S so that the first member encloses the second.

(Notice that while intersection is a symmetric relation, enclosure is a partial ordering relation.)

Again, we assume $\mathcal{R} = \{R^{(1)}, \ldots, R^{(N)}\}$. An earlier solution to the enclosure problem resorted to data structures as the segment and range trees [Vaishnavi–Wood (1980)]; the corresponding algorithm has running time $O(N \log^2 N + s)$ and uses space $O(N \log N)$. By transforming the rectangle enclosure problem to a point dominance problem, we shall now show that the same time bound can be achieved with optimal space $\theta(N)$ [Lee–Preparata (1982)].

Denoting as usual $R^{(i)} = [x_1^{(i)}, x_2^{(i)}] \times [y_1^{(i)}, y_2^{(i)}]$, we have that $R^{(i)} \subseteq R^{(j)}$ if and only if the following four conditions jointly hold:

$$x_1^{(i)} \leq x_1^{(j)}, \qquad x_2^{(j)} \leq x_2^{(i)}, \qquad y_1^{(i)} \leq y_1^{(j)}, \qquad y_2^{(j)} \leq y_2^{(i)}. \qquad (8.5)$$

These conditions are trivially equivalent to

$$-x_1^{(j)} \leq -x_1^{(i)}, \qquad x_2^{(j)} \leq x_2^{(i)}, \qquad -y_1^{(j)} \leq -y_1^{(i)}, \qquad y_2^{(j)} \leq y_2^{(i)}, \qquad (8.6)$$

which express the well-known relation "$\prec$" of dominance between two four-dimensional points, that is

$$(-x_1^{(j)}, x_2^{(j)}, -y_1^{(j)}, y_2^{(j)}) \prec (-x_1^{(i)}, x_2^{(i)}, -y_1^{(i)}, y_2^{(i)}). \qquad (8.7)$$

Thus, after mapping each $R^{(i)} \in \mathcal{R}$ into its corresponding four-dimensional point, the rectangle enclosure problem becomes the point dominance problem in 4-space.[7] Specifically

PROBLEM R.11 (DOMINANCE). Given a set $S = \{p_1, \ldots, p_n\}$ of points in d-space, for each point $p_i \in S$ find a subset D_i of S defined by $D_i = \{p: p \in S, p \prec p_i\}$.

The technique we shall describe to solve Problem R.11 is (not surprisingly) closely reminiscent of the method introduced earlier to solve the problem of finding the maxima of a set of vectors (Section 4.1.3). Again, for $d = 4$, we let u_1, u_2, u_3, and u_4 denote the coordinates of E^4. The first preliminary step consists of transforming each $R^{(i)} \in \mathcal{R}$ to a point $p(R^{(i)})$ of E^4, where the

---

[7] This correspondence has also been noted in [Edelsbrunner–Overmars (1981)].

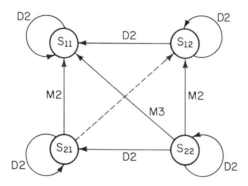

Figure 8.34   Directed-graph representation of the divide-and-conquer algorithm for DOMINANCE MERGE. Each subset is a node and each subproblem is an arc.

function $p(\ )$ is described by (8.6) and (8.7) above. Thus we obtain the set S = $\{p(R^{(i)}): R^i \in \mathcal{R}\} = \{p_1, \ldots, p_N\}$, whose elements have been reindexed so that $(i < j) \Rightarrow (u_1(p_i) \le u_1(p_j))$. We then have:

procedure DOMINANCE

D1. (Divide) Partition S into (S_1, S_2) where $S_1 = \{p_1, \ldots, p_{\lfloor N/2 \rfloor}\}$ and $S_2 = \{p_{\lfloor N/2 \rfloor+1}, \ldots, p_N\}$.

D2. (Recur) Solve the point-dominance problem on S_1 and S_2, separately.

D3. (Merge) Find all the pairs $p_i \prec p_j$, where $p_i \in S_1$ and $p_j \in S_2$.

We shall now discuss the implementation of step D3. Note that this step solves Problem R.9, "dominance merge." For $p_i \in S_1$ and $p_j \in S_2$, since $u_1(p_i) \le u_1(p_j)$ by construction, we have $p_i \prec p_j$ if and only if $u_l(p_i) \le u_l(p_j)$ for $l = 2, 3, 4$. Thus Step D3 is, in effect, a *three-dimensional problem*. Once again, we use divide-and-conquer, and denote by $\bar{u}_2$ the median of $\{u_2(p_i): p_i \in S_2\}$.

procedure DOMINANCE MERGE

M1. (Divide) Partition S_1 into (S_{11}, S_{12}) and S_2 into (S_{21}, S_{22}), so that $S_{11} = \{p: p \in S_1, u_2(p) \le \bar{u}_2\}$, $S_{21} = \{p: p \in S_2, u_2(p) \le \bar{u}_2\}$, and $S_{12} = S_1 - S_{11}, S_{22} = S_2 - S_{21}$.

M2. (Recur) Solve the merge problem on the pairs of set (S_{11}, S_{21}) and (S_{12}, S_{22}).

M3. (Combine) Find all pairs $p_i \prec p_j$ such that $p_i \in S_{11}$ and $p_j \in S_{22}$.

To establish the correctness of the technique outlined above, we note at first that S is partitioned into sets S_{11}, S_{12}, S_{21}, and S_{22}. Referring to Figure 8.34 (where each subproblem is represented by an arc), within each of these four subsets the point-dominance problem is solved in D2; it remains to be solved between pairs of subsets. Of the six pairs, (S_{11}, S_{12}) and (S_{21}, S_{22}) are also processed in D2; (S_{11}, S_{21}) and (S_{12}, S_{22}) are processed in M2; (S_{11}, S_{22}) is processed in M3, while (S_{12}, S_{21}) need not be considered, because for each

$p_i \in S_{12}$ and $p_j \in S_{21}$ we have $u_1(p_i) \le u_1(p_j)$ and $u_2(p_i) > u_2(p_j)$. Also note that Step M3 (Combine) is a two-dimensional merge problem (in u_3 and u_4).

It is obvious that the key operation of the entire task is therefore the implementation of step M3, the *two-dimensional* Merge (or Combine). Indeed the entire computation reduces to the careful sequencing of steps like M3; therefore, in what follows we shall concentrate on devising an efficient implementation of "Combine." We shall show that "Combine" can be implemented in time linear in the input size, with $O(n \log n)$ time for a preliminary sort, which can be charged to the overall procedure. We shall later see how this result affects the entire task.

The sets S_{11} and S_{22}, appearing in Step M3, are collections of two-dimensional points in the plane (u_3, u_4). We represent each of these two sets as a doubly-threaded bidirectional list as follows: for each point p there is a node containing the information $(u_3(p), u_4(p))$; in addition, there are two pointers, NEXT3 and NEXT4, respectively describing the ordering on u_3 and u_4. Bidirectional links are established by two additional pointers PRED3 and PRED4. We temporarily ignore the cost of constructing the doubly-threaded lists.

By BEG31 and BEG32 we denote pointers to the first elements of the u_3-coordinate lists for S_{11} and S_{22}, respectively. The algorithm has the general form of a plane sweep, whose event points are the u_3-coordinates of $S_{11} \cup S_{22}$, and whose sweep-line status is represented by a list L. This list contains a sequence, sorted by increasing u_4-coordinates of a subset of S_{11} (specifically the u_4-coordinates of the points of S_{11} whose u_3-coordinate is no larger than the current scan value). Temporarily, we use NEXTL and BEGL to denote the forward and initial pointers for L, although—as we shall see below—NEXT4 can be used in place of NEXTL. We propose the following algorithm:

```
      procedure COMBINE
 1    begin j₁ := BEG31; j₂ := BEG32;
 2          while (j₂ ≠ Λ) do
 3                begin if (u₃[j₁] ≤ u₃[j₂]) then
 4                      begin insert u₄[j₁] into L maintaining sorted order;
 5                            j₁ := NEXT3[j₁]
                      end
 6                else begin l := BEGL;
 7                      while (l ≠ Λ) and (u₄[j₂] ≥ u₄[l]) do
 8                            begin print (j₂, l);
 9                                  l := NEXTL[l]
                            end;
10                      j₂ := NEXT3[j₂]
                end
          end
      end.
```

The above algorithm has obviously the structure of a merge technique. On line

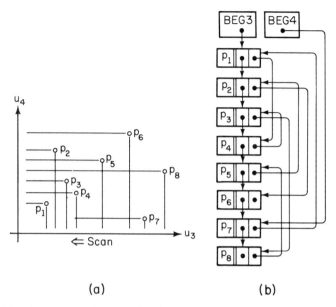

(a)   (b)

Figure 8.35 An example of set $S_1^1 = \{p_1,\ldots,p_8\}$ and of the associated doubly threaded list. NEXT3 links are shown by broken lines; NEXT4 links by solid lines.

3 we test whether we should advance on S_{11} or on S_{22}. In the first case we must insert $u_4[j_1]$ into L (line 4). In the second case (lines 6–9) we scan the list L from its smallest element, thereby determining all the points dominated by p_{j_2}; this part of the procedure is straightforward and runs in time proportional to the number of pairs (j_2,l) that are printed.

The crucial task of the procedure occurs on line 4: "Insert $u_4[j_1]$ into L maintaining sorted order." Indeed, at first sight, it appears to globally require time proportional to $|S_{11}|^2$, since each insertion may require a full scan of L, while a more sophisticated implementation of L with a rearrangeable tree (either a height-balanced or a weight-balanced tree) would cut the global time down to $(|S_{11}|\log|S_{11}|)$. However, there is an interesting way to organize this task so that its global time requirement is $O(|S_{11}|)$. Our objective is the generation of the schedule of insertions into L of the elements of the u_4-list of S_{11}. Referring to Figure 8.35 this is accomplished as follows. Observe that the u_4-coordinate of the rightmost element of S_{11}, i.e., p_8, must be inserted into L (the sorted sequence of the u_4-coordinates of the points to the left of p_8) between $u_4[3]$ and $u_4[5]$. But $u_4[3]$ is exactly PRED4[8]; therefore, PRED4[8] gives the position where to insert $u_4[8]$ into L. Now, deleting p_8 from the u_4-list and repeating the process on the resulting list, gives the insertion position of a new element, $u_4[7]$. In this manner, by scanning the u_3-list in reverse order, we obtain the insertion schedule for each element of the u_4-list, as shown by the following algorithm:

begin $l := \text{LAST}(u_3\text{-list})$;
    while $(\text{PRED3}[l] \neq \text{BEG3})$ **do**
        begin $\text{NEXT4}[\text{PRED4}[l]] := \text{NEXT4}[l]$;
              $\text{PRED4}[\text{NEXT4}[l]] := \text{PRED4}[l]$;
              $l := \text{PRED3}[l]$
        end
end.

EXAMPLE. Given the set S_{11} depicted in Figure 8.35(a), in Figure 8.35(b) we illustrate the initial configuration of the u_3- and u_4-list. The initial configuration of the array PRED4 is

$$j \quad : \quad 1 \quad 2 \quad 3 \quad 4 \quad 5 \quad 6 \quad 7 \quad 8$$

PRED4: | 7 | 5 | 4 | 1 | 8 | 2 | BEG | 3 |

The evolution of this array when executing the above scan is shown compactly below (entries being updated are encircled).

j:	1	2	3	4	5	6	7	8	After Scanning
Initial PRED4	7	5	4	1	8	2	BEG	3	—
	7	5	4	1	③	2	BEG	3	P_8
	(BEG)	5	4	1	3	2	BEG	3	P_7
	BEG	5	4	1	3	2	BEG	3	P_6
	BEG	③	4	1	3	2	BEG	3	P_5
	BEG	3	①	1	3	2	BEG	3	P_4
	BEG	①	1	1	3	2	BEG	3	P_3
Final	BEG	1	1	1	3	2	BEG	3	P_2

(insertion schedule)

Therefore, the final configuration of the array PRED4 completely specifies the insertion schedule into the L-list (which becomes the u_4-list when the scan is complete) and line 4 of COMBINE can be executed in constant time. This shows that the entire COMBINE procedure runs in time linear in $|S_{11}| + |S_{22}|$ and in the number of pairs (point dominances) obtained. With the subroutine COMBINE at our disposal, we can see how to organize the entire procedure DOMINANCE.

First we consider the data structure. After sorting the set S on each of the four coordinates, we set up a quadruply-threaded list (QTL). As for the doubly-threaded list mentioned earlier, all links are bidirectional and pointers NEXTj and PREDj are used for the u_j-list. Obviously the construction of the

QTL for S uses $O(N \log N)$ time. The QTL lends itself, very naturally, to the linear-time implementation of the set-splitting operations specified by steps D1 and M1 of the preceding algorithms. Indeed, suppose we wish to split S into (S_1, S_2) and mark the elements of S_1. Then, by traversing the QTL for each selected pointer NEXTi, for $i = 1, 2, 3, 4$, the list corresponding to this pointer can be split into two lists, corresponding to the two sets S_1 and S_2 of the partition. Analogously given S_1 and S_2 we can merge the two corresponding lists using "natural merge" [Knuth (1973)] in linear time. Note that the splitting and merging operations simply involve modification of the pointers and use no additional space for storing data.

To analyze the performance of the proposed technique we note:

(1) All processing occurs in place, uses the QTL arrays, and reduces to transformations of the pointers' values. Thus the space used is $\theta(N)$.
(2) As regards processing time, each dominance pair (i.e., each nested pair of rectangles) is found exactly once and in constant time by the *while*-loop (7)–(9) of COMBINE. Thus, if s is the number of pairs, $\theta(s)$ optimal time is used for this activity. The remaining computing time depends exclusively on the size N of S: denote it by $D(N)$. Also denote by $M_d(r, s)$ the running time of Algorithm DOMINANCE MERGE on two sets with r and s d-dimensional points, where $d = 2, 3$. Assuming, for simplicity, that N is even, we have

$$D(N) = 2D(N/2) + M_3(N/2, N/2) + O(N), \qquad (8.8)$$

where $O(N)$ is the time used by the "divide" step D1. Analogously, we have (assuming that $|S_{21}| = m$ and r is even)

$$M_3(r, s) = M_3(r/2, m) + M_3(r/2, s - m) + M_2(r/2, \max(m, s - m)) \\ + O(r + s), \qquad (8.9)$$

where, again, $O(r + s)$ time is needed to perform the set splitting. An upper bound to $M_3(r, s)$ is obtained by maximizing the right side of (8.9) with respect to m. Since $M_2(r', s')$ is $O(r' + s')$, arguing as in [Kung–Luccio–Preparata (1975)], we obtain $M_3(r, s) = O((r + s) \log (r + s))$ and, consequently, $D(N) = O(N(\log N)^2)$.

This proves the following theorem.

Theorem 8.10. *The rectangle enclosure problem on N rectangles (and its equivalent* DOMINANCE *problem in four dimensions) can be solved in time* $O(N \log^2 N + s)$ *and optimal space* $\theta(N)$, *where s is the number of dominance pairs.*

Of course, it remains an open question whether a better algorithm can be found for the above two equivalent problems. On the other hand, the establishment of an $\Omega(N \log^2 N)$ lower bound is a very improbable prospect.

8.9 Notes and Comments

Several of the problems discussed in this chapter lend themselves to interesting extensions or generalizations. For example, the problem of reporting the intersections of N rectangles in the plane can be generalized to an arbitrary number d of dimensions. If K is the number of intersecting pairs of huperrectangles, Six and Wood obtained an $O(N\log^{d-1} N + K)$ time and $O(N\log^{d-1} N)$ space algorithm [Six–Wood (1982)], whose time bound was later improved by Edelsbrunner (1983) to $O(N\log^{d-2} N)$. Within the same time bound Chazelle and Incerpi (1983) and Edelsbrunner and Overmars (1985) have reduced the space bound down to $\theta(N)$, which is obviously optimal. The corresponding counting problem, unlike for the case of segment intersection (see Section 7.2.3) can be solved in $O(N\log^{d-1} N)$ time and $O(N\log^{d-2} N)$ space by slightly modifying the intersection reporting algorithm of Six and Wood, although the algorithm of Chazelle and Incerpi cannot be adapted to solve this problem efficiently.

An interesting generalization of the rectangle intersection problem is the class conveniently referred to as ORTHOGONAL INTERSECTION SEARCHING, where, given N isothetic objects, we must find all those that intersect a given isothetic query object. (Of course, the canonical form of an isothetic object in E^d is the cartesian product of d intervals, each of which may reduce to a single value. For instance, a point, an isothetic rectangle, and a horizontal or vertical line segment are orthogonal objects.) Examples of such problems, discussed in Chapter 2, are multidimensional range searching, both in its "reporting" and in its "counting" form. The INVERSE RANGE SEARCHING, or POINT ENCLOSURE problem i.e., given N isothetic rectangles, find the rectangles that enclose a query point, also belongs to the class. Vaishnavi (1982) gives a data structure that supports searching in $O(\log N + K)$ query time with space $= O(N\log N)$. Chazelle (1983c) gives an optimal algorithm for this problem, i.e., $\theta(N)$ space and $\theta(\log N + K)$ time. Another instance of such problems is the ORTHOGONAL SEGMENT INTERSECTION SEARCHING problem, which requests, for given N horizontal and vertical line segments to find all segments intersecting a given orthogonal query segment; this problem was investigated by Vaishnavi and Wood (1982), who gave an $O(\log N + K)$ query time algorithm with $O(N\log N)$ preprocessing and space. The space bound can be reduced to $\theta(N)$ [Chazelle (1983c)]. If the set of distinct coordinate values is of cardinality $O(N)$ and when insertions or deletions of segments are allowed, McCreight presents a dynamic data structure [McCreight (1981)] for this problem which requires $O(N)$ space and $O(\log^2 N + K)$ time; and Lipski and Papadimitriou (1981) provide an algorithm with $O(\log N \cdot \log\log N + K)$ time and $O(N\log N)$ space. Edelsbrunner and Maurer (1981) have recently unified several of the previous approaches to solving this class of intersection searching problems and obtained the following results. For the static problem, there exists a data structure achieving the following performance: query time $O(\log^{d-1} N + K)$ with preprocessing time and storage both $O(N\log^d N)$. The space bound has been later improved to $O(N\log^{d-1} N\log\log N)$ [Chazelle–Guibas (1984)]. For the dynamic problem, there exists a dynamic data structure achieving the following performance: query time $O(\log^d N + K)$, with space $O(N\log^d N)$ and update time $O(\log^d N)$ [Edelsbrunner (1980)].

8.10 Exercises

1. Given the set of nontrivial circuits of the union F of N isothetic rectangles, design an algorithm that selects the circuits forming the nontrivial contour. (The target time of this algorithm is $O(N \log N)$.)

2. *3-D DOMINANCE.* Given a set of N points in E^3, design an algorithm that constructs the s dominance pairs in time $O(N \log N + s)$ and space $\theta(N)$.

3. *3-D DOMINANCE.* Prove that the time required to compute the s dominance pairs in a set of N points in E^3 is lower-bounded by $\Omega(N \log N + s)$.

4. *Enclosure of Squares.* The ENCLOSURE-OF-SQUARES problem is obtained by replacing the world "square" to the word "rectangle" in Problem R.10. Show that ENCLOSURE-OF-SQUARES is equivalent to 3-D DOMINANCE.

5. Show that 4-D DOMINANCE MERGE for a set of N points can be solved in time $O(N \log^2 N + s)$, where s is the size of the reported set of pairs.

6. *Route-in-a-Maze.* A maze is defined by two collections of segments (N in total), each of which consists of parallel segments (for simplicity, the two respective directions are orthogonal). A maze partitions the plane into connected regions. Given two points s and t in the plane, a route is a curve connecting s and t without crossing any segment.
   (a) Design an algorithm to test if there is a route between s and t.
   (b) Design an algorithm to construct the route if there exists one.

References

A. V. Aho, J. E. Hopcroft and J. D. Ullman, *The Design and Analysis of Computer Algorithms*, Addison-Wesley, Reading, Mass., 1974.

S. B. Akers, Routing, in *Design Automation of Digital Systems*, M. Breuer, ed., Prentice-Hall, Englewood Cliffs, NJ, 1972.

S. G. Akl, Two remarks on a convex hull algorithm, *Info. Proc. Lett.* **8**, 108–109 (1979).

S. G. Akl and G. T. Toussaint, Efficient convex hull algorithms for pattern recognition applications, *Proc. 4th Int'l Joint Conf. on Pattern Recognition*, Kyoto, Japan, pp. 483–487 (1978).

A. M. Andrew, Another efficient algorithm for convex hulls in two dimensions, *Info. Proc. Lett.* **9**, 216–219 (1979).

H. C. Andrews, *Introduction to Mathematical Techniques in Pattern Recognition*, Wiley-Interscience, New York, 1972.

M. J. Atallah, A linear time algorithm for the Hausdorff distance between convex polygons, *Info. Proc. Lett.* **8**, 207–209 (Nov. 1983).

D. Avis, On the complexity of finding the convex hull of a set of points, McGill University, School of Computer Science; Report SOCS 79.2, 1979

D. Avis and B. K. Bhattacharya, Algorithms for computing *d*-dimensional Voronoi diagrams and their duals, in *Advances in Computing Research*. Edited by F. P. Preparata. **1**, JAI Press, 159–180 (1983).

H. S. Baird, Fast algorithms for LSI artwork analysis, *Design Automation and Fault-Tolerant Computing*, **2**, 179–209 (1978).

R. E. Barlow, D. J. Bartholomew, J. M. Bremner and H. D. Brunk, *Statistical Inference under Order Restrictions*, Wiley, New York, 1972.

M. Ben-Or, Lower bounds for algebraic computation trees, *Proc. 15th ACM Annual Symp. on Theory of Comput.*, pp. 80–86 (May 1983).

R. V. Benson, *Euclidean Geometry and Convexity*, McGraw-Hill, New York, 1966.

J. L. Bentley, Multidimensional binary search trees used for associative searching, *Communications of the ACM* **18**, 509–517 (1975).

J. L. Bentley, Algorithms for Klee's rectangle problems, Carnegie-Mellon University, Pittsburgh, Penn., Department of Computer Science, unpublished notes, 1977.

J. L. Bentley, Decomposable searching problems, *Info. Proc. Lett.* **8**, 244–251 (1979).

J. L. Bentley, Multidimensional divide and conquer, *Comm. ACM* **23**(4), 214–229 (1980).

J. L. Bentley, G. M. Faust and F. P. Preparata, Approximation algorithms for convex hulls, *Comm. ACM* **25**, 64–68 (1982).

J. L. Bentley, H. T. Kung, M. Schkolnick and C. D. Thompson, On the average number of maxima in a set of vectors, *J. ACM* **25**, 536–543 (1978).

J. L. Bentley and H. A. Maurer, A note on Euclidean near neighbor searching in the plane, *Inform. Processing Lett.* **8**, 133–136 (1979).

J. L. Bentley and H. A. Maurer, Efficient worst-case data structures for range searching, *Acta Informatica* **13**, 155–168 (1980).

J. L. Bentley and T. A. Ottmann, Algorithms for reporting and counting geometric intersections, *IEEE Transactions on Computers* **28**, 643–647 (1979).

J. L. Bentley and M. I. Shamos, Divide-and-Conquer in Multidimensional Space, *Proc. Eighth ACM Annual Symp. on Theory of Comput.*, pp. 220–230 (May 1976).

J. L. Bentley and M. I. Shamos, A problem in multivariate statistics: Algorithms, data structure, and applications, *Proceedings of the 15th Annual Allerton Conference on Communication, Control, and Computing*, pp. 193–201 (1977).

J. L. Bentley and M. I. Shamos, Divide and conquer for linear expected time, *Info. Proc. Lett.* **7**, 87–91 (1978).

J. L. Bentley and D. F. Stanat, Analysis of range searches in quad trees, *Info. Proc. Lett.* **3**, 170–173 (1975).

J. L. Bentley and D. Wood, An optimal worst case algorithm for reporting intersection of rectangles, *IEEE Transactions on Computers* **29**, 571–577 (1980).

P. Bézier, *Numerical Control—Mathematics and Applications*. Translated by A. R. Forrest. Wiley, New York, 1972.

B. Bhattacharya, Worst-case analysis of a convex hull algorithm, unpublished manuscript, Dept. of Computer Science, Simon Fraser University, February 1982.

G. Bilardi and F. P. Preparata, Probabilistic analysis of a new geometric searching technique, unpublished manuscript, 1981.

G. Birkhoff and S. MacLane, *A Survey of Modern Algebra*, McMillan, New York, 1965.

M. Blum, R. W. Floyd, V. Pratt, R. L. Rivest and R. E. Tarjan, Time bounds for selection, *Jour. Compt. Sys. Sci.* **7**, 448–461 (1973).

B. Bollobás, *Graph Theory, An Introductory Course*, Springer-Verlag, New York, 1979.

K. Q. Brown, Fast intersection of half spaces, Carnegie-Mellon University, Pittsburgh, Pennsylvania, Department of Computer Science; Report CMU-CS-78-129, 1978.

K. Q. Brown, Geometric transformations for fast geometric algorithms, Ph. D. thesis, Dept. of Computer Science, Carnegie Mellon Univ., Dec. 1979a.

K. Q. Brown, Voronoi diagrams from convex hulls, *Info. Proc. Lett.* **9**, 223–228 (1979b).

W. A. Burkhard and R. M. Keller, Some approaches to best match file searching, *Comm. ACM* **16**, 230–236 (1973).

A. Bykat, Convex hull of a finite set of points in two dimensions, *Info. Proc. Lett.* **7**, 296–298 (1978).

J. C. Cavendish, Automatic triangulation of arbitrary planar domains for the finite element method, *Int'l J. Numerical Methods in Engineering* **8**, 679–696 (1974).

D. R. Chand and S. S. Kapur, An algorithm for convex polytopes, *JACM* **17**(1), 78–86 (Jan. 1970).

B. M. Chazelle, Reporting and counting arbitrary planar intersections, Rep. CS-83-16, Dept. of Comp. Sci., Brown University, Providence, RI, 1983a.

B. M. Chazelle, Optimal algorithms for computing depths and layers, *Proc. 21st Allerton Conference on Comm., Control and Comput.*, pp. 427–436 (Oct. 1983b).

B. M. Chazelle, Filtering search: A new approach to query answering, *Proc. 24th IEEE*

Symp. on Foundations of Comp. Sci., Tucson, AZ, 122–132 Nov. 1983c.

B. M. Chazelle, R. Cole, F. P. Preparata, and C. K. Yap, New upper bounds for neighbor searching, submitted for publication (1984).

B. M. Chazelle and D. P. Dobkin, Detection is easier than computation, *Proc. 12th ACM Annual Symp. on Theory of Comput.*, pp. 146–153, (May 1980).

B. M. Chazelle and H. Edelsbrunner, Optimal solutions for a class of point retrieval problems, Tech. Rep. CS-84-16, Dept. of Computer Science, Brown University, July 1984.

B. M. Chazelle and L. J. Guibas, Fractional cascading: A data structuring technique with geometric applications, manuscript, 1984.

B. M. Chazelle, L. J. Guibas and D. T. Lee, The power of geometric duality, *Proc. 24th IEEE Annual Symp. on Foundations of Comput. Sci.*, pp. 217–225 (Nov. 1983).

B. M. Chazelle and J. Incerpi, Triangulating a polygon by divide-and-conquer, *Proc. 21st Allerton Conference on Comm. Control and Comput.*, pp. 447–456 (Oct. 1983).

D. Cheriton and R. E. Tarjan, Finding minimum spanning trees, *SIAM J. Comput.* **5**(4), 724–742 (Dec. 1976).

F. Chin and C. A. Wang, Optimal algorithms for the intersection and the minimum distance problems between planar polygons, *IEEE Trans. Comput.* **C-32**(12), 1203–1207 (1983).

F. Chin and C. A. Wang, Minimum vertex distance between separable convex polygons, *Info. Proc. Lett.* **18**, 41–45 (1984).

N. Christofides, Worst-case analysis of a new heuristic for the travelling salesman problem, Symposium on Algorithms and Complexity. Department of Computer Science, Carnegie-Mellon University, Apr. 1976.

V. Chvátal, A combinatorial theorem in plane geometry, *J. Comb. Theory B.* **18**, 39–41 (1975).

J. L. Coolidge, *A Treatise on the Circle and the Sphere*, Oxford University Press, Oxford, England, 1916. Reprinted 1971 by Chelsea.

B. Dasarathy and L. J. White, On some maximin location and classifier problems, Computer Science Conference, Washington, D.C., 1975 (Unpublished lecture).

P. J. Davis, *Interpolation and Approximation*, Blaisdell, NY (1963). Reprinted 1975 by Dover, New York.

B. Delaunay, Sur la sphère vide, *Bull. Acad. Sci. USSR(VII)*, Classe Sci. Mat. Nat., 793–800 (1934).

R. B. Desens, Computer processing for display of three-dimensional structures, Technical Report CFSTI AD-7006010, Naval Postgraduate School, Oct. 1969.

E. W. Dijkstra, A note on two problems in connection with graphs, *Numer. Math.* **1**(5), 269–271 (Oct. 1959).

D. P. Dobkin and D. G. Kirkpatrick, Fast algorithms for preprocessed polyhedral intersection detection, submitted for publication (1984).

David Dobkin and Richard Lipton, Multidimensional searching problems, *SIAM J. Comput.* **5**(2), 181–186 (June 1976).

David Dobkin and Richard Lipton, On the complexity of computations under varying set of primitives, *Journal of Computer and Systems Sciences* **18**, 86–91 (1979).

R. L. Drysdale, III, Generalized Voronoi diagrams and geometric searching, Ph.D. Thesis, Dept. Comp. Sci., Stanford University, Tech. Rep. STAN-CS-79-705 (1979).

R. O. Duda and P. E. Hart, *Pattern Classification and Scene Analysis*, Wiley-Interscience, New York, 1973.

R. D. Düppe and H. J. Gottschalk, Automatische Interpolation von Isolinien bei willkürlich verteilten Stützpunkten, *Allgemeine Vermessungsnachrichten* **77**, 423–426 (1970).

M. E. Dyer, A simplified O(nlogn) algorithm for the intersection of 3-polyhedra, Teesside Polytechnic, Middlesbrough, United Kingdom, Department of Mathematics and Statistics; Report TPMR 80-5, 1980.

M. E. Dyer, Linear time algorithms for two- and three-variable linear programs, *SIAM J. Comp.* **13**(1), 31–45 (Feb. 1984).

M. I. Edahiro, I Kokubo and T. Asano, A new point-location algorithm and its practical efficiency—comparison with existing algorithms, Res. Memo. RMI 83-04, Dept. Math. Eng. B Instrumentation Physics, Univ. of Tokyo, Oct. 1983.

W. Eddy, A new convex hull algorithm for planar sets, *ACM Trans. Math. Software* **3**(4), 398–403 (1977).

H. Edelsbrunner, Dynamic data structures for orthogonal intersection queries, Rep. F59, Tech. Univ. Graz, Institute für Informationsverarbeitung 1980.

H. Edelsbrunner, Intersection problems in computational geometry, Ph. D. Thesis, Rep. 93, IIG, Technische Universität Graz, Austria, 1982.

H. Edelsbrunner, A new approach to rectangle intersections, Part II, *Int'l. J. Comput. Math.* **13**, 221–229 (1983).

E. Edelsbrunner, L. J. Guibas and J. Stolfi, Optimal point location in a monotone subdivision, *SIAM J. Comput.*, to appear (1985).

H. Edelsbrunner, J. van Leeuwen, T. A. Ottman, and D. Wood, Connected components of orthogonal geometric objects, McMaster University, Hamilton, Ontario, Unit for Computer Science, Report 81-CS-04, 1981.

H. Edelsbrunner and H. A. Maurer; On the intersection of orthogonal objects, *Info. Proc. Lett.* **13**, 177–181 (1981).

H. Edelsbrunner and M. H. Overmars, On the equivalence of some rectangle problems, *Information Processing Letters* **14**(3), 124–127 (May 1982).

H. Edelsbrunner and M. H. Overmars, Batched dynamic solutions to decomposable searching problems, *J. Algorithms*, to appear (1985).

H. Edelsbrunner, M. H. Overmars and D. Wood, Graphics in flatland: A case study, in *Advances in Computing Research*. Edited by F. P. Preparata. Vol. 1, JAI Press, 35–59 (1983).

H. Edelsbrunner and R. Seidel, Arrangements of planes and Voronoi diagrams, in preparation (1985).

H. Edelsbrunner and E. Welzl, Halfplanar range search in linear space and O(n** 0.695) query time, Rep. 111, IIG, Technische Universität Graz, Austria, 1983.

J. Edmonds, Maximum matching and a polyhedron with 0,1 vertices, *J. Res. NBS*, 69B, 125–130 (Apr.–June 1965).

B. Efron, The convex hull of a random set of points, *Biometrika* **52**, 331–343 (1965).

H. El Gindy and D. Avis, A linear algorithm for computing the visibility polygon from a point, *J. Algorithms* **2**(2), 186–197 (1981).

D. J. Elzinga and D. W. Hearn, Geometrical solutions for some minimax location problems, *Transportation Science* **6**, 379–394 (1972a).

D. J. Elzinga and D. W. Hearn, The minimum covering sphere problem, *Mgmt. Sci.* **19**(1), 96–104 (Sept. 1972b).

P. Erdös, On sets of distances of *n* points, *Amer. Math. Monthly* **53**, 248–250 (1946).

P. Erdös, On sets of distances of *n* points in Euclidean space, *Magy. Tud. Akad. Mat. Kut. Int. Kozi.* **5**, 165–169 (1960).

K. P. Eswaran, J. N. Gray, R. A. Lorie and I. L. Traiger, The notions of consistency, and predicate locks in a database system, *Comm. ACM* **19**, 624–633 (1976).

S. Even, *Graph Algorithms*, Computer Science Press, Potomac, MD, 1979.

H. Eves, *A Survey of Geometry*, Allyn and Bacon, Newton, Mass., 1972.

G. Ewald, *Geometry: An Introduction*, Wadsworth, Belmont, Calif., 1971.

I. Fáry, On straight-line representation of planar graphs, *Acta Sci. Math. Szeged.* **11**, 229–233 (1948).

R. A. Finkel and J. L. Bentley, Quad-trees; a data structure for retrieval on composite keys, *Acta Inform.* **4**, 1–9 (1974).

A. R. Forrest, Computational Geometry, *Proc. Royal Society London*, **321** Series 4,

187–195 (1971).

R. L. Francis and J. A. White, *Facility Layout and Location: An Analytical Approach*, Prentice-Hall, Englewood Cliffs, NJ, 1974.

M. L. Fredman, A lower bound of the complexity of orthogonal range queries, *J. ACM* **28**, 696–705 (1981).

M. L. Fredman and B. Weide, The complexity of computing the measure of $\cup[a_i, b_i]$, *Comm. ACM* **21**(7), 540–544 (July 1978).

H. Freeman, Computer processing of line-drawing images, *Comput. Surveys* **6**, 57–97 (1974).

H. Freeman and P. P. Loutrel, An algorithm for the solution of the two-dimensional hidden-line problem, *IEEE Trans. Elec. Comp.* **EC-16**(6), 784–790 (1967).

H. Freeman and R. Shapira, Determining the minimum-area encasing rectangle for an arbitrary closed curve, *Comm. ACM.* **18**(7), 409–413 (1975).

J. H. Friedman, J. L. Bentley and R. A. Finkel, An algorithm for finding best match in logarithmic expected time, *ACM Trans. Math. Software* **3**(3), 209–226 (1977).

H. Gabow, An efficient implementation of Edmond's maximum matching algorithm, Technical Report 31. Computer Science Department, Stanford Univ., 1972.

K. R. Gabriel and R. R. Sokal, A new statistical approach to geographic variation analysis, *Systematic Zoology* **18**, 259–278 (1969).

R. Galimberti and U. Montanari, An algorithm for hidden-line elimination, *CACM* **12**(4), 206–211 (1969).

M. R. Garey, R. L. Graham and D. S. Johnson, Some NP-complete geometric problems, *Eighth Annual Symp. on Theory of Comput.*, pp. 10–22 (May 1976).

M. R. Garey and D. S. Johnson, *Computers and Intractability. A Guide to the Theory of NP-Completeness*. W. H. Freeman, San Francisco, 1979.

M. R. Garey, D. S. Johnson and L. Stockmeyer, Some simplified NP-complete graph problems, *Theor. Comp. Sci.* **1**, 237–267 (1976).

M. Garey, D. S. Johnson, F. P. Preparata, and R. E. Tarjan, Triangulating a simple polygon, *Info. Proc. Lett.* **7**(4), 175–180 (1978).

S. I. Gass, *Linear Programming*, McGraw-Hill, New York, 1969.

J. Gastwirth, On robust procedures, *J. Amer. Stat. Assn.* **65**, 929–948 (1966).

J. A. George, Computer implementation of the finite element method, Technical Report STAN-CS-71-208. Computer Science Department, Stanford University, 1971.

P. N. Gilbert, New results on planar triangulations, Tech. Rep. ACT-15, Coord. Sci. Lab., University of Illinois at Urbana, July 1979.

T. Gonzalez, Algorithms on sets and related problems, Technical Report. Department of Computer Science, University of Oklahoma, 1975.

J. C. Gower and G. J. S. Ross, Minimum spanning trees and single linkage cluster analysis, *Appl. Stat.* **18**(1), 54–64 (1969).

R. L. Graham, An efficient algorithm for determining the convex hull of a finite planar set, *Info. Proc. Lett.* **1**, 132–133 (1972).

R. L. Graham and F. F. Yao, Finding the convex hull of a simple polygon, Tech. Rep. No. STAN-CS-81-887, Stanford University, 1981; also *J. Algorithms* **4**(4), 324–331 (1983).

P. J. Green and B. W. Silverman, Constructing the convex hull of a set of points in the plane, *Computer Journal* **22**, 262–266 (1979).

B. Grünbaum, A proof of Vazsonyi's conjecture, *Bull. Res. Council Israel* **6**(A), 77–78 (1956).

B. Grünbaum, *Convex Polytopes*, Wiley-Interscience, New York, 1967.

R. H. Güting, An optimal contour algorithm for isooriented rectangles, *Jour. of Algorithms* **5**(3), 303–326 (Sept. 1984).

H. Hadwiger and H. Debrunner, *Combinatorial Geometry in the Plane*, Holt, Rinebart and Winston, New York, 1964.

M. Hanan and J. M. Kurtzberg, Placement techniques, in *Design Automation of Digital Systems*. Edited by M. Breuer. Prentice-Hall, Englewood Cliffs, NJ, 1972.

M. Hanan, Layout, interconnection, and placement, *Networks* **5**, 85–88 (1975).

J. A. Hartigan, *Clustering Algorithms*, Wiley, New York, 1975.

T. L. Heath, *A History of Greek Mathematics*, Oxford University Press, Oxford, England, 1921.

S. Hertel, K. Mehlhorn, M. Mäntylä and J. Nievergelt, Space sweep solves intersection of two convex polyhedron elegantly, *Acta Informatica* (to appear) (1985).

D. Hilbert, *Foundations of Geometry*, (1899). Repr. 1971 by Open Court, La Salle, IL.

J. G. Hocking and G. S. Young, *Topology*, Addison-Wesley, Reading, MA, 1961.

C. A. R. Hoare, Quicksort, *Computer Journal* **5**, 10–15 (1962).

I. Hodder and C. Orton, *Spatial Analysis in Archaeology*, Cambridge University Press, Cambridge, England, 1976.

P. G. Hoel, *Introduction to Mathematical Statistics*, Wiley, New York, 1971.

P. J. Huber, Robust statistics: A review, *Ann. Math. Stat.* **43**(3), 1041–1067 (1972).

F. K. Hwang, An $O(n \log n)$ algorithm for rectilinear minimal spanning tree, *J. ACM* **26**, 177–182 (1979).

J. W. Jaromczyk, Lower bounds for polygon simplicity testing and other problems, *Proc. MFCS-84*, Prague (Springer-Verlag), 339–347 (1984).

R. A. Jarvis, On the identification of the convex hull of a finite set of points in the plane, *Info. Proc. Lett.* **2**, 18–21 (1973).

S. C. Johnson, Hierarchical clustering schemes, *Psychometricka* **32**, 241–254 (1967).

M. Kallay, Convex hull algorithms in higher dimensions, unpublished manuscript, Dept. Mathematics, Univ. of Oklahoma, Norman, Oklahoma, 1981.

R. M. Karp, Reducibility among combinatorial problems, in *Complexity of Computer Computations*. Edited by R. E. Miller and J. W. Thatcher. Plenum Press, New York, 1972.

J. M. Keil and J. R. Sack, Minimum decomposition of polygonal objects, in *Computational Geometry*. Edited by G. T. Toussaint. North-Holland, Amsterdam, The Netherlands (1985).

M. G. Kendall and P. A. P. Moran, *Geometrical Probability*, Hafner, New York, 1963.

D. G. Kirkpatrick, Efficient computation of continuous skeletons, *Proc. 20th IEEE Annual Symp. on Foundations of Comput. Sci.*, pp. 18–27 (Oct. 1979).

D. G. Kirkpatrick, Optimal search in planar subdivisions, *SIAM J. Comput.* **12**(1), 28–35 (1983).

D. G. Kirkpatrick and R. Seidel, The ultimate planar convex hull algorithm? Tech. Rep. 83–577, Dept. Comput. Sci., Cornell Univ., Oct. 1983.

V. Klee, Convex polytopes and linear programming, *Proc. IBM Sci. Comput. Symp.: Combinatiorial Problems*, IBM, White Plains,NY, pp. 123–158 (1966).

V. Klee, Can the measure of $\cup [a_i, b_i]$ be computed in less than $O(n \log n)$ steps?, *Amer. Math. Monthly*, **84**(4), 284–285 (April 1977).

V. Klee, On the complexity of d-dimensional Voronoi diagrams, *Archiv der Mathematik*, **34**, 75–80 (1980).

F. Klein, *Das Erlangen Programm: vergleichende Betrachtungen über neuere geometrisches Forschungen*, 1872. Reprinted by Akademische VerlagsGesellschaft Greest und Portig, Leipzig, 1974.

D. E. Knuth, *The Art of Computer Programming. Volume I: Fundamental Algorithms*, Addison-Wesley, Reading, MA, 1968.

D. E. Knuth, *The Art of Computer Programming. Volume III: Sorting and Searching*, Addison-Wesley, Reading, Mass., 1973.

D. E. Knuth, Big omicron and big omega and big theta, *SIGACT News* **8**(2), 18–24 (April–June 1976).

J. F. Kolars and J. D. Nystuen, *Human Geography*, McGraw-Hill, New York, 1974.

J. B. Kruskal, On the shortest spanning subtree of a graph and the traveling salesman problem, *Proc. AMS* **7**, 48–50 (1956).

H. T. Kung, F. Luccio and F. P. Preparata, On finding the maxima of a set of vectors, *JACM* **22**(4), 469–476 (Oct. 1975).

U. Lauther, 4-dimensional binary search trees as a means to speed up associative searches in the design verification of integrated circuits, *Jour. of Design Automation and Fault Tolerant Computing*, **2**(3), 241–247 (July 1978).

C. L. Lawson, The smallest covering cone or sphere, *SIAM Rev.* **7**(3), 415–417 (1965).

C. L. Lawson, Software for C^1 surface interpolation, JPL Publication, pp. 77–30 (Aug. 1977).

D. T. Lee, On finding k nearest neighbors in the plane, Technical Report. Department of Computer Science, University of Illinois, 1976.

D. T. Lee, Proximity and reachability in the plane, Tech. Rep. No. R-831, Coordinated Sci. Lab., Univ. of Illinois at Urbana, IL, 1978.

D. T. Lee, On k-nearest neighbor Voronoi diagrams in the plane, *IEEE Trans. Comput.* **C-31**(6), 478–487 (June 1982).

D. T. Lee, Two dimensional Voronoi diagram in the L_p-metric, *J. ACM* **27**, 604–618 (1980a).

D. T. Lee, Farthest neighbor Voronoi diagrams and applications, Tech. Rep. No. 80-11-FC-04, Dpt. EE/CS, Northwestern Univ., 1980b.

D. T. Lee, On finding the convex hull of a simple polygon, Tech. Rep. No. 80-03-FC-01, EE/CS Dept., Northwestern Univ., 1980. See also *Int'l J. Comput. and Infor. Sci.* **12**(2), 87–98 (1983a).

D. T. Lee, Visibility of a simple polygon, *Computer Vision, Graphics and Image Processing* **22**, 207–221 (1983b).

D. T. Lee and I. M. Chen, Display of visible edges of a set of convex polygons, in *Computational Geometry.* Edited by G. T. Toussaint. North Holland, Amsterdam, to appear (1985).

D. T. Lee and R. L. Drysdale III, Generalized Voronoi diagrams in the plane, *SIAM J. Comput.* **10**, 1, 73–87 (Feb. 1981).

D. T. Lee and F. P. Preparata, Location of a point in a planar subdivision and its applications, *SIAM Journal on Computing* **6**(3), 594–606 (Sept. 1977).

D. T. Lee and F. P. Preparata, The all nearest neighbor problem for convex polygons, *Info. Proc. Lett.* **7**, 189–192 (1978).

D. T. Lee and F. P. Preparata, An optimal algorithm for finding the kernel of a polygon, *Journal of the ACM* **26**, 415–421 (1979).

D. T. Lee and F. P. Preparata, An improved algorithm for the rectangle enclosure problem, *Journal of Algorithms*, **3**(3), 218–224 (1982).

D. T. Lee and B. Schachter, Two algorithms for constructing Delaunay triangulations, *Int'l J. Comput. and Info. Sci.* **9**(3), 219–242 (1980).

D. T. Lee and C. K. Wong, Worst-case analysis for region and partial region searches in multidimensional binary search trees and balanced quad trees, *Acta Informatica* **9**, 23–29 (1977).

D. T. Lee and C. K. Wong, Voronoi diagrams in L_1-(L_∞-) metrics with 2-dimensional storage applications, *SIAM J. Comput.* **9**(1), 200–211 (1980).

J. van Leeuwen and D. Wood, Dynamization of decomposable searching problems, *Info. Proc. Lett.* **10**, 51–56 (1980).

J. van Leeuwen and D. Wood, The measure problem for rectangular ranges in d-space, *J. Algorithms* **2**, 282–300 (1981).

E. Lemoine, *Géométrographie*, C. Naud, Paris, 1902.

W. Lipski, Jr. and C. H. Papadimitriou, A fast algorithm for testing for safety and detecting deadlocks in locked transaction systems, *Journal of Algorithms* **2**, 211–226 (1981).

W. Lipski, Jr. and F. P. Preparata, Finding the contour of a union of iso-oriented rectangles, *Journal of Algorithms* **1**, 235–246 (1980).

W. Lipski, Jr. and F. P. Preparata, Segments, rectangles, contours, *Journal of Algorithms* **2**, 63–76 (1981).

R. J. Lipton and R. E. Tarjan, A separator theorem for planar graphs, *Conference on Theoretical Computer Science*, Waterloo, Ont., pp. 1–10 (August 1977a).

R. J. Lipton and R. E. Tarjan, Applications of a planar separator theorem, *Proc. 18th IEEE Annual Symp. on Found. of Comp. Sci.*, Providence, RI, pp. 162–170 (October 1977b).

R. J. Lipton and R. E. Tarjan, Applications of a planar separator theorem, *SIAM J. Comput.* **9**(3), 615–627 (1980).

C. L. Liu, *Introduction to Combinatorial Mathematics*, McGraw-Hill, New York, 1968.

E. L. Lloyd, On triangulations of a set of points in the plane, *Proc. 18th IEEE Annual Symp. on Foundations of Comput. Sci.*, pp. 228–240 (Oct. 1977).

H. Loberman and A. Weinberger, Formal procedures for connecting terminals with a minimum total wire length, *JACM* **4**, 428–437 (1957).

D. O. Loftsgaarden and C. P. Queensberry, A nonparametric density function, *Ann. Math. Stat.* **36**, 1049–1051 (1965).

P. P. Loutrel, A solution to the hidden-line problem for computer-drawn polyhedra, *IEEE Trans. Comp.* **C-19**(3), 205–215 (March 1970).

G. S. Lueker, A data structure for orthogonal range queries, *Proceedings of the 19th Annual IEEE Symposium on Foundations of Computer Science*, pp. 28–34 (1978).

U. Manber and M. Tompa, Probabilistic, nondeterministic and alternating decision trees, Tech. Rep. No. 82-03-01, Univ. Washington, 1982.

H. G. Mairson and J. Stolfi, Reporting line segment intersections in the plane, Tech. Rep., Dept. of Computer Sci., Stanford University, 1983.

Y. Matsushita, A solution to the hidden line problem, Technical Report 335. Department of Computer Science, University of Illinois, 1969.

D. W. Matula and R. R. Sokal, Properties of Gabriel graphs relevant to geographic variation research and the clustering of points in the plane, *Geographical Analysis* **12**, 205–222 (July 1980).

D. McCallum and D. Avis, A linear algorithm for finding the convex hull of a simple polygon, *Info. Proc. Lett.* **9**, 201–206 (1979).

E. M. McCreight, Priority search trees, Tech. Rep. Xerox PARC CSL-81-5, 1981.

P. McMullen and G. C. Shephard, *Convex Polytopes and the Upper Bound Conjecture*, Cambridge University Press, Cambridge, England, 1971.

C. A. Mead and L. Conway, *Introduction to VLSI Systems*, Addison-Wesley, Reading, Mass., 1979.

N. Megiddo, Linear time algorithm for linear programming in R^3 and related problems, *SIAM J. Comput.* **12**(4), 759–776 (Nov. 1983).

W. S. Meisel, *Computer-Oriented Approaches to Pattern Recognition*, Academic Press, New York, 1972.

Z. A. Melzak, *Companion to Concrete Mathematics*, Wiley-Interscience, New York, 1973.

J. Milnor, On the Betti numbers of real algebraic varieties, *Proc. AMS* **15**, 2745–280 (1964).

J. Milnor, *Singular Points of Complex Hypersurfaces*, Princeton Univ. Press, Princeton, NJ, 1968.

M. I. Minski and S. Papert, *Perceptrons*, MIT Press, Amherst, Mass., 1969.

J. W. Moon, Various proofs of Cayley's formula for counting trees, in *A Seminar on Graph Theory*. Edited by F. Harary. Holt, Rinehart and Winston, New York, 1976.

D. E. Muller and F. P. Preparata, Finding the intersection of two convex polyhedra, *Theoretical Computer Science* **7**(2), 217–236 (Oct. 1978).

K. P. K. Nair and R. Chandrasekaran, Optimal location of a single service center of certain types, *Nav. Res. Log. Quart* **18**, 503–510 (1971).

W. M. Newman and R. F. Sproull, *Principles of Interactive Computer Graphics*, McGraw-Hill, New York, 1973.

J. Nievergelt and F. Preparata, Plane-sweep algorithms for intersecting geometric figures, *Comm. ACM* **25**(10), 739–747 (1982).

A Nijenhuis and H. S. Wilf, *Combinatorial Algorithms*, Academic Press, New York, 1975.

J. O'Rourke, C.-B Chien, T. Olson and D. Naddor, A new linear algorithm for intersecting convex polygons, *Computer Graphics and Image Processing* **19**, 384–391 (1982).

R. E. Osteen and P. P. Lin, Picture skeletons based on eccentricities of points of minimum spanning trees, *SIAM J. Comput.* **3**(1), 23–40 (March 1974).

M. H. Overmars, The design of dynamic data structures, Ph. D. Thesis, University of Utrecht, The Netherlands (1983).

M. H. Overmars and J. van Leeuwen, Maintenance of configurations in the plane, *J. Comput. and Syst. Sci.* **23**, 166–204 (1981).

C. H. Papadimitriou, The Euclidean traveling salesman problem is NP-complete, *Theoret. Comput. Sci.* **4**, 237–244 (1977).

C. H. Papadimitriou and K. Steiglitz, Some complexity results for the traveling salesman problem, *Eighth ACM Annual Symp. on Theory of Comput.*, pp. 1–9 (May 1976).

E. Pielou, *Mathematical Ecology*, Wiley-Interscience, New York, 1977.

F. P. Preparata, Steps into computational geometry, Technical Report. Coordinated Science Laboratory, University of Illinois, 1977.

F. P. Preparata, An optimal real time algorithm for planar convex hulls, *Comm. ACM* **22**, 402–405 (1979).

F. P. Preparata, A new approach to planar point location, *SIAM J. Comput.* **10**(3), 473–482 (1981).

F. P. Preparata and S. J. Hong, Convex hulls of finite sets of points in two and three dimensions, *Comm. ACM* **2**(20), 87–93 (Feb. 1977).

F. P. Preparata and D. E. Muller, Finding the intersection of n half-spaces in time $O(n \log n)$, *Theoretical Computer Science* **8**(1), 45–55 (Jan. 1979).

F. P. Preparata and K. J. Supowit, Testing a simple polygon for monotonicity, *Info. Proc. Lett.* **12**(4), 161–164 (Aug. 1981).

R. C. Prim, Shortest connecting networks and some generalizations, *BSTJ* **36**, 1389–1401 (1957).

M. O. Rabin, Proving simultaneous positivity of linear forms, *Journal of Computer and System Sciences* **6**, 639–650 (1972).

H. Rademacher and O. Toeplitz, *The Enjoyment of Mathematics*, Princeton University Press, Princeton, NJ, 1957.

H. Raynaud, Sur l'enveloppe convexe des nuages des points aléatoires dans R^n, I, *J. Appl. Prob.* **7**, 35–48 (1970).

E. M. Reingold, On the optimality of some set algorithms, *Journal of the ACM* **19**, 649–659 (1972).

E. M. Reingold, J. Nievergelt, and N. Deo, *Combinatorial Algorithms: Theory and Practice*, Prentice-Hall, Englewood Cliffs, NJ, 1977.

A. Rényi and R. Sulanke, Ueber die konvexe Hulle von n zufallig gewahlten Punkten, I, *Z. Wahrschein.* **2**, 75–84 (1963).

R. Riesenfeld, Applications of b-spline approximation to geometric problems of computer-aided design, Technical Report UTEC-CSc-73-126. Department of Computer Science, University of Utah, 1973.

R. T. Rockafellar, *Convex Analysis*, Princeton University Press, Princeton, NJ 1970.

C. A. Rogers, *Packing and Covering*, Cambridge University Press, Cambridge, England, 1964.

A. Rosenfeld, *Picture Processing by Computers*, Academic Press, New York, 1969.

D. J. Rosenkrantz, R. E. Stearns and P. M. Lewis, Approximate algorithms for the traveling salesperson problem, *Fifteenth Annual IEEE Symposium on Switching and Automata Theory*, pp. 33–42 (May 1974).

T. L. Saaty, *Optimization in Integers and Related Extremal Problems*, McGraw-Hill, New York, 1970.

L. A. Santaló, *Integral Geometry and Geometric Probability*, Encyclopedia of Mathematics and Its Applications, Vol. 1. Addison-Wesley, Reading, Mass., 1976.

J. B. Saxe, On the number of range queries in k-space, *Discrete Applied Mathematics* **1**, 217–225 (1979).

J. B. Saxe and J. L. Bentley, Transforming static data structures into dynamic structures, *Proc. 20th IEEE Annual Symp. on Foundations of Comput. Sci.*, pp. 148–168 (1979).

A. M. Schönhage, M. Paterson and N. Pippenger, Finding the median, *J. Comput. and Syst. Sci.* **13**, 184–199 (1976).

J. T. Schwartz, Finding the minimum distance between two convex polygons, *Info. Proc. Lett.* **13**(4), 168–170 (1981).

D. H. Schwartzmann and J. J. Vidal, An algorithm for determining the topological dimensionality of point clusters, *IEEE Trans. Comp.* **C-24**(12), 1175–1182 (Dec. 1975).

R. Seidel, A convex hull algorithm optimal for points in even dimensions, M.S. thesis, Tech. Rep. 81-14, Dept. of Comput. Sci., Univ. of British Columbia, Vancouver, Canada, 1981.

R. Seidel, The complexity of Voronoi diagrams in higher dimensions, *Proc. 20th Allerton Conference on Comm., Control and Comput.*, pp. 94–95 (1982). See also Tech. Rep. No. F94, Technische Universität Graz, Austria, 1982.

M. I. Shamos, Problems in computational geometry, unpublished manuscript, 1975a.

M. I. Shamos, Geometric Complexity, *Seventh ACM Annual Symp. on Theory of Comput.*, pp. 224–233 (May 1975b).

M. I. Shamos, Geometry and statistics: Problems at the interface, in *Recent Results and New Directions in Algorithms and Complexity*. Edited by J. F. Traub. Academic Press (1976).

M. I. Shamos, Computational geometry, Ph.D. thesis, Dept. of Comput. Sci., Yale Univ., 1978.

M. I. Shamos and D. Hoey, Closest-point problems, *Sixteenth Annual IEEE Symposium on Foundations of Computer Science*, pp. 151–162 (Oct. 1975).

M. I. Shamos and D. Hoey, Geometric intersection problems, *Seventeenth Annual IEEE Symposium on Foundations of Computer Science* , pp. 208–215 (Oct. 1976).

H. W. Six and D. Wood, The rectangle intersection problem revisited, *BIT* **20**, 426–433 (1980).

H. W. Six and D. Wood, Counting and reporting intersections of d-ranges, *IEEE Trans. Comput.* **C-31**, 181–187 (1982).

J. Sklansky, Measuring concavity on a rectangular mosaic, *IEEE Trans. Comp.* **C-21**, 1355–1364 (1972).

E. Soisalon-Soininen and D. Wood, An optimal algorithm for testing for safety and detecting deadlock in locked transaction systems, *Proc. of ACM Symposium on Principles of Data Bases*, Los Angeles, March 1982; pp. 108–116.

J. M. Steele and A. C. Yao, Lower bounds for algebraic decision trees, *J. Algorithms* **3**, 1–8 (1982).

J. Stoer and C. Witzgall, *Convexity and Optimization in Finite Dimensions I*, Springer-Verlag, New York, 1970.

G. Strang and G. Fix, *An Analysis of the Finite Element Method*, Prentice-Hall, Englewood Cliffs, NJ, 1973.

K. J. Supowit, The relative neighborhood graph with an application to minimum spanning trees, *J. ACM* **30**(3), 428–447 (July 1983).

I. E. Sutherland, Computer graphics: ten unsolved problems, *Datamation* **12**(5), 22–27 (May 1966).

J. J. Sylvester, On Poncelet's approximate linear valuation of surd forms, *Phil. Mag.* Ser. 4, **20**, 203–222 (1860).

R. Thom, Sur l'homologie des variétés algébraïques réelles, *Differential and Combinatorial Topology*. Edited by S. S. Cairns. Princeton Univ. Press, Princeton, NJ, 1965.

C. Toregas, R. Swain, C. Revelle and L. Bergman, The location of emergency service facilities, *Operations Research* **19**, 1363–1373 (1971).

G. T. Toussaint, Pattern recognition and geometrical complexity, *Proc. 5th Int'l Conference on Pattern Recognition*, pp. 1324–1347 (Dec. 1980a).

G. T. Toussaint, The relative neighborhood graph of a finite planar set, *Pattern Recognition* **12**(4), 261–268 (1980b).

G. T. Toussaint, An optimal algorithm for computing the minimum vertex distance between two crossing convex polygons, *Proc. 21st Allerton Conference on Comm.,Control and Comput.*, pp. 457–458 (1983a).

G. T. Toussaint, Computing largest empty circles with location constraints, *Int'l J. Computer and Info. Sci.* **12**(5), 347–358 (1983b).

V. Vaishnavi, Computing point enclosures, *Pattern Recog.* **15**, 22–29 (1982).

V. Vaishnavi and D. Wood, Data structures for the rectangle containment and enclosure problems, *Computer Graphics and Image Processing* **13**, 372–384 (1980).

V. Vaishnavi and D. Wood, Rectilinear line segment intersection, layered segment trees, and dynamization, *J. Algorithms* **3**, 160–176 (1982).

F. A. Valentine, *Convex Sets,* McGraw-Hill, New York, 1964.

G. Voronoi, Nouvelles applications des parametres continus à la theorie des formes quadratiques. Deuxième Mémoire: Recherches sur les paralléloèdres primitifs, *J. reine angew. Math.* **134**, 198–287 (1908).

J. E. Warnock, A hidden-surface algorithm for computer generated halftone pictures, Technical Report TR 4-15. Computer Science Department, University of Utah, 1969.

G. S. Watkins, A real-time visible surface algorithm, Technical Report UTECH-CSc-70-101. Computer Science Department, University of Utah, June 1970.

D. E. Willard, Predicate-oriented database search algorithms, Harvard University, Cambridge, MA, Aiken Computation Laboratory, Ph.D. Thesis, Report TR-20-78, 1978.

D. E. Willard, Polygon retrieval, *SIAM J. Comput.* **11**, 149–165 (1982).

N. Wirth, *Algorithms + Data Structures = Programs*, Prentice-Hall, Englewood Cliffs, NJ, 1976.

I. M. Yaglom and V. G. Boltyanskii, *Convex Figures*, Holt, Rinehart and Winston, New York, 1961.

A. C. Yao, An $O(E \log \log V)$ algorithm for finding minimum spanning trees, *Info. Proc. Lett* **4**, 21–23 (1975).

A. C. Yao, A lower bound to finding convex hulls, *J. ACM* **28**, 780–787 (1981).

A. C. Yao, On contructing minimum spanning trees in k-dimensional space and related problems, *SIAM J. Comput.* **11**(4), 721–736 (1982).

M. Z. Yannakakis, C. H. Papadimitriou and H. T. Kung, Locking policies: safety and freedom for deadlock, *Proceedings 20th Annual Symposium on Foundations of Computer Science*, pp. 286–297 (1979).

C. T. Zahn, Graph-theoretical methods for detecting and describing gestalt clusters, *IEEE Trans. Comp.* **C-20**(1), 68–86 (Jan. 1971).

S. I. Zhukhovitsky and L. I. Avdeyeva, *Linear and Convex Programming*, Saunders, Philadelphia, 1966.

Author Index

Subject Index*

---

*Page numbers in italics refer to formal
definitions.

Texts and Monographs in Computer Science

continued